Handbook of Experimental Pharmacology

Volume 98

Pharmacology of Asthma

Contributors

A.G. Alexander, P.J. Barnes, K.F. Chung, R.J. Flower, L.G. Garland
R.G. Goldie, S.T. Holgate, A.B. Kay, A. Laitinen, L.A. Laitinen
K.M. Lulich, C.P. Page, J.W. Paterson, R. Pauwels, C.G.A. Persson
I.W. Rodger, M. Schmitz-Schumann, R.C. Small, S.F. Smith
A. Szczeklik, C. Virchow, J.W. Wilson

Editors

Clive P. Page and Peter J. Barnes

Springer-Verlag
Berlin Heidelberg New York
London Paris Tokyo
Hong Kong Barcelona

Clive P. Page, B.Sc.,Ph.D.
Reader in Pharmacology
Department of Pharmacology
King's College
University of London
Manresa Road
London SW3 6LX, Great Britain

Peter J. Barnes, M.A.,D.M.,D.Sc.,FRCP
Professor of Thoracic Medicine
Department of Thoracic Medicine
National Heart and Lung Institute
Dovehouse Street
London SW3 6LY, Great Britain

With 45 Figures

ISBN-13:978-3-642-75857-7 e-ISBN-13:978-3-642-75855-3
DOI: 10.1007/978-3-642-75855-3

Library of Congress Cataloging-in-Publication Data. Pharmacology of asthma / contributors, A.G. Alexander ... [et al.]; editors, Clive P. Page and Peter J. Barnes. p. cm. — (Handbook of experimental pharmacology; v. 98) Includes bibliographical references. Includes index. ISBN-13:978-3-642-75-857-7 1. Asthma—Pathophysiology. 2. Antiasthmatic agents—Testing. 3. Bronchodilator agents—Testing. 4. Asthma—Chemotherapy. I. Alexander, A.G. II. Page, C.P. III. Barnes, Peter J., IV. Series. [DNLM: 1. Asthma—drug therapy. 2. Asthma—pathology. 3. Inflammation—physiopathology. W1 HA51L v. 98 / WF 553 P5359] QP905.H3 vol. 98 [RC591] 615'.1 s—dc20 [616.2'38] DNLM/DLC for Library of Congress.

Softcover reprint the hardcover 1st edition 1991

Typesetting: Best-set Typesetter Ltd., Hong Kong
27/3130-543210 – Printed on acid-free paper

List of Contributors

A. G. Alexander, Department of Allergy and Clinical Immunology, National Heart & Lung Institute, Dovehouse Street, London SW3 6LY, Great Britain

P. J. Barnes, Department of Thoracic Medicine, National Heart and Lung Institute, Dovehouse Street, London SW3 6LY, Great Britain

K. F. Chung, Department of Thoracic Medicine, National Heart and Lung Institute, Dovehouse Street, London SW3 6LY, Great Britain

R. J. Flower, Department of Biochemical Pharmacology, St. Bartholomew's Hospital Medical College, Charterhouse Square, London EC1M 6BQ, Great Britain

L. G. Garland, Anti-Inflammatory Research: Research Division, The Wellcome Research Laboratories, Langley Court, South Eden Park Road, Beckenham, Kent BR3 3BS, Great Britain

R. G. Goldie, Department of Pharmacology, University of Western Australia, Perth, Nedlands, 6009, Western Australia

S. T. Holgate, Immunopharmacology Group, Medicine 1 and Clinical Pharmacology, Level D, Centre Block, Southampton General Hospital, Tremona Road, Southampton S09 4XY, Great Britain

A. B. Kay, Department of Allergy and Clinical Immunology, National Heart & Lung Institute, Dovehouse Street, London SW3 6LY, Great Britain

A. Laitinen, Department of Electron-Microscopy, University of Helsinki, Mannerheimintie 172, SF-00280 Helsinki, Finland, and Department of Medical and Physiological Chemistry, University of Lund, Box 94, S-22100 Lund, Sweden

L. A. Laitinen, Department of Lung Medicine, University of Lund, University Hospital of Lund, S-22185 Lund, and Explorative Clinical Research, AB Draco, Box 34, S-22100 Lund, Sweden

K. M. Lulich, Department of Pharmacology, University of Western Australia, Perth, Nedlands, 6009, Western Australia

C.P. PAGE, Department of Pharmacology, King's College, University of London, Manresa Road, London SW3 6LX, Great Britain

J.W. PATERSON, Department of Pharmacology, University of Western Australia, Perth, Nedlands, 6009, Western Australia

R. PAUWELS, Department of Respiratory Diseases, State University, Academic Hospital, De Pintelaan 185, B-9000 Gent, Belgium

C.G.A. PERSSON, AB Draco, Box 34, S-22100 Lund, Sweden

I.W. RODGER, Department of Physiology and Pharmacology, Royal College, University of Strathclyde, 204 George Street, Glasgow G1 1XW, Great Britain

M. SCHMITZ-SCHUMANN, Hochgebirgsklinik Davos-Wolfgang, Asthma- und Allergieklinik, CH-7265 Davos-Wolfgang, Switzerland

R.C. SMALL, Department of Physiological Sciences, Medical School, University of Manchester, Stopford Building, Oxford Road, Manchester M13 9PT, Great Britain

S.F. SMITH, Department of Medicine, Charing Cross and Westminster Medical School, Fulham Palace Road, London W6 8RF, Great Britain

A. SZCZEKLIK, Copernicus Academy of Medicine, Department of Medicine, ul. Skawinska 8, 31-066 Krakow, Poland

C. VIRCHOW, Hochgebirgsklinik Davos-Wolfgang, Asthma- und Allergieklinik, CH-7265 Davos-Wolfgang, Switzerland

J.W. WILSON, Immunopharmacology Group, Medicine 1 and Clinical Pharmacology, Southampton General Hospital, Tremona Road, Southampton S09 4XY, Great Britain

Preface

This volume forms part of a prestigious series and covers the latest advances in our understanding of the pathophysiology and treatment of asthma. Our understanding of asthma has changed dramatically in recent years, and much of this new information is brought together in this volume written by internationally recognised authorities.

The aim of the book is to review in depth the changing concepts of inflammatory processes in asthma and to discuss the implications for research and future therapy of this common chronic disease. Many of the advances in our understanding of asthma have originated from a pharmacological approach, and this volume highlights the promising new options for pharmacological intervention.

It is hoped this book will be invaluable for research scientists and clinicians involved in asthma research and will be a major reference resource for chest physicians and those involved in the development of novel pharmaceutical entities.

Each chapter is extensively referenced, generously illustrated with clear diagrams and photographs, and represents a state-of-the-art review of this growing area.

C.P. Page
P.J. Barnes

Contents

CHAPTER 3

Inflammatory Mediators

CHAPTER 9

Pharmacology of Prophylactic Anti-Asthma Drugs

L.G. Garland ... 261

CHAPTER 10

Pathophysiology and Pharmacology of Aspirin-Induced Asthma

A. Szczeklik, C. Virchow, and M. Schmitz-Schumann.
With 6 Figures ... 291

CHAPTER 1

The Pathology of Asthma: An Overview

L.A. LAITINEN and A. LAITINEN

A. Introduction

Classical dogma equates asthma with mucus plugs, bronchial smooth muscle hyperplasia, eosinophilia and thickening of the epithelial basement membrane. However, these characteristics are those seen at autopsy of lungs obtained from asthmatics who died from status asthmaticus and neither reflect the changes in early disease nor how they relate to disease severity (HOUSTON et al. 1953; CARDELL 1956; DUNNILL et al. 1969; THURLBECK et al. 1970). So far the fatal cases have contributed the most to our understanding of the pathological changes in asthma. In fact, until recent years it has been very uncommon to see biopsy material from asthmatics, and thus the histology of the airways of living stable asthmatics was unknown. The hallmark of asthma is the functional change associated with airways obstruction. At least in the early stages, bronchospasm is a function of smooth muscle tone, and it has been suggested that there may even not be any early morphological changes. Developments in bronchoscopic and electron microscopic techniques have been essential in obtaining knowledge of airway morphology in living asthmatics. In the future, along with the introduction of new techniques, special attention must be paid to careful patient characterization.

B. Methods to Investigate the Pathology of Human Asthma

There have been many indirect approaches to examine airway structure, especially with respect to airway inflammation and hyperresponsiveness (LEE et al. 1977; HINSON et al. 1984; MURLAS and ROUM 1985a,b; HULBERT et al. 1985), but few studies have been conducted in living patients. One difficulty lies in obtaining representative specimens of human airways. There have been some attempts to overcome this problem.

Several investigators (COHEN and PRENTICE 1959; NAYLOR 1962; SANERKIN and EVANS 1965; FRIGAS et al. 1981) studied the sputum of asthmatic patients. The presence of respiratory epithelial cells in sputum was one of the first pathological abnormalities observed in asthma and was described by CURSHMANN in 1885 (see NAYLOR 1962). NAYLOR (1962) noted the tendency of bronchial epithelium to exfoliate during acute asthmatic attacks in studies of the sputum of asthmatics. He introduced the term "Creola body" to describe large sheets of exfoliated epithelium which condense to form spherical or elongated massed in the overlying mucus layer. Sputum from asthmatics frequently contains eosinophils and occasionally Charcot-Leyden crystals derived from eosinophils (DOR et al. 1984), but neither are specific for asthma and may also be seen in allergic bronchopulmonary aspergillosis, chronic eosinophilic pneumonia, some drug reactions and parasitic infestations (SCHATZ et al. 1981). Recent studies have concentrated on measuring immunoglobulins (TURNBULL et al. 1978; ACKERMANN et al. 1981) and eosinophil derived proteins (DOR et al. 1984) in the sputum.

The histological structure of the bronchial mucosa has been studied in patients who died from an asthmatic attack or from other natural or violent causes (HUBER and KOESSLER 1922; DUNNILL 1975; CUTZ et al. 1978; SOBONYA 1984). Asthmatic patients seldom have lung surgery; thus, fresh specimens of lung tissue can usually be obtained with bronchoscopy. However, until recently it has been uncommon to see biopsy material from asthmatics (GLYNN and MICHAELS 1960; CUTZ et al. 1978; L.A. LAITINEN et al. 1985; LUNDGREN et al. 1988; BEASLEY et al. 1989). Additional information on the bronchial airways can be obtained by bronchoalveolar lavage (BAL) (WARDLAW et al. 1988).

For quantitative studies, bronchial biopsy specimens should be taken and processed using a standardized method (L.A. LAITINEN et al. 1985). An electron microscopic method has been developed to study greater areas of airway mucosa by using slot grids without bars and by making photomontages of adjacent areas. This method allows one to examine and photograph the whole thin section, which is approximately 1 mm × 1 mm. With this method, the number of mast cells, neutrophils and eosinophils in the airway mucosa can be quantitated per mm^2 with ultrastructural recognition of cells, including those in different stages of degranulation (LAITINEN 1989).

There is a great lack of information concerning the pathology of the airway mucosa in living asthmatic patients. Thus, a review of bronchial biopsy findings is presented with special reference to changes in the airway epithelium in relation to our present knowledge of airway structure. Other airway structures, i.e. smooth muscle, bronchial glands, nerves and the bronchial circulation are also described.

C. Bronchial Epithelium and Inflammatory Cells in Asthmatic Patients Between Attacks

The epithelium of the mammalian tracheobronchial tree is of the ciliated pseudostratified columnar type in the larger airways (trachea and proximal bronchi) and of the simple cuboidal type inside the lung (distal bronchi and bronchioli). The cell types comprising the bronchial epithelium differ somewhat in different species. In the normal human bronchial epithelium, four main cell types, which all rest on a basement membrane, are most often seen (RHODIN 1974; REID and JONES 1979; HEINO 1987): (1) ciliated cells, (2) basal cells, (3) secretory (mucous or goblet) cells and (4) Kulchitsky (neuroendocrine, amine-containing, APUD or K) cells. Ciliated, secretory and some of the K cells reach the lumen. There are roughly three to five ciliated cells for every mucous cell. The epithelium also contains a few lymphocytes and

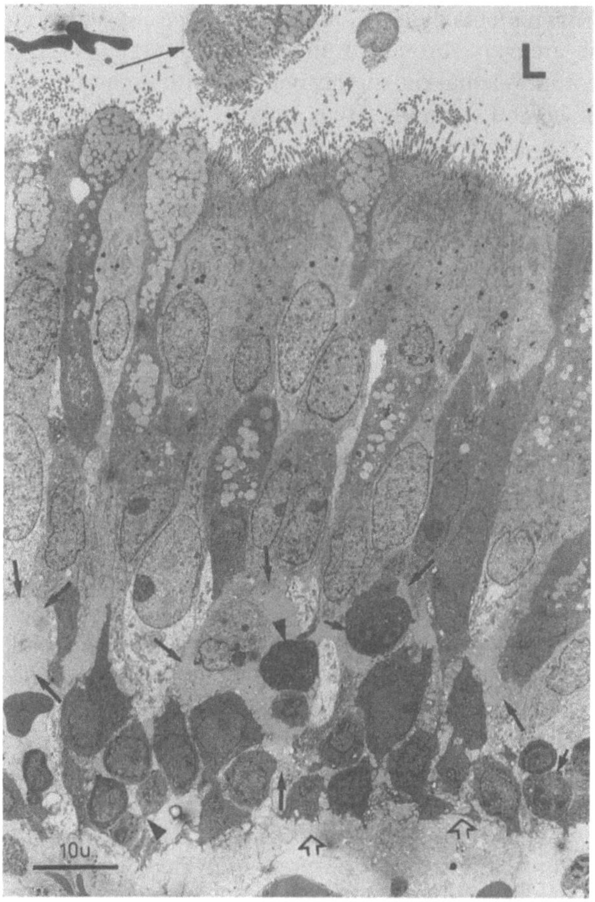

Fig. 1. Transmission electron microscopy. The specimen is from a clinically severe asthmatic patient who had the disease for several years. The epithelium shows fragility and shedding of ciliated and goblet cells. The shedding is caused by the presence of a homogeneous mass, probably representing oedematous fluid, separating the cells at the base of the epithelium. At the lumen, the cells are firmly joined together. The epithelium is highly infiltrated with neutrophils (*small arrow heads*). Shed epithelial cells are seen in the lumen at top of the picture (*long thin arrow*). *L*, lumen; *open arrows*, basement membrane. Original magnification ×2000; *bar*, 10μm

nerve profiles near the lumen and basement membrane (HEINO et al. 1982; A. LAITINEN 1985). The so-called intermediate cell type referred to in the literature is actually either a pre-ciliated or pre-secretory cell according to ultrastructural criteria, although an indeterminate cell type may exist.

Examination of airway epithelial structure is especially important as it serves as a primary target organ for exogenous irritants such as allergens. Structural changes, e.g. shedding, may greatly modify passage of irritants to deeper mucosa.

According to reports in the literature, at the end stage of asthma, in specimens from patients dying in status asthmaticus, it is extremely difficult to find normal areas of bronchial mucosa (DUNNILL 1982). A prominent described feature is a marked airway oedema with separation of the epithelial cells leaving behind in many areas only a layer of basal or reserve cells (DUNNILL 1960). More recently, similar changes have been described in light and electron microscopic studies of bronchial biopsies from living stable asthmatics (L.A. LAITINEN et al. 1985; LUNDGREN et al. 1988).

The airway epithelium in asthmatics can show a clearly fragile appearance caused either by shedding of the columnar epithelial cells or by intracellular signs of destruction, including vacuolization of the endoplasmic reticulum. In areas of shedding, there is accumulation of a homogenous mass, probably oedema fluid, in the widened intercellular spaces at the base of the eptihelium (Fig. 1). The homogenous mass, which pushes the uppermost columnar cells away, can often be seen close to areas of extensive epithelial destruction, e.g. in areas where only the basal cells are present or where the basement membrane is totally denuded. Despite this structurally destructive separation process, the columnar epithelial cells may have quite a normal appearance and may still be attached to each other at the luminal side by tight junctions (L.A. LAITINEN et al. 1985).

Epithelial shedding seems to be a commonly described feature in asthma and is also one of the first features described in the airway pathology of asthma (NAYLOR 1962). Other, less specific changes in asthmatic airway epithelium are goblet cell hyperplasia (Fig. 2) and epithelial cell metaplasia. Hyperplasia and metaplasia may be related to airway epithelial degeneration and regeneration processes. The basement membrane has usually (Fig. 3), but not always, been described as thickened (DUNNILL 1960; GLYNN and MICHAELS 1960; CUTZ et al. 1978; L.A. LAITINEN and A. LAITINEN 1988; BEASLEY et al. 1989). In addition, asthmatics show great individual variability in the number of mucosal mast cells, eosinophils and neutrophils present in lung tissue.

I. Mast Cells

The regular occurrence of mast cells in the airway epithelium, even in mild asthmatics with short duration of the disease, supports the idea that these cells may be important in the initial stage of the disease (SALVATO 1968; L.A. LAITINEN and A. LAITINEN 1988). Degranulation of mast cells and destruction

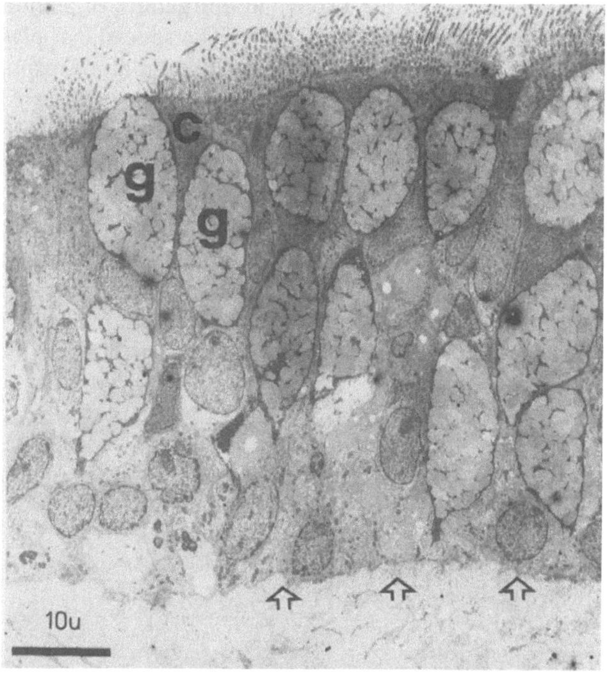

Fig. 2. A specimen from a clinically mild asthmatic patient who had the disease for only a couple of months. The epithelium shows goblet cell hyperplasia, the number of goblet cells having increased to a ratio of 1:1 with ciliated cells. *Open arrows*, basement membrane; *g*, goblet cell; *c*, ciliated cell. Original magnification ×2000; *Bar*, 10 μm

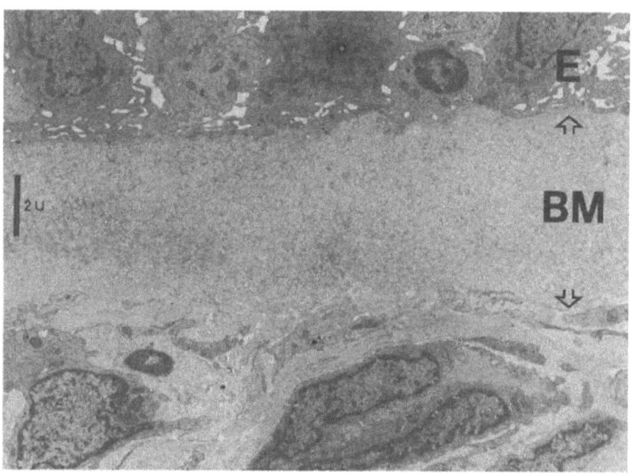

Fig. 3. The electron micrograph shows the thickened basement membrane in a patient with newly detected asthma. *E*, epithelium; *BM*, basement membrane. Original magnification ×7800; *bar*, 2 μm

of the surrounding tissue in the bronchial epithelium of asthmatics point to the activated status of the cells (Figs. 4–6). The special cytoplasmic granules of the mast cell are associated with initating inflammatory responses to specific stimuli. The ability of the activated mast cell to cause allergic inflammation has been studied extensively in the skin (SOLLEY et al. 1976; TANNENBAUM et al. 1980; OERTEL and KALINER 1981). In the nasal mucosa, mast cells are related to allergic but not to infectious inflammation (MELEN and Pipkorn 1985). There are several mediators in the granules which have been shown experimentally to generate chemotactic factors and to recruit inflammatory cells, Mast cell mediators with chemotactic activity for neutrophils have been described (PHILLIPS et al. 1982; GOETZL et al. 1983). Following mast cell degranulation, chemotactic factors are released, such as eosinophil chemo-

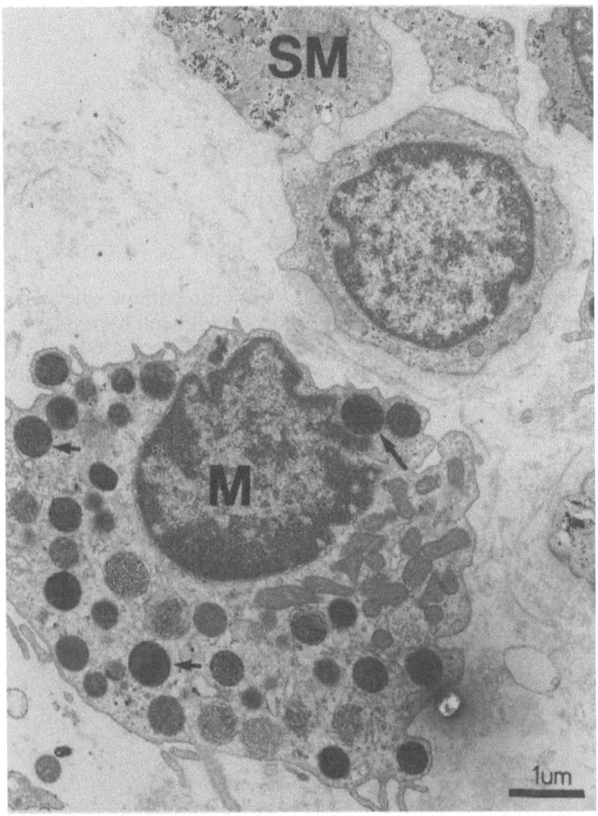

Fig. 4. The electron micrograph shows a mast cell in the smooth muscle layer of a segmental bronchus from a mild asthmatic patient. Some of the mast cell granules contain typical scroll-like substructures. Most of the granules contain either only digitated material or a combination of mainly longitudinally cut scroll and finger print-like substructures (*arrow heads*). *SM*, smooth muscle; *M*, mast cell. Original magnification ×15600; *bar*, 1 μm

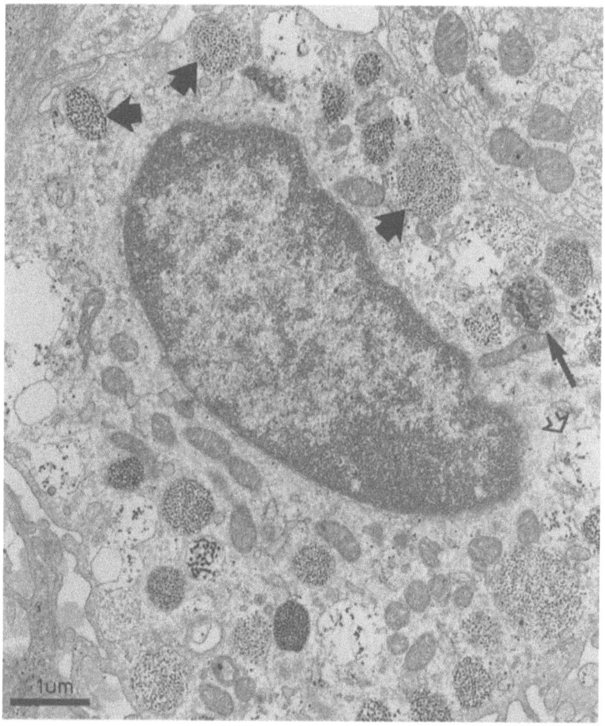

Fig. 5. The micrograph shows a partly degranulated mast cell in the airway epithelium in a patient with newly detected asthma. Most of the granules contain only digitated material (*thick arrows*). Only a few granules show typical scroll-type substructures (*thin, long arrow*); many empty granules (*open arrows*) can also be seen. This type of mast cell is commonly seen in the asthmatic airway epithelium together with the nearly completely degranulated mast cell seen in Fig. 6. Original magnification ×19500; *bar*, 1 μm

tactic factor of anaphylaxis (GOETZL and AUSTEN 1975) and intermediate molecular weight eosinophil chemotactic factor (BOSWELL et al. 1978).

The findings from BAL studies are in agreement with biopsy observations. In mild and even asymptomatic asthma, there is an increased influx of mast cells into the bronchial wall. WARDLAW et al. (1988) examined patients with mild atopic asthma and found that they had a five- to six-fold increase in the percentage of mast cells compared with control subjects. WARDLAW et al. (1988) suggested that mast cells found in asthmatic airways are activated and thus release more mediators. This has also been shown in atopic non-asthmatics (WENZEL et al. 1988).

II. Eosinophils

Even though eosinophil infiltration is a characteristic feature of asthmatic airways in necropsy specimens (FLINT et al. 1985), eosinophils occur in the

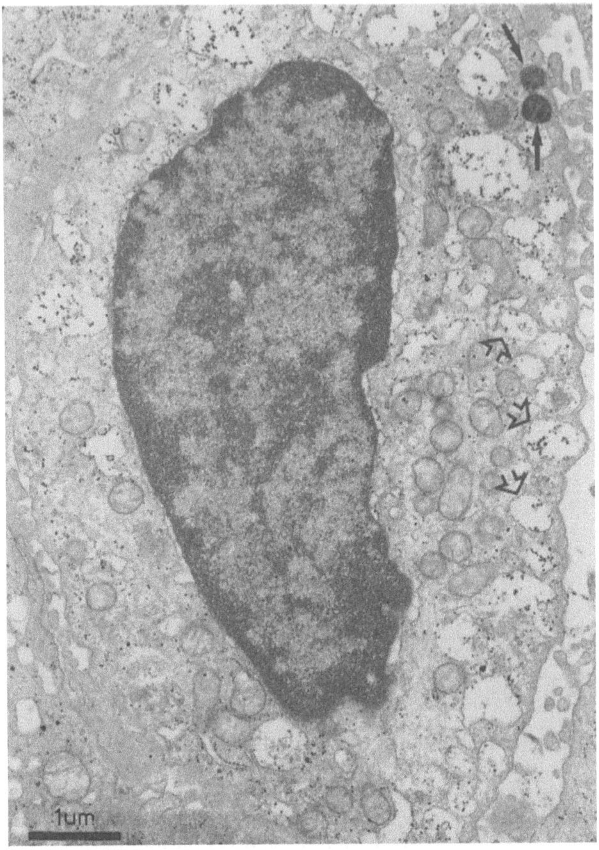

Fig. 6. A nearly completely degranulated mast cell in the airway epithelium of a newly detected asthmatic patient. Mainly enlarged empty granules are seen (*open arrows*). The cell can still be identified by two granules containing longitudinally cut scroll substructures (*black arrows* at the right upper corner of the picture). Original magnification ×25600; *bar*, 1 μm

airway epithelium of only a few stable asthmatic patients (LAITINEN 1989). Influx of eosinophils into the airway mucosa (Fig. 7) beneath the epithelium may be a characteristic feature of asthma (GLYNN and MICHAELS 1960; SALVATO 1968; CUTZ et al. 1978; L.A. LAITINEN and A. LAITINEN 1988; BEASLEY et al. 1989). Allergen inhalation results in a marked increase in eosinophils in the BAL at the time of the late reaction (FRIGAS and GLEICH 1986). An increased number of eosinophils was found only in asthmatics with evidence of disease activity (WARDLAW et al. 1988). Increases in the number of eosinophils have been shown to correlate with the severity of change in lung function (FRIGAS et al. 1981). This is in agreement with the finding that an increase in the number of eosinophils in the bronchial mucosa is related to worsening of asthma symptoms (L.A. LAITINEN, A. LAITINEN, M. HEINO, T.

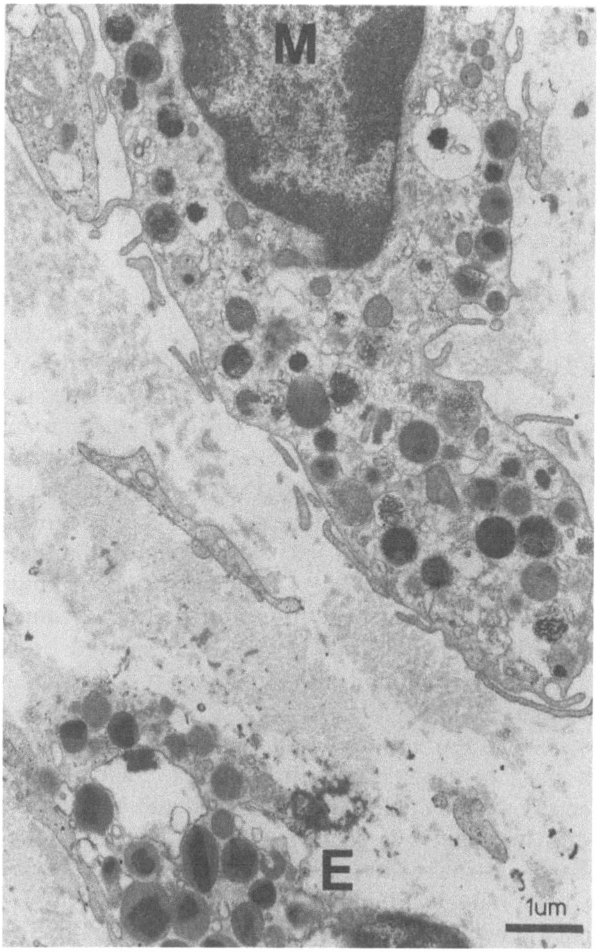

Fig. 7. A segmental bronchus of a mild asthmatic. A mast cell and part of an eosinophil are seen in the lamina propria and the difference in the granular morphology between these two cell types is seen. The eosinophil contains typical crystalloid core bars in the granule; the mast cell contains granules filled with digitated material and with scroll-type substructures. *M*, mast cell; *E*, eosinophil. Original magnification ×15600; *bar*, 1 μm

HAAHTELA, unpublished observations). Eosinophils may thus contribute to the pathophysiological process responsible for the change of stable asthma into an asthma attack. However, several lavage studies have shown that eosinophils are also present in BAL of patients with day-to-day asthma (GODARD et al. 1982; TOMIOKA et al. 1984; FLINT et al. 1985). WARDLAW et al. (1988) measured the amount of eosinophilic major basic protein and suggested that it is a more specific indicator of increased eosinophil activity than eosinophil count. Eosinophils release not only a variety of mediators, including leukotriene C_4 (WELLER et al. 1983) and platelet activating factor, but also basic proteins such as major basic protein (FILLEY et al. 1982) and eosinophil cationic protein, which are toxic to airway epithelium. Recent studies have associated eosinophils and, especially, eosinophilic major basic protein with epithelial damage in asthma (FILLEY et al. 1982).

III. Neutrophils

Neutrophils have been associated with acute (transient) bronchial hyperresponsiveness in many studies (NADEL 1984; BOUSHEY and HOLTZMANN 1985). According to HOLTZMANN et al. (1983), neutrophils are active participants in epithelial damage. The conclusion is based on a study in dogs, in which ozone was used to recruit neutrophils to destroy airway epithelium and simultaneously induce bronchial hyperresponsiveness; however, other finding do not agree. MURLAS and ROUM (1985a) described tracheal epithelial changes in guinea pigs that developed hyperreactive airways 2 h after ozone exposure. In a further study (MURLAS and ROUM 1985b), the same authors showed that the pathogenesis of ozone-induced bronchial hyperreactivity in guinea pigs is not dependent on inflammatory cell infiltration of the airway mucosa, which is in agreement with HULBERT et al. (1985), who showed bronchial hyperreactivity but not neutrophil infiltration in guinea pigs 30 min after exposure to cigarette smoke.

The presence of a few neutrophils in the epithelium is probably a normal phenomenon since neutrophils may be found even in the lavage fluid of normal subjects (CRYSTAL et al. 1986). Increased numbers of both neutrophils and eosinophils have been associated with the late asthmatic response in allergen challenge tests (DE MONCHY et al. 1985; FABBRI et al. 1987). The late response has been thought to mimic symptomatic day-to-day asthma.

In bronchial specimens from three out of ten asthmatics, no neutrophils were found in the epithelium (LAITINEN 1989). However, in patients with severe asthma of long duration, the neutrophil counts were very high (Fig. 8). Thus, it seems that chronic asthma of long duration and with lung function impairment is associated with epithelial and mucosal neutrophil influx, probably reflecting a far advanced, severe stage of the inflammatory process in the airways.

D. Bronchial Epithelial Inflammation During an Asthma Attack

Data on bronchial epithelial changes during an asthma attack are extremely limited. It is known that allergen inhalation results in a marked increase in eosinophils in BAL fluids obtained at the time of late reaction. In addition, a

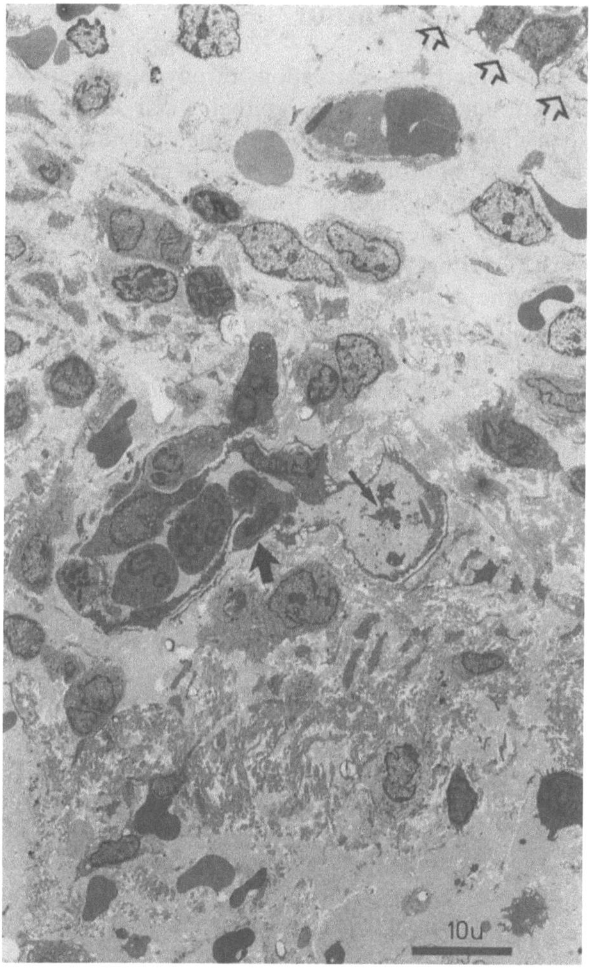

Fig. 8. Same specimen as in Fig. 1. The electron micrograph shows part of the lamina propria beneath the epithelium and basement membrane. The lamina propria is infiltrated with inflammatory cells, especially neutrophils and lymphocytes. A small vessel (*centre*) contains several neutrophils, one of which is penetrating the vessel wall (*thick arrow*). There are also platelets in the vessel (*thin arrow*). *Open arrows*, basement membrane. Original magnification ×2000; *bar*, 10 μm

close relationship exists between peripheral eosinophils and bronchial hyperresponsiveness (Frigas and Gleich 1986).

Bronchial biopsy specimens taken from one asthmatic patient during both a stable phase and an asthma attack showed during a spontaneous asthma attack a significant increase in the number of eosinophils in the bronchial mucosa (L.A. Laitinen, A. Laitinen, M. Heino, T. Haahtela, unpublished observations).

E. Epithelial Regeneration

The mechanisms behind the destruction, turnover and repair of the airway epithelium are not understood. After epithelial cell detachment, the basal or reserve cells which are left on the basement membrane may serve as a source and site of regenerating bronchial mucosa. Much of the metaplastic epithelium seen in asthmatics has been of the simple stratified non-ciliated variety exhibiting mitoses (DUNNILL 1982). This change has commonly been found in patients who died during an asthmatic attack. However, the appearance is very similar to epithelia found in the healing wounds of experimental animals after experimentally-induced trauma (WILHELM 1953).

Non-ciliated stratified epithelium has also been described in living asthmatics in bronchial biopsies analysed by light and electron microscopy (GLYNN and MICHAELS 1960; L.A. LAITINEN et al. 1985; LUNDGREN et al. 1988). There is no direct evidence in asthmatics whether this kind of epithelial change represents regeneration or damage.

When the tracheobronchial epithelium is damaged, mucous cells and basal cells proliferate in response to the injury. It is widely assumed that basal cells are the progenitor cells responsible for regeneration of the tracheobronchial epithelium following injury. However, according to the recent literature mucous cells may play a dominant proliferative role. As a result of a marked increase in their mitotic rate, pathological lesions are produced which include "goblet" cell hyperplasia, stratification and non-cornifying and cornifying epidermoid (squamous) metaplasias. Although morphologically dissimilar, these lesions are brought about by the wide and varied spectrum of phenotypic expression of mucous cells. Depending on the nature and extent of the injury, one or more of these lesions may be expressed simultaneously in the same specimen, and one lesion may change into another. Mitotic activity is very low in undisturbed tracheobronchial epithelium. Basal cells and mucous cells synthesize DNA and undergo mitosis; however, the extent to which each cell type normally participates in replacing effete cells, thereby maintaining the mucociliary state, is not yet resolved in humans or animals (see MCDOWELL and BEALS 1987).

Ciliated cells are easily damaged. When an epithelial injury persists or is severe, ciliated cells are lost from the epithelium. Ciliated cells are end stage cells; they neither synthesize DNA nor divide. They are not replaced in the regenerative process before the adverse situation is terminated. Loss of ciliated cells is an early response to many forms of injury (MCDOWELL et al. 1979; KEENAN et al. 1982a and b). Stimulated ciliogenesis may be one step in the regeneration of the epithelium. The number of cells with fibrogranular areas, which are precursors of cilia formation, is increased in asthmatics treated with oral prednisolone (HEINO et al. 1988).

It has also been suggested that fibronectin (FN) may play an important role in the healing process of some epithelial ulcers. FN refers to a group of structurally and immunologically related high molecular weight glycoproteins that are present in plasma and extracellular matrix. FN is important in cell-to-cell and cell-to-matrix interactions (YAMADA and OLDEN 1978; HYNES and YAMADA 1982; VAHERI et al. 1985). Cellular secretion of FN mediates adhesion of cultured cells to non-biological materials (YAMADA and OLDEN 1978). After tissue injury FN appears at the site of injury (CLARK et al. 1981; GRINNEL et al. 1981; SAKSELA et al. 1981; VAHERI et al. 1983). FN and fibrin are temporary components of the basement membrane extracellular matrix during healing of epithelial wounds in vivo. By immuno-fluorescence techniques the FN location in the mucosa of non-smoking asthmatic patients and controls has been studied (A. LAITINEN, K. TERVO, T. HAAHTELA, T. TERVO, L.A. LAITINEN, unpublished observation). In specimens from asthmatics, the basement membrane had a thickened structure. FN showed a bright reaction in the basement membrane at the bottom of an ulcer in the epithelium or adjacent to the edge of ulcerated epithelium. In controls the FN reaction was negative. Detection of FN in asthmatics at the possible zone of bronchial epithelial regeneration, which is identical to the FN location in other ulcerated human epithelia, is evidence of a wound healing process in the airways.

F. Airway Smooth Muscle

Some investigators have reported airway smooth muscle hypertrophy and/or hyperplasia in asthmatics (DUNNILL et al. 1969; HEARD and HASSAIN 1973; TAKIZAWA and THURLBECK 1971). However, it has been suggested that smooth muscle hypertrophy is a consequence rather than a cause of bronchial hyperreactivity since several studies have failed to show a relationship between airway responsiveness in vivo and airway smooth muscle quantity in airway specimens (ARMOUR et al. 1984a,b; MULLEN et al. 1986).

G. Bronchial Glands

Secretory glands are found in those bronchi which have cartilage in their walls. In trachea and larger bronchi they are abundant but become sparse in smaller bronchi. By definition bronchioli are found distal to the bronchi beyond the last plate of cartilage and proximal to the alveolar region of the lung. The submucosal glands extend only to the bifurcation of small bronchi (HAYWARD and REID 1952). Some studies have shown that the tracheobronchial mucous glands may be enlarged in asthmatic (EARLE 1953; GLYNN and MICHAELS 1960; DUNNILL et al. 1969) but not to the same degree as seen in chronic bronchitis (TAKIZAWA and THURLBECK 1971). Increases in gland size and goblet cell number have been reported in fatal cases of status asthmaticus, but it has been difficult to determine if the patients had chronic bronchitis and asthma or only asthma (HOUSTON et al. 1953). Airway secretions can be derived from the special secretions of the serous and mucous cells of the submucosal glands, from the secretory cells of the surface epithelium or from tissue fluid as transudate or exudate. Plugging of the small airways with mucous, epithelial and inflammatory cells can seriously impair respiratory function in asthmatics. It has been suggested that even if early in an attack the movement of mucus is in the normal direction with respect to the larynx, it may in later stages of the attack be reversed. This can be due either to reduced mucociliary clearance or to aspiration (REID 1974). There is still very little known about possible pathological changes in airway mucus secreting tissues, especially in living asthmatics.

H. Neural Pathways

The general pattern of innervation of human airways has been known for a long time (LARSELL and DOW 1933; GAYLOR 1934; RICHARDSON 1979). The vagus nerves and fibres from the upper four to five thoracic sympathetic ganglia form anterior and posterior plexuses at the hilum, from which the two main nerve networks arise: the peribronchial and periarterial plexuses. Nerves supply airways down to the level of respiratory bronchioles and also bronchial vessels (SPENCER and LEOF 1964). In addition to classical cholinergic and adrenergic mechanisms (PARTANEN et al. 1982; NADEL and BARNES 1984; WIDDICOMBE 1985; A. LAITINEN et al. 1985a), there exists a third component of neural control, which is neither adrenergic nor cholinergic and is thus called the nonadrenergic noncholinergic (NANC) nervous system (RICHARDSON and BELAND 1976; DOIDGE and SATCHELL 1982). Current evidence has implicated neuropeptides as possible neurotransmitters for the NANC system in airways (SAID et al. 1974; BARNES 1984; A. LAITINEN et al. 1985b).

The autonomic nervous system may influence, in addition to regulation of airway smooth muscle tone, secretion from submucosal glands, transport of fluid across airway epithelium, permeability and blood flow in the bronchial circulation and release of mediators from inflammatory cells (NADEL and BARNES 1984; BARNES 1986a).

Neural control of human airways is complex. Since changes in bronchomotor tone in asthma are rapid, it has been proposed that neurally mediated mechanisms are involved in the pathogenesis of asthma.

The precise relationship among neural pathological changes, bronchial hyperresponsiveness and clinical symptoms of asthma is not known. Therefore, this chapter will only deal in detail with afferent innervation of the airway epithelium of which there is some information in relation to the pathology of asthma.

I. Epithelium

Usually nerves are seen in the epithelium near the basal lamina (REID 1974). Only a few reports are available on the superficial location of nerves (DAS et al. 1978; FILLENZ and WOODS 1970; REID 1974; RICHARDSON 1979; HEINO et al. 1982; A. LAITINEN 1985; L.A. LAITINEN et al. 1985). The possible functional role of nerves cannot be judged on the morphological data only. On the basis of suggested ultrastructural criteria, nerve profiles containing many mitochondria resemble nerve endings classically considered afferent (KING et al. 1974). The exposure of superficially located nerves by sloughing of epithelial cells in asthma is possible. At places where the bronchial epithelium is missing because of shedding of the airway epithelium, a direct passage exists between the lumen and the subepithelial tissue; the latter also harbors nerves. Consequently, the exposure of mucosal afferent nerves to nonspecific stimuli or released mediators could partly explain bronchial hyperresponsiveness in patients with asthma.

II. Neuroepithelial Bodies and Neuroendocrine-Like (Granule-Containing) Cells

The airway epithelium contains both individual granule-containing cells and groups of cells defind as neuroepithelial bodies (NEB) (LAUWERYNS and PEUSKENS 1972). Various names have been given to these cells based on their morphological and cytochemical characteristics. The single cells have been referred to as Feyrter cells, Kulchitsky cells, AFG (argyrophil, fluorescent and granulated) cells, neuroendorcrine-like cells and APUD cells (DIAUGUSTINE and SONSTEGARD 1984). Both solitary cells and neuroepithelial bodies have been identified in many species, including humans (LAUWERYNS and PEUSKENS 1972; DIAUGUSTINE and SONSTEGARD 1984; L.A. LAITINEN 1988).

The cells usually have a triangular shape and rest on a basement membrane. The larger basal part of the cells contains many dense core vesicles. Often a fenestrated capillary in the lamina propria opposes the basement membrane at the point where the granule-containing cell is located (L.A. LAITINEN 1988).

Nerve profiles have been observed close to single granule-containing cells (L.A. LAITINEN 1988) and NEBs (LAUWERYNS et al. 1985). The function of these nerves has not been physiologically established, but experimental degeneration studies show that the majority of them are afferent (LAUWERYNS et al. 1985); more detailed investigations indicate that the NEBs may be stimulated to release their mediators in response to hypoxia (LAUWERYNS et al. 1983).

The apical poles of the cells contact the airway lumen (L.A. LAITINEN 1988). Granule-containing cells, by releasing their dense core vesicles, may produce their effects either via the nervous system or the blood circulation. Thus, either neural stimulation or a stimulus from the airway lumen to the amine-containing cells may release amines and possibly also neuropeptides into the local bloodstream to be carried into the deeper mucosa and smooth muscle layer.

Control of airway smooth muscle and glands may depend on diverse interactions of sensory stimuli from afferent nerve receptors, by mediators released from the neuroendocrine cells or mediators released from non-fixed cells, such as leukocytes and mast cells. These mechanisms may be disturbed in disease states like asthma. Both intraepithelial nerves and neuroendocrine cells have been found in asthmatic airways (L.A. LAITINEN et al. 1985; L.A. LAITINEN 1988), but no quantitative studies of their number exist.

I. Nerves and Other Airway Structures

Eosinophils and mast cells can be seen under the electron microscope in the smooth muscle layer between the cells. These cells are also seen close to nerve fibers in the mucosa (L.A. LAITINEN and A. Laitinen 1988), both in the normal state and in diseases such as bronchial asthma.

J. Bronchial Circulation

In humans and several other species, such as dog, bronchial arteries arise from multiple sites (SUSLOV 1895; DALY and HEBB 1966; MCLAUGHLIN 1983). They have been described as originating from the aorta, intercostal arteries and internal mammary or pericardial arteries (CUDKOWICZ and ARMSTRONG 1951; FLORANGE 1960). The bronchial arteries supply not only the walls of the bronchi, but also the adventitia or large pulmonary vessels, the nerves and some of the pleura (DALY and HEBB 1966; CUDKOWICZ 1979). The arteries in the peribronchial connective tissue give rise to smaller branches that reach the submucosa and form venous plexuses there (PIETRA et al. 1971).

Venous return from the larger bronchi is via bronchial vein to the azygos and hemiazygos veins. From more peripheral airways, bronchial blood is drained by tributaries of the pulmonary veins (MILLER 1947; LIEBOW 1960; PIETRA and MAGNO 1978; CHARAN et al. 1984; DEFFEBACH et al. 1987). The structure of the bronchial circulation in humans (Fig. 9) is similar to that in the dog. A rich network of capillaries runs close to the epithelium at all levels of the lower airways. The capillaries converge to form venules extending to a deeper plexus of larger venules and arterioles. The networks are continuous around and along the airways. Capillovenular structures are most frequent at the luminal edge of the mucosa, with only a few arterioles in the mucosa. Transmission electron microscopy shows capillaries very close to the epithelial basement membrane but not in the epithelium itself (L.A. LAITINEN and A. LAITINEN 1990).

The function of the airway vascular bed is not limited to providing nutrition to the structures it invests (DALY and HEBB 1966). Rather, the bronchial circulation may play a significant role in controlling the clearance of chemical mediators from the airways, regulating the interstitial fluid

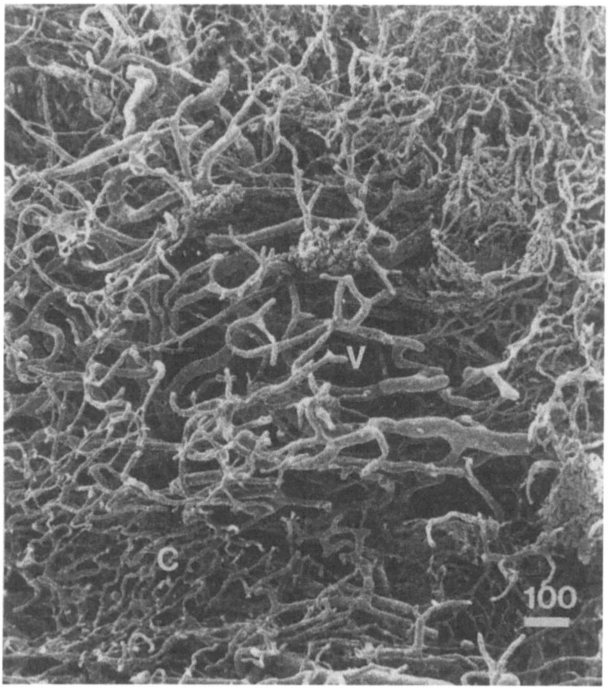

Fig. 9. A scanning electron micrograph showing a vascular cast of human airway microcirculation. The cast was prepared after lobectomy by injecting the bronchial artery with a polymerising substance; other airway structures were subsequently dissolved with potassium hydroxide. The capillary network lies beneath the airway epithelium in the lamina propria. Most of the vessels seen are capillaries (*c*) being approximately 10 micrometers in diameter; some larger postcapillary venules (*v*) are also seen; *bar*, 100μm. (For details on method, see A. LAITINEN et al. 1989a)

volume of the airway wall and participating in airway heat exchange (BAIER et al. 1985; PERSSON 1986, 1988), which has been proposed to be important in exercise-induced asthma.

K. Leakage of the Bronchial Vascular Bed in Patients with Asthma

The increase in the number of inflammatory cells, oedema formation and cell destruction found in the bronchial mucosa of asthma patients is only a part of the whole inflammatory process. In experimental animal studies it has been previously shown that drugs and inflammatory cell mediators, including histamine, methacholine, bradykinin, prostaglandins (D_2, E_1, $F_2\alpha$), platelet activating factor and some neuropeptides, e.g. substance P, vasoactive intestinal peptide, neurokinin A, calcitonin gene-related peptide, peptide histidine isoleucine and tyrosine, can cause dose-related decreases in airway vascular resistance indicating an increase in the blood flow (L.A. LAITINEN et al. 1987a,b). After injection of inflammatory mediators into the airway vascular bed in dogs, the histological appearance of the extravascular mucosal space is similar to that seen in some asthmatics, with extravasation of erythrocytes and neutrophils (L.A. LAITINEN et al. 1986, 1987b).

It has been proposed that airway epithelial damage follows neurogenic inflammation, which also, due to antidromic nerve stimulation, causes smooth muscle contraction, leakage of airway blood vessels and stimulation of bronchial glands (BARNES 1986b). Histologically the pathways of the axon reflexes have not been established. The mediator for axon reflexes has been suggested to be substance P (LUNDBERG and SARIA 1983; LUNDBERG et al. 1979), and these reflexes have been shown to cause contraction of airway smooth muscle. McDONALD (1987, 1988) has studied the effects of respiratory tract infection on neurogenic inflammation in rat trachea. He found that respiratory infections by certain viruses and, to a lesser extent, direct vagal

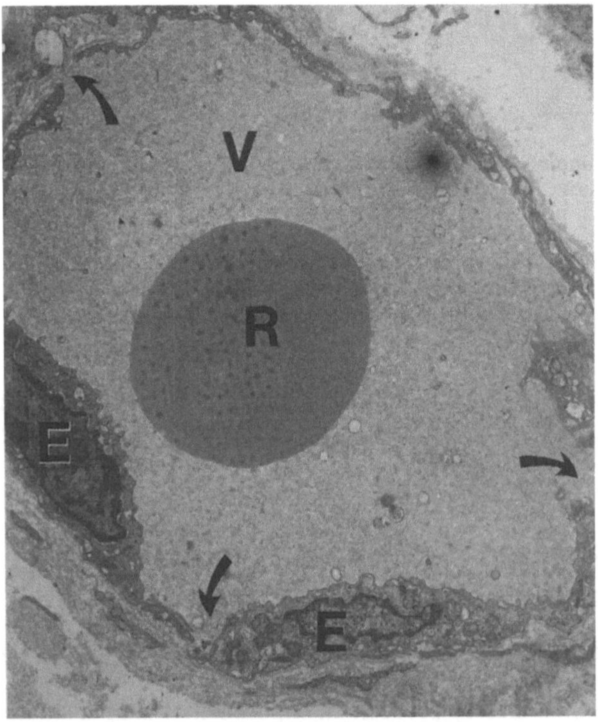

Fig. 10. Micrograph of a venule in the lamina propria of an asthmatic patient. In the vessel lumen a red blood cell is seen. Disruption of the vascular endothelium occur at three different sites (*arrows*). The endothelial gaps are about 1 micrometer wide. *V*, venule; *R*, red blood cell; *E*, endothelial cell. Original magnification ×7800

stimulation caused leakage of postcapillary venules and increased leukocyte adherence in the vessel wall. The airway epithelium showed changes, such as increased epithelial cell height, widening of intercellular spaces and an increase in the number of goblet cells. No counts of inflammatory cells in the airway epithelium were made. The increase in epithelial cell height and widening of intercellular spaces have also been found in asthmatics (L.A. LAITINEN et al. 1985; L.A. LAITINEN and A. LAITINEN 1988). These findings may result from increased exudation (PERSSON 1986). The endothelium of the bronchial vessels may have wide gaps in asthmatics (Fig. 10) (L.A. LAITINEN and A. LAITINEN 1988).

L. Conclusion

In experimental animal studies direct injection of the tracheobronchial arteries with inflammatory mediators, neural stimulation or induced respiratory tract infection results in similar histopathological features as observed in human asthma. These experimental models may help us to understand the mechanisms and sequence of events in the progress of asthma. Viral upper respiratory tract infection causes destructive epithelial changes in humans (HERS 1966), and otherwise healthy subjects can develop transient bronchial hyperresponsiveness (L.A. LAITINEN 1974; EMPEY et al. 1976). Asthmatics, in whom bronchial hyperreactivity is a constant feature, can have profound epithelial changes in the airways (L.A. LAITINEN et al. 1985). In especially severe asthmatics, features of the classical inflammatory reaction, including destruction of the tissue, vascular permeability changes with mucosal oedema and inflammatory cell influx, can be seen.

Bronchial epithelial damage may cause bronchial hyperreactivity. When, why, and how the disease in the airways proceeds to the condition of a fully developed, irreversible inflammatory reaction, as described at the beginning of this chapter, i.e. with eosinophilia, mucus plugs and oedema at autopsy, is not known. Could the process be interrupted with treatment? So far, the disease has mainly been monitored by clinical observation and measurement of respiratory function. With recent bronchoscopic studies in living asthmatics we have evidence that there are morphological changes even at a clinically early stage of the disease. Morphological studies may give important information about functional abnormalities in asthmatic airways and about the progress of changes leading to the fully developed inflammatory process.

Acknowledgement. The authors thank Mrs. Eva Engberg for her excellent secretarial assistance in the preparation of the manuscript.

References

Ackerman SJ, Gleich GJ, Weller PF, Ottesen EA (1981) Eosinophilia and elevated serum levels of eosinophil major basic protein and Charcot-Leyden crystal protein (lysophospholipase) after treatment of patients with Bancroft's filariasis. J Immunol 127:1093–1098

Armour CL, Black JL, Berend N, Woolcock AJ (1984a) The relationship between bronchial hyperresponsiveness to methacholine and airway smooth muscle structure and reactivity. Respir Physiol 58:223–233

Armour CL, Lazar NM, Schellenberg RR, Taylor SM, Chan N, Hogg JC, Pare PD (1984b) Comparison of in vivo and in vitro human airway reactivity to histamine. Am Rev Respir Dis 129:907–910

Baier H, Long WM, Wanner A (1985) Bronchial circulation in asthma. Respiration 48:199–205

Barnes PJ (1984) The third nervous system in the lung: physiology and clinical perspectives. Thorax 39:561–567

Barnes PJ (1986a) Airway inflammation and autonomic control. Eur J Respir Dis 69 [Suppl 147]:80–87

Barnes PJ (1986b) Asthma as an axon reflex. Lancet 1:242–244

Beasley R, Roche WR, Roberts JA, Holgate ST (1989) Cellular events in the bronchi in mild asthma and after bronchial provocation. Am Rev Respir Dis 139:806–817

Boswell RN, Austen KF, Goetzl EJ (1978) Intermediate molecular weight eosinophil factor in rat peritoneal mast cells; immunologic release, granule association, and demonstration of structural heterogeneity. J Immunol 120:15–20

Boushey HA, Holtzman MJ (1985) Experimental airway inflammation and hyperreactivity. Am Rev Respir Dis 13:312–313

Cardell BS (1956) Pathological findings in deaths from asthma. Int Arch Allergy 9:189

Charan NB, Turk G, Dhand R (1984) Gross and subgross anatomy of bronchial circulation in sheep. J Appl Physiol 57:648–664

Clark RAF, Dvorak HF, Colvin RB (1981) Fibronectin in delayed-type hypersensitivity skin reactions. J Immunol 262:497–500

Cohen RC, Prentice AID (1959) Metaplastic cells in sputum of patients with pulmonary eosinophilia. Tubercle 40:44–46

Crystal RG, Reynolds HY, Kalica AR (1986) Bronchoalveolar lavage. The report of an international congress. Chest 90:122–128

Cudkowicz L (1979) Bronchial arterial circulation in man: normal anatomy and response to disease. In: Moser KM (ed) Lung biology in health and disease, vol 14: pulmonary vascular diseases. Dekker, New York, pp 111–232

Cudkowicz L, Armstrong JB (1951) Observations on the normal anatomy of the bronchial arteries. Thorax 6:343–358

Curschmann H (1885) Einige Bemerkungen über die im Bronchialsecret vorkommenden Spiralen. Dtsch Arch Klim Med 36:578–585

Cutz E, Levison H, Cooper DM (1978) Ultrastructure of airways in children with asthma. Histopathology 2:407–421

Daly M de B, Hebb C (1966) Pulmonary and bronchial vascular systems. Arnold, London

Das RM, Jefferey PK, Widdicombe JG (1978) The epithelial innervation of the lower respiratory tract of the cat. J Anat 126:123–131

De Monchy JGR, Kauffman HF, Venge P et al. (1985) Bronchoalveolar eosinophilia during allergen-induced late asthmatic reactions. Am Rev Respir Dis 131:373–377

Deffebach ME, Charan N, Lakshminarayan F, Butler J (1987) The bronchial circulation; small, but a vital attribute of the lung. Am Rev Respir Dis 135:463–481

DiAugustine RP, Sonstegard KS (1984) Neuroendocrine-like (small granule) epithelial cells of the lung. Environ Health Perspect 55:271–295

Doidge JM, Satchell DG (1982) Adrenergic and nonadrenergic inhibitory nerves in mammalian airways. J Auton Nerv Syst 5:83–89
Dor PJ, Ackermann SJ, Gleich GJ (1984) Charcot-Leyden crystal protein and eosinophil granule major basic protein in sputum of patients with respiratory diseases. Am Rev Respir Dis 130:1072–1077
Dunnill MA (1975) The morphology of the airways in bronchial asthma. American College of Chest Physicians, Park Ridge, Illinois, pp 213–321
Dunnill MS (1960) The pathology of asthma with special reference to changes in the bronchial mucosa. J Clin Pathol 13:27–33
Dunnill MS (1982) Pulmonary pathology. Churchill Livingstone, Edinburgh
Dunnill MS, Massarella GR, Anderson JA (1969) A comparison of the quantitative anatomy of the bronchi in normal subjects, in status asthmaticus, in chronic bronchitis, and in emphysema. Thorax 24:176–179
Earle BV (1953) Fatal bronchial asthma; series of 15 cases with review of literature. Thorax 8:195–206
Empey DW, Laitinen LA, Jacobs L, Gold WM, Nadel JA (1976) Mechanism of bronchial hyperreactivity in normal subjects after upper respiratory tract infections. Am Rev Respir Dis 113:131–139
Fabbri LM, Boshetto P, Zocca E et al. (1987) Bronchoalveolar neutrophilia during late asthmatic reactions induced by toluene di-isocyanate. Am Rev Respir Dis 136:36–42
Fillenz M, Woods RI (1970) Sensory innervation of the airways. In: Porter R (ed) CIBA Foundation Symposium: breathing. Hering-Breuer Centenary Symposium. Churchill, London, pp 101–107
Filley WC, Holley KE, Kephart GM, Gleich GJ (1982) Identification by immunofluorescence of eosinophil granule major basic protein in lung tissues of patients with bronchial asthma. Lancet 2:11–15
Flint KC, Leung KBP, Hudspith BN et al. (1985) Bronchoalveolar mast cells in extrinsic asthma: a mechanism for the initiation of antigen specific bronchoconstriction. Br Med J 291:923
Florange W (1960) Anatomie und Pathologie der Arteria bronchialis. Ergeb Pathol Anat 39:152–213
Frigas E, Gleich GJ (1986) The eosinophil and the pathology of asthma. J Allergy Clin Immunol 77:527–537
Frigas E, Loegering DA, Solley GO, Farrow GM, Gleich GJ (1981) Elevated levels of eosinophil granule major basic protein in the sputum of patients with bronchial asthma. Proc Mayo Clin 56:345–353
Gaylor JB (1943) The intrinsic nervous mechanism of the human lung. Brain 57: 143–160
Glynn AA, Michaels L (1960) Bronchial biopsy in chronic bronchitis and asthma. Thorax 15:142–153
Godard P, Chaintreuil J, Damon M et al. (1982) Functional asessement of alveolar macrophages: comparison of cells from asthmatics and normal subjects. J Allergy Clin Immunol 70:88–94
Goetzl EJ, Austen KF (1975) Purification and synthesis of eosinophilotactic tetrapeptides of human lung tissue: identification as eosinophil chemotactic factor of anaphylaxis. Proc Natl Acad Sci USA 72:4123–4127
Goetzl EJ, Phillips MJ, Gold WM (1983) Stimulus specificity of the generation of leukotrienes by dog mastocytoma cells. J Exp Med 158:731–737
Grinnel F, Billingham RE, Burgess L (1981) Distribution of fibronectin during wound healing in vivo. J Invest Dermatol 76:181–189
Hayward J, Reid L (1952) Observations on the anatomy of the intrasegmental bronchial tree. Thorax 7:89–37
Heard BE, Hassain S (1973) Hyperplasia of bronchial muscle in asthma. J Pathol 110:319–331
Heino M (1987) Morphological changes related to ciliogenesis in the bronchial

epithelium in experimental conditions and clinical course of disease. Eur J Respir Dis 151 [Suppl]:1–39
Heino M, Mönkäre S, Haahtela T, Laitinen LA (1982) An electronmicroscopic study of the airways in patients with farmer's lung. Eur J Respir Dis 36:52–61
Heino M, Karjalainen J, Ylikoski J et al. (1988) Bronchial ciliogenesis and oral steroid treatment in patients with asthma. Br J Dis Chest 82:175–178
Hers JFPH (1966) Disturbances of the ciliated epithelium due to influenza virus. Am Rev Respir Dis 93:162–171
Hinson JM, Hutchinson AA, Brigham KL, Meyrick BO, Snapper JR (1984) Effects of granulocyte depletion on pulmonary responsiveness to aerosol histamine. J Appl Physiol 56:411–417
Holtzman MJ, Fabbri LM, O'Byrne PM et al. (1983) Importance of airway inflammation for hyperresponsiveness induced by ozone. Am Rev Respir Dis 127:686–690
Houston JC, DeNavasquez S, Trouce JR (1953) A clinical and pathologic study of fatal cases of status asthmaticus. Thorax 8:207–213
Huber HL, Koessler KK (1922) The pathology of bronchial asthma. Arch Intern Med 30:687–760
Hulbert VM, MeLean T, Hogg JC (1985) The effect of acute airway inflammation on bronchial reactivity in guinea-pigs. Am Rev Respir Dis 132:7–11
Hynes RO, Yamada KM (1982) Fibronectins: multifunctional modular glycoproteins. J Cell Biol 95:369–377
Keenan KP, Combs JW, McDowell EM (1982a) Regeneration of hamster tracheal epithelium after mechanical injury. I. Focal lesions: quantitative morphologic study of cell proliferation. Virchows Arch [B] 41:193–214
Keenan KP, Combs JW, McDowell EM (1982b) Regeneration of hamster tracheal epithelium after mechanical injury. II. Multifocal lesions: stathmokinetic and autoradiographic studies of cell proliferation. Virchows Arch [B] 41:215–229
King AS, McLelland J, Cook RD, King DZ, Walsh C (1974) The ultrastructure of afferent nerve endings in the avain lung. Respir Physiol 22:21–40
Laitinen A (1985) Ultrastructural organization of intraepithelial nerves in the human airway tract. Thorax 40:488–492
Laitinen A, Partanen M, Hervonen A, Laitinen LA (1985a) Electron microscopic study of the innervation of the human lower respiratory tract. Evidence of adrenergic nerves. Eur J Respir Dis 67:209–215
Laitinen A, Partanen M, Hervonen A, Pelto-Huikko M, Laitinen LA (1985b) VIP-like immunoreactive nerves in human respiratory tract. Histochemistry 82:313–319
Laitinen A, Laitinen LA, Moss R, Widdicombe JG (1989) Organization and structure of the tracheal and bronchial blood vessels in the dog. J Anat 165:133–140
Laitinen LA (1974) Histamine and methacholine challenge in the testing of bronchial reactivity. Scand J Respir Dis 86 [Suppl]:1–48
Laitinen LA (1988) Detailed analysis of neural elements in human airways. In: Kaliner M, Barnes P (eds) Neural regulation of the airways in health and disease. Dekker, New York pp 35–36
Laitinen LA (1989) Epithelial Damage. In: Hargreave FE, Hogg JC, Malo J-L, Toogood JH (Eds) Glucocorticoids and mechanisms of asthma. Excerpta Medica, pp 215–229
Laitinen LA, Laitinen A (1988) Mucosal inflammation and bronchial hyperreactivity. Eur Respir J1 (5):488–489
Laitinen LA, Laitinen A (1990) Histology and electron microscopy. In: Butler J (Ed) The bronchial circulation (The lung in health and disease-series). Dekker, New York (in press)
Laitinen LA, Heino M, Laitinen A, Kava T, Haahtela T (1985) Damage of the airway epithelium and bronchial reactivity in patients with asthma. Am Rev Respir Dis 131:599–606

Laitinen LA, Robinson NP, Laitinen A, Widdicombe JG (1986) Relationship between mucosal thickness and vascular resistance in dogs. J Appl Physiol 61(6):2186–2193
Laitinen LA, Laitinen A, Salonen RO, Widdicombe JG (1987a) Vascular actions of airway neuropeptides. Am Rev Respir Dis 136:59–64
Laitinen LA, Laitinen A, Widdicombe JG (1987b) Effects of inflammatory and other mediators on airway vascular beds. Am Rev Respir Dis 135 (6):67–70
Larsell G, Dow LS (1933) The innervation of the human lung. Am J Anat 52:125–146
Lauweryns JM, Peuskens JC (1972) Neuro-epithelial bodies (neuroreceptor or secretory organs?) in human infant bronchial and bronchiolar epithelium. Anat Rec 172:471–482
Lauweryns JM, De Boc V, Guelinckx P, Decramer M (1983) Effects of unilateral hypoxia on neuroepithelial bodies in rabbit lungs. J Appl Physiol 55:1665–1668
Lauweryns JM, Van Lommer AT, Dom RJ (1985) Innervation of rabbit intrapulmonary neuroepithelial bodies. Quantitative and qualitative ultrastructural study after vagotomy. J Neuro Sci 67:81–92
Lee L-Y, Bleeker ER, Nadel JA (1977) Effect of ozone on bronchomotor response to inhaled histamine aerosol in dog. J Appl Physiol 43:626–631
Liebow AA (1960) The bronchopulmonary venous collateral circulation with special reference to emphysema. Am J Pathol 37:361–380
Lundberg JM, Saria A (1983) Capsaicin-induced desensitisation of airway mucosa to cigarette smoke, mechanical and chemical irritants. Nature 302:251–253
Lundberg JM, Hökfelt T, Kewenter J, Pettersson G, Ahlman H, Edin R, Dahlström A, Nilsson G, Terenius L, Uvnäs-Wallensten K, Said S (1979) Substance P-, VIP- and enkephalin-like immunoreactivity in the human vagus nerve. Gastroenterology 77:469–471
Lundgren R, Söderberg M, Hörstedt, Stenling R (1988) Morphological studies of bronchial mucosal biopsies from asthmatics before and after ten years of treatment with inhaled steroids. Eur Respir J 1:883–889
McDonald DM (1987) Neurogenic inflammation in the respiratory tract: actions of sensory nerve mediators on blood vessels and epithelium of the airway mucosa. Am Rev Respir Dis 136:65–72
McDonald DM (1988) Respiratory tract infections increase susceptibility to neurogenic inflammation in the rat trachea. Am Rev Respir Dis 137:1432–1440
McDowell EM, Beals TF (1987) Biopsy pathology of the bronchi. Saunders, Philadelphia
McDowell EM, Becci PJ, Schürch W, Trump FF (1979) The respiratory epithelium. VII. Epidermoid metaplasia of hamster tracheal epithelium during regeneration following mechanical injury. JNCI 62:995–1008
McLaughlin RF Jr (1983) Bronchial artery distribution in various mammals and humans. Am Rev Respir Dis 128:S 57–58
Melen J, Pipkorn U (1985) Mast cells on the surface of the mucous membrane – a general features in inflammatory reactions in the nose? Rhinology 23:187
Miller SW (1947) The lung, 2nd edn. Thomas, Springfield
Mullen JBM, Wiggs BR, Wright JL, Hogg JC, Pare PD (1986) Nonspecific airway reactivity in cigarette smokers. Am Rev Respir Dis 133:120–125
Murlas C, Roum JH (1985a) Bronchial hyperreactivity occurs in steroid treated guineapigs depleted of leukocytes by cyclophosphamide. J Appl Physiol 58: 1630–1637
Murlas C, Roum JH (1985b) Sequence of pathological change in the airway mucosa of guinea-pigs during ozone induced bronchial hyperreactivity. Am Rev Respir Dis 131:314–320
Nadel J (1984) Inflammation and asthma. J Allergy Clin Immunol 73:651–653
Nadel JA, Barnes PJ (1984) Autonomic regulation of the airways. Annu Rev Med 35:451–467

Naylor B (1962) The shedding of the mucosa of the bronchial tree in asthma. Thorax 17:69–72
Oertel HL, Kaliner MA (1981) The biological activity of mast cell granules. III. Purification of inflammatory factors of anaphylaxis (IF-A) responsible for causing late-phase reactions. J Immunol 112:1398–1402
Partanen M, Laitinen A, Hervonen A, Toivanen M, Laitinen LA (1982) Catecholamine- and acetylcholinesterase-containing nerves in human lower respiratory tract. Histochemistry 76:175–188
Persson CGA (1986) Role of plasma exudation in asthmatic airways. Lancet 2: 1126–1128
Persson CGA (1988) Plasma exudation and asthma. Lung 166:1–23
Phillips MJ, Calonico L, Gold WM (1982) Morphological and pharmacological characterization of dog mastocytoma cells. Am Rev Respir Dis 125:63A
Pietra GG, Magno M (1978) Pharmacologic factors influencing permeability of the bronchial microcirculation. Fed Proc 37:2466–2470
Pietra GG, Szidon JP, Leventhal MM, Fisherman AP (1971) Histamine and interstitial pulmonary edema in the dog. Circ Res 29:323–337
Reid L (1974) Histological aspects of bronchial secretion. Experimental studies on bronchial secretion and therapeutical aspects of pathological bronchial secretion. Scand J Respir Dis 90 [Suppl]:9–15
Reid L, Jones R (1979) Bronchial mucosal cells. Fed Proc 38:191–196
Rhodin JAG (1974) Respiratory system. In: Histology. Oxford University Press, New York, pp 607–645
Richardson JB (1979) Nerve supply to the lungs. Am Rev Respir Dis 119:785–802
Richardson JB, Beland J (1976) Non-adrenergic inhibitory nervous system in human airways. J Appl Physiol 41:764–771
Said SI, Kitamura S, Yoshida T, Preskitt J, Holden LD (1974) Humoral control of airways. Ann NY Acad Sci 221:103–114
Saksela O, Alitalo K, Kiistala U, Vaheri A (1981) Basal lamina components in experimentally induced skin blisters. J Invest Dermatol 77:283–286
Salvato G (1986) Asthma and mast cells of bronchial connective tissue. Experientia 18:330–331
Sanerkin NG, Evans DMD (1965) The sputum in bronchial asthma: pathognomic patterns. J Pathol 89:535–541
Schatz M, Wasserman S, Patterson R (1981) Eosinophils and immunologic lung disease. Symposium on clinical allergy. Med Clin North Am 65:1055–1071
Sobonoya RE (1984) Concise clinical study: quantitative structural alterations in long standing allergic asthma. Am Rev Respir Dis 130:289
Solley GO, Gleich GJ, Jordan Scroeter AL (1976) Late phase of the immediate wheal and flare skin reaction: its dependence on IgE antibodies. J Clin Invest 58:408–420
Spencer H, Leof D (1964) The innervation of the human lung. J Anat 98:599–609
Suslov KI (1895) Some investigations on the anatomy of the bronchial arteries in man (in Russian) Thesis, St Petersburgh
Takizawa T, Thurlbeck WM (1971) Muscle and mucous gland size in the major bronchi of patients with chronic bronchitis, asthma and asthmatic bronchitis. Am Rev Respir Dis 104:331–336
Tannenbaum S, Oertel H, Henderson WR, Kaliner M (1980) The biologic activity of mast cell granules. I. Elicitation of inflammatory responses in rat skin. J Immunol 125:325–335
Thurlbeck WM, Henderson JA, Fraser RG, Bates DV (1970) Chronic obstructive lung disease. A comparison between clinical, roentgenologic, functional and morphologic criteria in chronic bronchitis, emphysema, asthma and bronchiectasis. Med 49:81–147
Tomioka M, Ida S, Shondoh Y et al. (1984) Mast cells in bronchoalveolar lumen of patients with bronchial asthma. Am Rev Respir Dis 129:1000–1005

Turnbull LS, Turnbull LW, Crofton LW, Kay AB (1978) Immunoglobulins, complement and arylsulphatase in sputum from chronic bronchitis and other pulmonary diseases. Clin Exp Immunol 32:226–232
Vaheri A, Salonen E-M, Vartio T, Hedman K, Stenman S (1983) Fibronectin and tissue injury. In: Woolf N (ed) Biology and pathology of the vessel wall. Praeger, Eastbourne, p 161
Vaheri A, Salonen EM, Vartio T (1985) Fibronectin in formation and degradation of the pericellular matrix. In: Evered D, Whelan J (eds) Fibrosis, Ciba Foundation Symposium. Pitman, London, p 111
Wardlaw AJ, Dunnette S, Gleich GJ et al. (1988) Eosinophils and mast cells in bronchoalveolar lavage in subjects with mild asthma. Am Rev Respir Dis 137:62
Weller PF, Lee CW, Foster DW, Corey EJ, Austen KF, Lewis RA (1983) Generation and metabolism of 5-lipoxygenase pathway leukotrienes by human eosinophils: predominant production of leukotriene C_4. Proc Natl Acad Sci USA 80:7626–7630
Wenzel SE, Fowler AA, Schwartz LB (1988) Activation of pulmonary mast cells by bronchoalveolar allergen challenge. Am Rev Respir Dis 137:1002–1008
Widdicombe JG (1985) Control of airway caliber. Am Rev Respir Dis 131 [Suppl]: 33–35
Wilhelm DL (1953) Regeneration of tracheal epithelium. J Pathos Bat 65:543–550
Yamada KM, Olden K (1978) Fibronectin-adhesive glycoproteins of cell surface and blood. Nature 275:179–184

CHAPTER 2

The Contribution of Inflammatory Cells to the Pathogenesis of Asthma

A.G. Alexander, J.W. Wilson, S.T. Holgate, and A.B. Kay

A. Inflammation and Inflammatory Cells

Inflammation is the response of vascularised tissue to injury and serves to resolve and repair the effect of damage. The factors that initiate inflammation, like those of cell injury, are diverse and include infectious agents (bacteria, viruses and parasites), physical agents (burns, radiation and trauma), chemical agents (drugs, toxins and industrial agents), ischaemic injury to tissues and immunological reactions such as allergy and autoimmunity. The histopathological features of inflammation consist of changes in blood flow and calibre of small blood vessels followed by alterations in vascular permeability and accompanied by a series of changes involving white blood cells. Acute inflammation is of short duration and characterised by exudation of fluid and plasma proteins (oedema) and leucocyte emigration, with granulocytes being prominent. In contrast, chronic inflammation is of longer duration again generally with granulocyte emigration but also with a dense infiltration of lymphocytes and monocytes/macrophages together with proliferation of blood vessels and connective tissue. In certain specialised circumstances, such as allergy and asthma, eosinophils are often found in particularly large numbers. Both acute and chronic inflammation are associated with some degree of fibrin deposition leading to platelet adherence and the release of platelet products. Basophils are also inflammatory cells and are prominent in certain forms of delayed-type hypersensitivity. They are also found in the upper airways in allergic rhinitis, but at the present time there is no conclusive evidence that the basophil participates in pathological processes in the lung. Thus the cells that migrate from blood vessels and which have been clearly identified in airway inflammation include neutrophils, eosinophils, lymphocytes and monocytes. Platelets may contribute to the reaction, but their ability actively to emigrate from the bloodstream remains controversial. Certain resident tissue cells, such as mast cells and epithelial cells, probably participate, as does the fibroblast. Inflammation at mucosal surfaces has two additional important features: mucus hypersecretion and shedding or denudation of the airway epithelial surface.

These inflammatory cells produce a wide range of cytokines (Fig. 1) and mediators which contribute not only to the development of the inflammatory

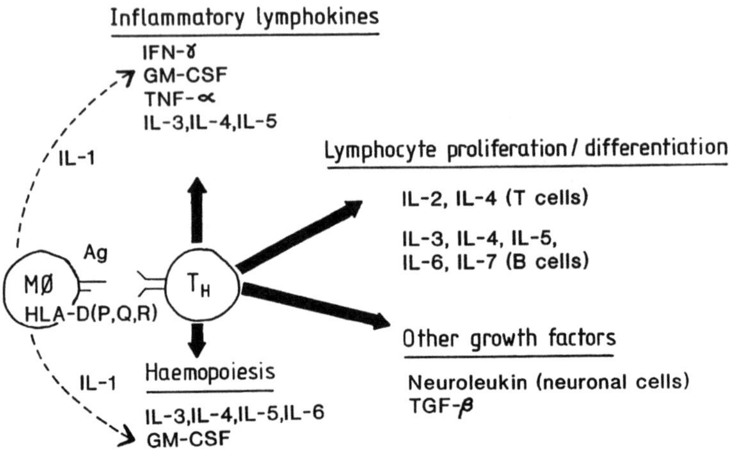

Fig. 1. T cell lymphokines grouped according to function. *IFN*-γ, interferon-gamma; *GM-CSF*, granulocyte/macrophage-colony stimulating factor, *TNF*-α, tumour necrosis factor-alpha; *IL*, interleukin; *MØ*, macrophage; *Ag*, antigen; *HLA*, human leucocyte antigen; T_H, T-helper lymphocyte; *TGF*-β, transforming growth factor-beta. (IL-8 might also be involved but unlike other interleukins it does not appear to be a growth factor (BAGGIOLINI et al. 1989))

response but also to its modulation by feedback mechanisms. Cellular events in the bronchial mucosa are also influenced by neuropeptides secreted by neuroendocrine tissue in the airways. The role of mediators and the neural control of lung function are addressed specifically in separate chapters.

It is very unlikely that one cell, or one mediator, will explain totally the mechanisms of asthma and attendant bronchial hyperreactivity, but rather that the *combined* effects of a number of cells and mediators are required for the observed effects.

B. Pathological Evidence for Inflammation in Asthma

Until recently, pathological studies in asthma were largely confined to tissues taken from patients who had died in status asthmaticus. With the advent of bronchoscopy and bronchial biopsy in asthma a number of reports of pathological findings in more moderate forms of the disease are beginning to appear. In asthma deaths certain features seem to be typical. These include intense infiltration of eosinophils and deposition of eosinophil products in and around the bronchial epithelium (DUNNILL 1960; DUNNILL et al. 1969). There are also large numbers of lymphocytes and macrophages. Neutrophils are also usually present. There is marked hyperaemia and dilatation of blood vessels and considerable mucus hypersecretion with plugging of the small airways. Shedding of epithelial cells is a common finding. These cells appear as clumps in the sputum and are known as Creola bodies.

It has been shown that asthmatics shed significantly greater numbers of epithelial cells into bronchial lavage fluid than non-asthmatics (Beasley et al. 1989). In this study a fivefold increase was demonstrated along with a significant correlation between epithelial cell count and bronchial responsiveness. The cells were adherent to each other and viable, suggesting that they had sustained selective damage to their basement membrane contacts.

Other important features include goblet cell hyperplasia and thickening of the basement membrane. Recently immunohistochemical and electron microscopic studies by Roche et al. (1989) have brought to our attention that this is not true thickening of the basement membrane but subepithelial fibrosis. These findings contradict the long-held notion of basement membrane thickening in asthma and indicate that this fibrosis is a result of fibroblast activation rather than bronchial epithelial cell dysfunction. Conspicuous in the pathological descriptions of asthmatic airways is the hyperplastic increase in smooth muscle (Dunnill 1960, 1978; Dunnill et al. 1969; Heard and Hossain 1973). Despite these structural changes, the occurrence of functional changes in smooth muscle contractility is controversial. Of interest in the study by Beasley et al. (1989) is that extensive epithelial shedding, subepithelial fibrosis and mucosal infiltration with inflammatory cells were present in the airways of particularly mild asthmatics suggesting that severe tissue disruption is present in mild disease and may persist even in the absence of clinical disease.

What is still significantly unclear is the dynamics of this process. For instance, what is the sequence of cell emigration and is it different from the sequence observed in inflammatory processes in other parts of the body? How does this inflammatory process resolve? What is the sequence of the structural changes? Do the airways of asthmatics all look the same or is there heterogeneity of the lesions?

C. Microvascular Leakage

It is virtually certain that microvascular leakage occurs in asthma of any severity not only because of the histopathological findings but because sputum (Ryley and Brogan 1968; Brogan et al. 1975) and bronchoalveolar lavage (BAL) fluid from asthmatics have been shown to contain elevated concentrations of albumin (Lam et al. 1985; Boschetto et al. 1987). The amount of albumin in lavage fluid is also increased after exposure to inhaled antigen (Fick et al. 1987). Many of the mediators implicated in asthma are known to cause microvascular leakage of postcapillary venules. These include histamine, bradykinin, sulphidopeptide leukotrienes (LT) and platelet activating factor (PAF) (Saria et al. 1983, Persson 1987; Hua et al. 1985; Evans et al. 1987). In addition, stimulation of the vagus nerve or injection of capsaicin, via release of sensory neuropeptides such as substance P, causes microvascular leakage in rodents (LUNDBERG et al. 1987). Further, mediators which increase bronchial blood flow might be expected to exaggerate leakage in asthmatic airways. Oedema, as a result of increased capillary permeability, may have several sequelae relevant to asthma. These include a contribution to narrowing of small airways, epithelial shedding, the formation of mucus plugs, inhibition of mucociliary clearance and, due to the rich source of plasma proteins, provision of substrates for complement-derived anaphylatoxins and kinins. To the extent that oedema represents an alteration in permeability, it may also contribute directly to alterations in responsiveness by allowing greater (or different) agonist penetration into the smooth muscle layer.

D. Neutrophils

The evidence that neutrophils by themselves play an important role in asthma and bronchial hyperresponsiveness is controversial. A number of studies in experimental animals as well as control models of asthma in humans indicate that neutrophils may play a part early on in the asthma process. Neutrophils appear to be "normal" residents of larger airways both in normoresponsive and hyperresponsive individuals. For instance, WARDLAW et al. (1988) found a large percentage of neutrophils in the bronchial wash of non-atopic controls and hayfever sufferers as well as mild asthmatics. The numbers were approximately equal in all groups and considerably higher than those observed in bronchoalveolar wash. In fact neutrophils accounted for almost 50% of the total cell count in bronchial washes in normal subjects. Is this a reflection of the local environment that the subjects were exposed to or is it a general finding? In any event it raises the question of whether the airway is poised for an inflammatory response if the right trigger factors and underlying disease susceptibility are present. A further question is whether the neutrophils of asthmatics demonstrate a different degree of activation or stimulation in comparison with those of normals. It is relevant that scrapings from the nasal mucosa of normal individuals also contained large numbers of neutrophils (A.J. WARDLAW, unpublished work) as did normal conjunctiva.

Numerous studies indicate that peripheral blood neutrophils become "activated" after allergen- and exercise-induced early- and late-phase asthmatic responses (PAPAGEORGIOU et al. 1983; MOQBEL et al. 1986; CARROLL et al. 1985; DURHAM et al. 1984). Activation was assessed by increased membrane expression of complement receptors and enhanced cytotoxicity for complement-coated targets. Such observations suggest that the circulating neutrophils have been exposed to some stimulus. However, many of the currently known stimuli for neutrophils are not specific for this cell type so that activation of other inflammatory cells may also result from the same circumstances.

Significant increases in the percentage of neutrophils in BAL and bronchial mucus have been observed in subjects experiencing late-phase reactions after bronchial challenge (DIAZ et al. 1989). Comparable observations were made in BAL during late-phase reactions in sensitised subjects challenged with toluene diisocyanate, in which it was found that increases in neutrophils, eosinophils and lymphocytes were inhibited by prior administration of oral prednisolone (BOSCHETTO et al. 1987). In contrast, BAL neutrophilia was not a feature of late reactions elicited by plicatic acid in red cedar asthma (LAM et al. 1987).

The elaboration of a high molecular weight neutrophil chemotactic activity (HMW-NCA) into the circulation of patients after allergen- or exercise-induced early- and late-phase reactions is well documented (ATKINS et al. 1977; NAGY et al. 1982; T.H. LEE et al. 1982a, 1983a). HMW-NCA was associated with molecules having a molecular weight of approximately 600 kD and a near neutral isoelectric point. This activity has recently been identified from the serum of asthmatics admitted to hospital with acute severe disease (status asthmaticus) (BUCHANAN et al. 1987). The molecular size of NCA in acute severe asthma was heterogeneous, i.e. 800, 600 and <20 kD. When peripheral blood mononuclear cells from patients with acute severe asthma were cultured in serum-free medium NCA could also be detected in the supernatant (NAGY et al. 1989). Present evidence suggests that HMW-NCA may be derived from lymphocytes and/or monocytes and that it is related to the 10 kD neutrophil chemotactic factor now fully characterised and sequenced by a number of groups (YOSHIMURA et al. 1987; VAN DAMME et al. 1988; GREGORY et al. 1988; MAESTRELLI et al. 1988a,b). The high molecular weight of the serum factor might be an artefact of heating to 56°C for 30 min.

The accumulation of neutrophils into the airways under conditions of inflammation has the potential to result in significant tissue damage. This cell type is associated with tissue injury in many inflammatory conditions (HENSON and JOHNSTON 1987). The mechanisms by which it may do this include the release of oxygen metabolites, proteases and cationic materials. Additionally, the neutrophil is a potential source of a wide variety of mediators, including potent lipid mediators such as prostaglandins (PG), thromboxanes, LTB_4 and PAF which may contribute to airway responses and/or exacerbation of the inflammatory response.

Convincing evidence suggesting that neutrophils are capable of altering airway function comes from animal experiments. The cell has been firmly implicated in ozone-induced and antigen-induced hyperreactivity in dogs (L.Y. LEE et al. 1977; CHUNG et al. 1985; MURPHY et al. 1986; HOLTZMAN et al. 1983; O'BYRNE et al. 1984), in antigen-induced late-phase and hyperresponsiveness reactions in rabbits (MARSH et al. 1985; MURPHY et al. 1986) and in sheep exposed to both endotoxin (HUTCHINSON et al. 1983; HINSON et al. 1984) and antigen (ABRAHAM et al. 1985). Rabbits demonstrated increased responsiveness when airway inflammation was induced with phlogistic fragments of C5, and this too was dependent on neutrophils (IRVIN et al. 1986). The requirement for granulocytes (especially neutrophils) in these experiments has been established by association (lavage and histology), depletion studies (e.g. nitrogen mustard) and, more convincingly, by depletion followed by selective repletion with enriched populations of purified neutrophils (MURPHY et al. 1986). Furthermore supernatants from phagocytosing neutrophils in vitro induced hyperreactivity when nebulised into the airways of rabbits (IRVIN et al. 1985), whereas the supernatants from eosinophils in vitro did not. The active agents in this model have yet to be identified. By contrast the guinea pig may be different (MURLAS and ROUM 1985). While there was also a sevenfold increase at 6 h and a 17-fold increase at 17 h in neutrophils in BAL in a guinea pig model of late-phase and "late-late-phase" bronchoconstriction (HUTSON et al. 1988), nedocromil sodium blocked the late reaction and subsequent eosinophil infiltration in BAL but did not affect the neutrophil infiltration. This suggested that in the late-phase and the "late-late-phase" neutrophil infiltration is less critical for the development of airways obstruction in this species (CHURCH et al. 1988).

Whether or not the neutrophil participates in the functional alterations associated with the asthmatic condition remains an open question. The frequent recovery of neutrophils from BAL of asthmatic patients (DIAZ et al. 1984; MARTIN et al. 1988), the presence of neutrophils in pathological sections of asthmatic airways (LAITINEN et al. 1985; SALVATO 1968) and the ability of human neutrophils to alter airway function suggests that they probably do.

E. Eosinophils

There is circumstantial evidence which suggests that eosinophils are important proinflammatory cells in the asthma process. It is well known that a blood and sputum eosinophilia is often, but not invariably, found in association with most forms of asthma. A blood eosinophilia accompanied late-phase but not single early asthmatic responses, and there was an inverse correlation between the blood eosinophil count and the degree of non-specific bronchial hyperreactivity as measured by the methacholine provocation concentration $(PC)_{20}$ (DURHAM and KAY 1985). FRICK et al. (1989) have extended these findings observing that the proportion of hypodense eosinophils in peripheral blood rises in such patients and better reflects the severity of asthma than does the total eosinophil count. The accumulation of eosinophils and eosinophil products (major basic protein, MBP; eosinophil cationic protein, ECP; and eosinophil-derived neurotoxin, EDN) have been observed in BAL during allergen-induced late-phase reactions (DE MONCHY et al. 1985a; DIAZ et al. 1989; METZGER et al. 1987). Similar observations were made in red cedar asthma in which it was observed that plicatic acid inhalation elicited a BAL eosinophilia together with sloughing of bronchial epithelial cells (LAM et al.

1987). In a placebo-controlled double-blind study it was shown that disodium cromoglycate (SCG) suppressed the local accumulation of eosinophils in bronchial mucus and BAL fluid and that these reductions in lung eosinophils were related to clinical improvement (DIAZ et al. 1984).

Eosinophils are also very prominent in many of the histopathological sections obtained from asthma deaths (DUNNILL 1960; DUNNILL et al. 1969). In fact, the pathology is often termed "chronic eosinophilic desquamative bronchitis". MBP was prominent in the bronchial wall and in the mucus plugs of virtually all of these patients, even though only a few intact eosinophils were observed by routine light microscopy (FILLEY et al. 1982). MBP concentrations are also elevated in the sputum from asthmatics.

Studies on atopic asthmatics have provided further information on the eosinophil. BEASLEY et al. (1989) found that in bronchial biopsies the presence of eosinophils in the lamina propria was confined to asthmatics, being absent in controls, and that these cells showed morphological features of activation. AZZAWI et al. (1989) demonstrated a significantly higher number of eosinophils in atopic asthmatics than in atopic and non-atopic controls and a significant increase in eosinophil activation in the asthmatic group.

ECP and MBP are both cytotoxic to the respiratory epithelium (GLEICH et al. 1979) in vitro and in experimental animals, and both may contribute to the denudation of the epithelium seen in asthma. Asthmatics with airway hyperreactivity (PC_{20} <4 mg/ml) had significant elevations in both the eosinophil count and the concentration of MBP in BAL fluid (WARDLAW et al. 1988). Furthermore, there were significant correlations between the amount of MBP recovered and the percentage of eosinophils. These changes were even more marked when asthmatics with airway hyperreactivity were compared with subjects with normoreactive airways. There were inverse correlations between the PC_{20} and the percentage of eosinophils and epithelial cells and the amount of MBP in BAL. This study and that by LAM et al. (1987) clearly suggest that bronchial hyperresponsiveness is associated with increased amounts of eosinophils and their products. It is also consistent with the suggestion that the responsiveness is secondary to epithelial cell damage mediated through eosinophil-derived granule products. However, this suggestion presupposes that the hyperresponsiveness is caused by epithelial damage and that eosinophils have a particular ability to induce this damage. Both of these may be the case but need to be clearly demonstrated. Evidence from FLAVAHAN et al. (1988) adds support to this hypothesis. The increase in reactivity of in vitro preparations of tracheal rings in the presence of MBP was shown to be dependent on the presence of the epithelium. This suggests that the association between non-specific bronchial hyperresponsiveness and MBP in BAL may be due to the effect of MBP on epithelial cells which may then affect the reactivity of the smooth muscle.

In addition to the basic proteins of the eosinophil granule, membrane phospholipid-derived mediators may also play a role in the pathogenesis of asthma. Eosinophils produced considerable quantities of LTC_4 after ionophore- (JORG et al. 1982; WELLER et al. 1983; SHAW et al. 1984), IgG- (SHAW et al. 1985) and IgE-dependent stimuli (MOQBEL et al. 1990) and small elevations in LTC_4 were noted in BAL in "baseline, resting" asthma (WARDLAW et al. 1989) and in late-phase reactions when fluid from diluent challenge was compared with that obtained from allergen challenge (DIAZ et al. 1989).

Eosinophils have the capacity to generate quantities of PAF (LEE et al. 1984; CHAMPION et al. 1988) comparable to those produced by neutrophils. In humans, PAF may be of particular relevance to asthma because of its ability to cause vasodilatation and increased vascular permeability, to induce eosinophil chemotaxis (WARDLAW et al. 1986) and adherence to endothelial cells (KIMANI et al. 1988) and to enhance mucus secretion and increase bronchial hyperresponsiveness after inhalation (CUSS et al. 1986). However, there are no convincing studies to date of PAF elaboration associated with clinical asthma.

The mechanism of recruitment of the eosinophil, in preference to the neutrophil, to the human asthmatic airway still needs to be explained. Is it merely a response to a larger stimulus, a less efficient removal process or are the stimuli (signals) somehow different or more prolonged in comparison with those in other inflammatory reactions? PAF is a potent chemotactic factor for eosinophils but is also effective in vitro at evoking directional neutrophil migration. It seems likely that in vitro chemotaxis is not a true model of cell accumulation in vivo since it does not take into account the requirement for the chemoattractant to cross the vessel wall, the special requirements of cell adhesion to endothelial cells, the process of emigration itself nor the metabolism that any chemotactic signal might be subjected to in the tissue.

It is now appreciated that T cell-derived products, which play a vital role in eosinophil maturation, also affect the mature cell. For instance, both granulocyte/macrophage-colony stimulating factor (GM-CSF) and interleukin (1L)-5 activate mature eosinophils, in terms of increased cytotoxicity and oxidative metabolism, and prolong the life of eosinophils in vitro (LOPEZ et al. 1988; YAMAGUCHI et al. 1988). Thus lymphocyte and macrophage products and/or PAF acting either alone, in combination or in sequence might alter the inflammatory cell in such a way as to cause its adherence to blood vessel walls. Other factors such as IL-1, may alter the endothelium to render it more adherent to granulocytes.

Recruitment and activation of eosinophils is strongly inhibited by corticosteroids, an effect which could explain the efficacy of these drugs in modifying late-phase bronchoconstriction. However, such an explanation must be tempered by the controversial effect of corticosteroids on bronchial hyperresponsiveness (McFADDEN 1988), in contrast to the effects on the late-phase asthmatic response, and the common clinical experience of asthma persisting long after the peripheral blood eosinophilia has been corrected.

F. Mast Cells and Basophils

In humans, mast cells are located in the lumen of the airways (where they can be recovered by BAL), in the bronchial epithelium, in the submucosa, and in the lung parenchyma. Basophils have not been identified in bronchial pathology or in any situation associated with hyperresponsiveness, although they probably exist in the lung, as they do in most other organs.

Current evidence suggests that the early allergic asthmatic reaction is predominantly mast cell-mediated. The immediate response to inhaled allergens in atopic subjects (accompanied by elevations in plasma histamine) is rapid in onset and easily prevented by prior inhalation of β2-adrenoceptor agonists (i.e. salbutamol) or SCG (Howarth et al. 1985). In these situations the drugs are assumed to act primarily on the mast cell, although as far as SCG is concerned in vitro data do not support this view (Holgate et al. 1984). Corticosteroids given over a period of time also attenuate the immediate reaction (Burge et al. 1982), possibly by depletion of mast cells in the mucosa (King et al. 1985; Otsuka et al. 1986). After allergen challenge in atopic individuals there was an elevation in plasma histamine (Lee et al. 1982b), in BAL histamine (Wardlaw et al. 1988; Casale et al. 1987) and both histamine and mast cell tryptase in BAL. This occurred within minutes of challenge. Over the following few hours increased urinary secretion of a major catabolite, *N*-methylhistamine, was identified (De Monchy et al. 1985b). Mast cells recovered from the airways by lavage within the first 15 min of allergen challenge have all the morphological features of non cytotoxic degranulation (Metzger et al. 1986, 1987). Human lung mast cells also elaborate LTC_4, PGD_2, PAF, various chemotactic peptides, proteolytic enzymes and proteoglycans. However, their pathophysiological role is as yet unknown.

Mast cells have been noted in various stages of degranulation in asthmatics but not in controls which supports the suggestion that degranulation and mediator release occur continuously within the bronchial mucosa of atopic asthmatics (Beasley et al. 1989). Activation of mast cells as a pathogenetic mechanism of immediate bronchoconstriction is not limited to allergen exposure since there is some evidence that it is also involved in asthma provoked by exercise, cold air and hyperventilation (Lee et al. 1983b). In ongoing, day-to-day asthma there is an inverse correlation between the methacholine PC_{20} and the percentage of mast cells in BAL (Wardlaw et al. 1988) and the amount of histamine in BAL fluid (Casale et al. 1987). Furthermore, asthmatics with airway hyperreactivity had significant increases in spontaneous histamine release from BAL mast cells.

What is the role of the mast cell in late-phase reactions (LPR) airways reactivity and ongoing asthma? Almost a decade ago it was shown that $F(ab)'_2$ anti-IgE, when injected into the skin, produced a LPR which had many of the histopathological features of allergen-induced LPR. For this reason it was hypothesised that mast cells were essential for the LPR and that mast cell-derived chemotactic factors accounted for the subsequent infiltration of eosinophils, neutrophils and basophils (Fig. 2). More recently, a similar dependency on IgE was shown for the induction of an aeroallergen-induced LPR in the airways of rabbits (inflammation and bronchoconstriction) (BEHRENS et al. 1984). However, these do not prove the involvement of mast cells since the anti-IgE or antigen may also have interacted with macrophages or lymphocytes through IgE on their FcεRII receptors and these cells, in turn, may have contributed to the LPR. The airway LPR was inducible in rabbits passively sensitised with IgE-containing serum, further suggesting a "type I allergic reaction". On the other hand, evidence is now accumulating that LPR (mainly from the skin and to a lesser extent in the lung) are associated with the presence of lymphocytes and may therefore include elements of delayed-type hypersensitivity (type IV reactions) (see Sect. I).

It has recently been reported that Fcε RI-mediated activation of murine mast cells results in the production of GM-CSF and IL-3 (WODNAR-FILIPOWICZ et al. 1989). Messenger ribonucleic acid (mRNA) for several other cytokines, including IL-5 and interferon (IFN)-γ, has also been identified in similarly activated mast cell clones (BURD et al. 1989). These cytokines, previously thought to be secreted only by lymphocytes and mononuclear phagocytes, have the ability to recruit, prime and activate neutrophils, monocytes and eosinophils. Their elaboration by mast cells in response to allergen would enable the local defence system to become self-regulated and could also provide a mechanism by which the mast cell could participate in the

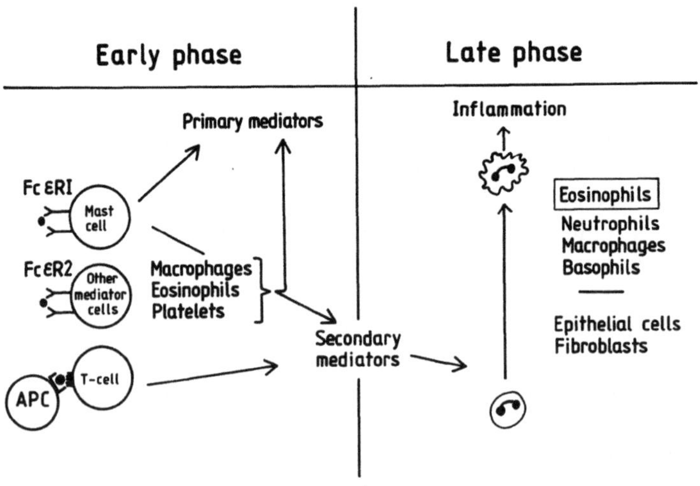

Fig. 2. Hypothetical scheme showing possible relationship between early- and late-phase asthmatic reactions. *FcεRI*, high affinity IgE receptor present on mast cells; *FcεRII*, low affinity IgE receptor present on other inflammatory cells (macrophages, eosinophils and platelets); *APC*, antigen-presenting cell (macrophage or dendritic cell); *filled circle*, antigen

LPR. Nevertheless, the evidence that mast cells are involved in immediate bronchoconstriction to immunological and non-immunological stimuli is still persuasive. The evidence that mast cells play a role in the LPR is debatable and there is little evidence for or against the suggestion that the cell plays a pivotal role in ongoing, day-to-day, chronic asthma. Mast cells may be involved in inflammation in general and are found in association with a variety of pulmonary pathologies. Indeed, the most dramatic increases in mast cell numbers in the lung are found in fibrotic conditions, although as noted above the presence of the cell does not necessarily indicate what role it may play in pathophysiological processes.

G. Monocytes and Macrophages (Mononuclear Phagocytes)

The majority of cells recovered from BAL in normal and asthmatic subjects are macrophages. These cells contain functional IgE receptors (FcεRII) (MELEWICZ et al. 1982; JOSEPH et al. 1983a) and the numbers of these IgE-bearing cells are substantially greater in atopic asthmatics than in normal controls (CAPRON et al. 1986). Macrophages have the capacity to produce a wide range of lipid mediators including eicosanoids from the lipoxygenase and cyclooxygenase pathways and PAF. Mononuclear phagocytes also make and secrete a vast array of proteins and peptides which have biological activities that may contribute to airway reactions. It would be very surprising if this cell type did not participate in airway responses to inhaled allergens. Appreciable amounts of free β-glucuronidase were recovered in BAL (TONNEL et al. 1983), although the macrophage is only one potential source of this marker for lysosomal enzyme secretion, and a substantial increase in the number of macrophages has been reported after allergen challenge (METZGER et al. 1987). This increase appears to be due to the migration of monocytes into the lung since most of the increase can be accounted for by peroxidase positive cells (a characteristic feature of monocytes rather than alveolar macrophages).

Lung macrophages are also activated in the LPR after allergen challenge as shown by an increase in the number of complement rosettes (DIAZ et al. 1989). Chronic severe asthmatics who are relatively unresponsive to corticosteroids have increased numbers of circulating activated monocytes (KAY et al. 1981). At one time this was thought to be a primary monocyte defect but this now appears to be secondary to T lymphocyte activation (GRANT et al. 1984).

These observations provide only circumstantial support for the mononuclear phagocyte as a player in the pathogenesis of asthma. Nevertheless, the cell type is pluripotential and is involved not only in the majority of inflammatory reactions but in the afferent arm of the immune response as well. Histologically the inflammatory lesions in asthma include mononuclear phagocytes and they may be presumed to participate in the process. However, it is also worth recognising the role that the mononuclear phagocyte series of cells plays in the resolution of inflammation (HASLETT and HENSON 1988). Thus it is unclear at present whether these cells contribute actively to the pathophysiology of asthma and/or are involved in removing and modulating the effects of other cells and the mediators they produce. Since macrophages can manifest many different functional phenotypes it is entirely possible that the airways of asthmatics contain macrophages that contribute to the disease as well as cells that serve to limit the extent of both the inflammation and the pathophysiology.

H. Platelets

Initial experiments in the rabbit and in the guinea pig suggested that platelet depletion prevented PAF-induced airway responses (MAZZONI et al. 1985; VARGAFTIG et al. 1980), although the presence of these cells may not always be necessary (STIMLER and O'FLAHERTY 1983; LEFORT et al. 1984). Platelets have been reported in the lavage fluid following allergen challenge (METZGER et al. 1985) and platelet-like structures have been identified histologically in the inflamed airways of guinea pigs (LELLOUCH-TUBIANA et al. 1985). Unfortunately, it is not always possible to identify platelets clearly by morphological means and specific tags (e.g. monoclonal antibodies) have not yet been used. The question of how platelets might get out of the blood vessels and through the tissues has not been addressed. The cells are only poorly chemotactic in vitro and have not yet been shown to have that property in vivo. Alternatively, the platelets might be present as a result of loss of vessel wall integrity.

In another approach to possible platelet involvement, platelet factor 4 was identified in the plasma of atopic subjects (KNAUER et al. 1981) but further studies have not provided confirmation (DURHAM et al. 1985). Platelets do bear the second IgE Fc receptor in a functionally active form, as shown by IgE-dependent oxidative metabolism and cytotoxicity (JOSEPH et al. 1983b). By and large, the evidence that platelets play a pivotal role in human asthma and are directly concerned in airway hyperresponsiveness is absent. The observations that suggest a role raise some important questions relating to the possible mechanisms by which platelets might become involved. Intravascular participation of platelets in inflammatory reactions in general and in airways inflammation in particular seems highly likely, and the cells can release many molecules and mediators that could contribute to the process. If they migrate into the tissues themselves, they would certainly represent a potentially important additional inflammatory cell. However, this process needs to be formally demonstrated. In any event, one intriguing role that could be played by this cell is as a major source of lipid mediators.

I. Lymphocytes

An area of considerable current interest is the role of the T lymphocyte in the regulation and expression of inflammation associated with allergy and asthma. The T cell-derived lymphokines IL-4, IL-5 and IFN-γ are intimately involved in the regulation of IgE production (LEUNG and GEHA 1987). Some lymphokines are active in the control of eosinophil production by the bone marrow (IL-5, GM-CSF, IL-3) and in the regulation of mast cell differentiation. Others have chemotactic activity for neutrophils, eosinophils, basophil granulocytes and monocytes and can activate or degranulate these effector cells. T lymphocytes also play a general role in the regulation of specific immune responses and are a possible target cell for desensitisation immunotherapy.

Direct evidence for T lymphocyte changes in asthma and allergy comes from a variety of sources. Post mortem examination of the airways of asthmatic patients revealed large numbers of lymphocytes (DUNNILL 1960; DUNNILL et al. 1969). Increased numbers of "atypical intraepithelial lymphocytes" were found in an ultrastructural morphological study of bronchial biopsies taken during life from subjects with mild asthma (JEFFERY et al. 1987, 1990). That these lymphocytes are activated is supported by recent analysis of the cellular infiltrate using monoclonal antibodies (Mabs) and immunocytochemical techniques in atopic asthmatics with rhinitic and non-atopic controls (AZZAWI et al. 1989). No significant differences were seen between the groups in total leucocyte or total lymphocyte numbers. There were, however, significantly higher numbers of activated T cells in the asthmatic group. Increased natural killer (NK) activity has been described in the peripheral blood of asthmatic patients (TIMONEN and STENIUS-AARNALA 1985). NK activity is an inducible property of T cells and of non-T, non-B lymphocytes and is thus a nonspecific indicator of lymphocyte activation.

Measurements in chronic asthmatics indicated that patients who were relatively refractory to treatment with corticosteroids had a relative decrease in the number of circulating T suppressor (CD8+) cells (POZNANSKY et al. 1985). These patients also had an abnormality of T cell growth in vitro (colony counts in soft agar) since, unlike T cells from normal individuals and corticosteroid-responsive subjects, cell proliferation from the refractory patients were not inhibited by optimal concentrations of methylprednisolone (POZNANSKY et al. 1984). A defect in concanavalin A-induced suppressor cell function has been described in asthma (HARPER et al. 1980; ROLA-PLESZCZYNSKI and BLANCHARD 1981; RIVLIN et al. 1981; HWANG et al. 1985; ILFELD et al. 1985) and successful immunotherapy was associated with an increase in the relative number of T suppressor cells (OKT8) (ROCKLIN et al. 1980).

T lymphocyte subset changes have also been studied using the model of bronchial allergen challenge in three separate studies (METZGER et al. 1987;

GERBLICH et al. 1984; GONZALEZ et al. 1987). The design of each of these investigations was different but taken together it appears that T lymphocytes bearing the CD4 marker (helper/inducer subset) were depleted in peripheral blood and selectively retained the lung after challenge. However, the kinetics and significance of this finding is as yet unclear. In addition it appears that there is a difference in the profile of regulatory T cell subsets present in BAL in asthmatics who respond with an early reaction alone as compared with dual (early- and late-phase) responders, suggesting that those who go on to develop a LPR to allergen challenge may have a relative inability to recruit CD8+ T lymphocytes to the lung.

A study of cell traffic and activation in human allergen-induced LPR in the skin revealed a lymphocyte infiltration which was almost exclusively CD4+ (helper/inducer subset) (FREW and KAY 1988). Some of these T cells were activated in that they stained positively for the presence of IL-2 receptors (Fig. 3). Further evidence of T lymphocyte activation was provided by the observation that endothelial cells in the allergen-challenged biopsies showed an increased density of HLA-DR expression compared with control sites. This indicated local secretion of the T lymphocyte-derived soluble inflammatory mediator IFN-γ (although as mentioned above it is possible that mast cells may also elaborate this). In the same study it was shown that eosinophil accumulation and activation were striking features of the late-phase skin reaction with numerous activated eosinophils (EG2+) present at 6 h after challenge and persisting in tissue for up to 48 h.

More recently the kinetics and phenotype of T lymphocytes infiltrating the airways of guinea pigs undergoing late-phase asthmatic reactions (LAR)

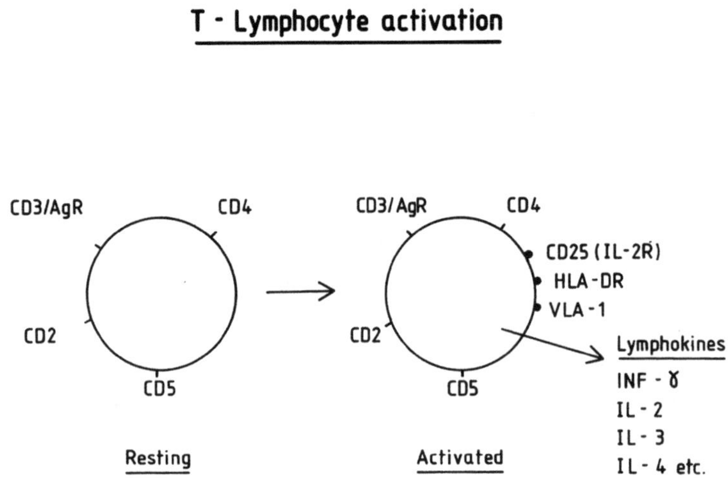

Fig. 3. Cell surface markers expressed on the resting and activated T lymphocyte. *IL-2R*, interleukin-2 receptor which is expressed on IL-2 secreting cells; *VLA-1*, very late activation antigen; *HLA-DR*, class II histocompatibility antigen

were studied with Mabs, cytofluorimetry and immunocyto chemistry (FREW et al. 1990). Challenge of sensitised animals with aerosolised ovalbumin was followed by early-phase (2 h) and late-phase (17 h) bronchoconstriction. The induction of hypersensitivity by aerosolised antigen was associated with an increase in mucosal T cell numbers, which consisted almost entirely of CD8+ T cells. Following allergen challenge of fully sensitised animals, a biphasic rise in total T cell (CD3+) numbers was observed in the bronchial mucosa, peaking at 17 and 48 h. A similar pattern of T cell accumulation was observed in the bronchial adventitia but with an extra, early peak at 2 h. In contrast to the T cell influx of the sensitisation phase, the post-challenge infiltrate consisted largely of CD3+ CD8− cells. Eosinophil numbers were elevated in both submucosa and adventitia, with a single broad peak between 17 and 48 h. Whereas correlations between T cell and eosinophil numbers varied over the 96 h of the experiment, strong associations were observed between CD8+ numbers and eosinophils in the adventitia at 6 h and between CD3+/CD8− numbers and eosinophils in the submucosa at 72 h. No significant changes were detected in T cell or eosinophil numbers in the lung parenchyma. There was a post-challenge increase in eosinophils (but not T cells) in BAL. In contrast, analysis of blood leucocytes showed no changes in T cell total or subset numbers during the progression of the LAR. These observations indicate that accumulation of T cells in the airways is a feature of the LAR in this model and are consistent with human BAL and skin studies of the allergen-induced LPR. This study also illustrates that evaluation of distant compartments (blood and BAL) may not fully reflect rapidly evolving tissue infiltration in the bronchial mucosa.

Further evidence to support the view that T cells are involved in the pathogenesis of chronic asthma (Fig. 4) comes from studies of acute severe asthma (ASA, status asthmaticus). CORRIGAN et al. (1988) were able to demonstrate significant elevation of the expression of three surface proteins associated with T lymphocyte activation (IL-2 receptor, IL-2R; class II histocompatibility antigen, HLA-DR; and very late activation antigen VLA-1) (Fig. 3) in ASA compared with control subjects (mild asthma, chronic obstructive airways disease and normal). Phenotypic analysis of the IL-2R positive T lymphocytes showed that these cells were exclusively of the CD4+ "helper-inducer" phenotype. In contrast, CD8+ cells were devoid of IL-2R and VLA-1 in both asthmatics and controls and the expression of HLA-DR on these cells was not increased above that of controls.

The percentages of IL-2R and HLA-DR positive (but not VLA-1 positive) lymphocytes tended to decrease as the patients were treated and improved clinically. In addition the serum concentrations of IFN-γ and soluble IL-2R were significantly elevated in patients with ASA compared with all the control groups (CORRIGAN and KAY 1990). Concentrations decreased as the patients improved clinically during the first 3 day period of hospital treatment. At 3 days a significant correlation was observed between the improvement in airways obstruction, as measured by the change in peak

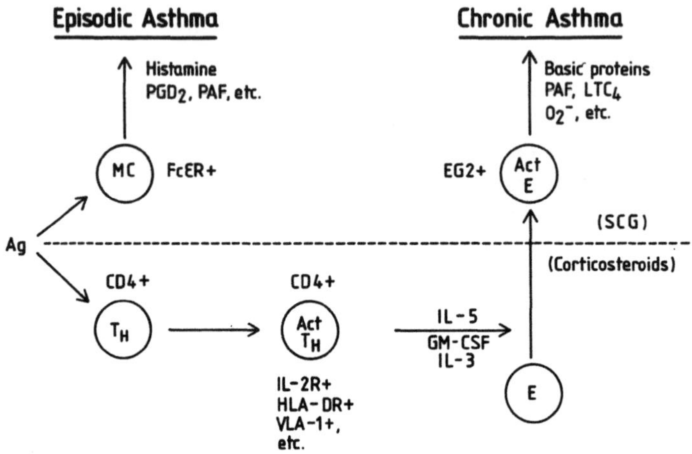

Fig. 4. Possible role of the T lymphocyte in chronic "ongoing" asthma. Episodic and chronic asthma are represented by the models of allergen-induced early and late asthmatic reactions. It should be noted that there is a possibility that the T lymphocyte could also be activated via mast cell cytokines, at least in mice. *MC*, mast cell; *Act* T_H, activated T helper lymphocyte; *E*, eosinophil; *SCG*, disodium cromoglycate; *Act E*, activated eosinophil; *EG2+*, cell surface marker of eosinophil activation (*EG2* is the Mab specific for the cleaved form of eosinophil cationic protein); O_2^-, oxygen radicals

expiratory flow rate, and both the decrease in the percentage of peripheral blood CD4+ T lymphocytes expressing IL-2R and the decrease in the serum concentration of soluble IL-2R.

Atopic ASA patients had significantly smaller numbers of peripheral blood CD4+ T cells on admission than non-atopic patients and also tended to have lower or normal serum concentrations of soluble T cell mediators, suggesting that additional factors may have been contributing to the acute bronchoconstriction in these patients. In those patients studied for 7 days a peak expiratory flow rate of less than 50% of the predicted value was associated with significantly higher numbers of activated CD4+ T cells and higher serum concentrations of IL-2R and IFN-γ.

Thus, taken together these observations provide further evidence that CD4+ T lymphocyte activation may contribute to the pathogenesis of acute severe asthma.

J. Concluding Comments

An attempt has been made to review the relative importance of various inflammatory cells in the pathogenesis of asthma. The possible participation of neutrophils, eosinophils, mast cells, basophils, mononuclear phagocytes, platelets and lymphocytes has been considered, particularly as products from these cell types have the potential for contributing directly to the features of bronchial asthma. Furthermore, the epithelium is a dynamic tissue which may play a role in defence of the airways. In asthma and in other inflammatory diseases of the airways, there are conspicuous alterations in the epithelium, many of which may be presumed to result from the inflammatory process itself. On the other hand, the epithelium is by no means passive and almost certainly contributes to and modulates inflammation.

It is not possible at the present time to judge the relative importance of each individual cell type. It seems far more likely that they act together in a complex cascade (Fig. 5). However, in the absence of knowledge of the dynamics of the bronchial pathology associated with the asthmatic process, anything more than speculation is unwarranted. Furthermore, the heterogeneity of asthma and associated pathology is still poorly understood and so the likelihood exists that different cells and their mediators may have importance in the various forms of asthma and at different stages in their natural history.

The aetiology of asthma remains unknown. One possibility is that, as with allergic asthma in which the inflammation is an inappropriate response to antigen, all asthma is antigen driven. The resident inflammatory cells of the bronchial mucosa, following inappropriate antigen recognition, might become misdirected and forsake their probable defensive role to cause tissue damage. The nature of the putative activating antigen in non-atopic asthma is unclear. Possibilities include infectious agents, altered (auto-

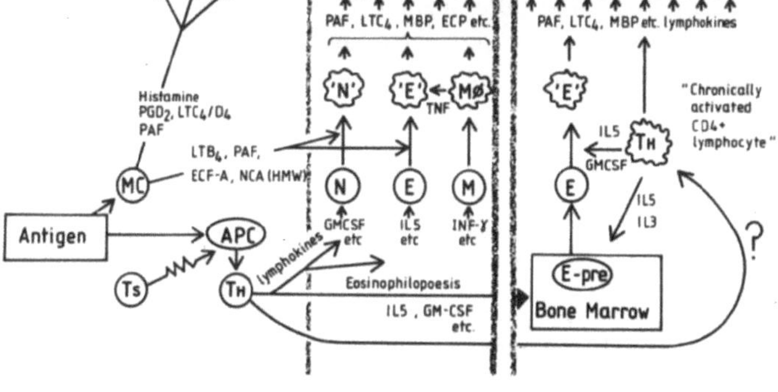

Fig. 5. Summary scheme showing how the various inflammatory cells may interact with the various cytokines and mediators to produce the pathological and physiological features of clinical and experimental asthma. *ECF-A*, eosinophil chemotactic factor of anaphylaxis; *Ts*, T suppressor lymphocyte; *E-pre*, eosinophil precursor

immune) antigens or a predominantly T cell response to an inhaled aeroallergen independent of the presence of an IgE response. An alternative hypothesis is that epithelial cell loss is the primary defect, exposing the submucosal cells to increased antigen penetration and other inflammatory stimuli.

There are a number of particular issues which need to be urgently addressed. These include:

1. Elucidation of the factors responsible for the persistence of inflammation in asthma. This might explain why, in the case of some occupational antigens, inflammation (and asthma) continue long after exposure has ceased.
2. The relationship of mucosal inflammation to both bronchial hyperresponsiveness and clinical asthma. This would explain how severe pathological changes may persist even in the absence of clinical disease.
3. The relationship between the immediate and the late-phase response in provoked asthma (and, in turn, the significance of these to the hyperresponsiveness of persistent asthma).
4. The relevance of clinical models of asthma and hyperresponsiveness to natural disease. The role of bronchial hyperresponsiveness in asthma has been the subject of much debate. Recent observations by JOSEPHS et al. (1989), that individual measurements were not consistently related to concurrent asthma severity, suggest that non specific bronchial reactivity is only one mechanism underlying airflow obstruction in asthma.

Therapeutic possibilities are based on clinical and pathological knowledge and include pharmacological manipulation of mediators and cytokines and the use of Mabs against cell surface molecules on various cell types, subsets or clones. Advances in technology and the ability to probe the airways by methods such as fibreoptic bronchosocopy and mucosal biopsy together with prospects for analysing complex biological fluids in the microenvironment of the bronchial mucosa may pave the way to greater understanding of the pathogenesis of asthma.

References

Abraham WM, Perruchoud AP, Sielczak MW, Yerger LD, Stevenson JS (1985) Airway inflammation during antigen-induced late bronchial obstruction. Prof Respir Res 19:48–88

Atkins PC, Norman M, Weiner H, Zweiman B (1977) Release of neutrophil chemotactic activity during immediate hypersensitivity reactions in humans. Ann Intern Med 86:415–418

Azzawi M, Bradley B, Jeffery PK, Frew AJ, Assoufi B, Collins JV, Durham SR, Kay AB (1989) Activated T lymphocytes and eosinophils in bronchial biopsy specimens in asthma: association with airway hyperresponsiveness. Thorax 44:882P–883P (abstr)

Baggiolini M, Walz A, Kunkel SL (1989) Neutrophil-activating peptide-1/interleukin 8, a novel cytokine that activates neutrophils. J Clin Invest 84:1045–1049
Beasley R, Roche WR, Roberts JA, Holgate ST (1989) Cellular events in the bronchi in mild asthma and after bronchial provocation. Am Rev Respir Dis 139:806–817
Behrens BL, Clark RAF, Marsh W, Larsen GL (1984) Modulation of the late asthmatic response by antigen-specific immunoglublin G in an animal model. Am Rev Respir Dis 130:1134–1139
Boschetto P, Fabbri LM, Zocca E, Milani G, Pivirotto F, Dal Vecchio A, Plevani M, Mapp CE (1987) Prednisone inhibits late asthmatic reactions and airway inflammation induced by toluene diisocyanate in sensitized subjects. J Allergy Clin Immunol 80:261–267
Brogan TD, Ryley HC, Neale L, Yassa L (1975) Soluble proteins of bronchopulmonary secretions from patients with cystic fibrosis, asthma and bronchitis. Thorax 30:72–79
Buchanan DR, Cromwell O, Kay AB (1987) Neutrophil chemotactic activity in acute severe asthma ("status asthmaticus"). Am Rev Respir Dis 136:1397–1402
Burd PR, Rogers HW, Gordon JR, Martin CA, Jayayaman S, Wilson SD, Dvorak AM, Galli SJ, Dorf ME (1989) Interleukin 3-dependent and -independent mast cells stimulated with IgE and antigen express multiple cytokines. J Exp Med 170:245–257
Burge PS, Efthimiou J, Turner-Warwick M, Nelmes PJ (1982) Double-blind trials of inhaled beclomethasone dipropionate and fluocortin butyl ester in allergen-induced immediate and late asthmatic reactions. Clin Allergy 12:523–531
Capron M, Jouault T, Prin C, Joseph M, Ameisen JC, Butterworth AE, Papin JP, Kusneirz JP, Capron A (1986) Functional study of a monoclonal antibody to IgE-Fc receptor of eosinophils, platelets and macrophages ($Fc\varepsilon R_2$). J Exp Med 164:72–89
Carroll M, Durham SR, Walsh GM, Kay AB (1985) Activation of neutrophils and monocytes after allergen- and histamine-induced bronchoconstriction. J Allergy Clin Immunol 75:290–296
Casale TB, Wood D, Richerson HB, Zehr B, Zavala D, Hunninghake GW (1987) Direct evidence of a role for mast cells in the pathogenesis of antigen-induced bronchoconstriction. J Clin Invest 80:1507–1511
Champion A, Wardlaw AJ, Moqbel R, Cromwell O, Shepherd D, Kay AB (1988) IgG-dependent generation of platelet-activating factor by normal and 'low density' human eosinophils. J Allergy Clin Immunol 81:207 (abstr no 157)
Chung KF, Becker AB, Lazarus SC, Frick OL, Nadel JA, Gold WM (1985) Antigen-induced airway hyperresponsiveness and pulmonary inflammation in allergic dogs. J Appl Physiol 558:1347–1353
Church MK, Hutson PA, Holgate ST (1988) Comparison of nedocromil sodium and albuterol against late phase bronchoconstriction and cellular infiltration in guinea pigs. Am Rev Respir Dis 137:136 (abstr)
Corrigan CJ, Kay AB (1990) CD4 T-lymphocyte activation in acute severe asthma: relationship to disease severity and atopic status. Am Rev Respir Dis 141:970–977
Corrigan CJ, Hartnell A, Kay AB (1988) T lymphocyte activation in acute severe asthma. Lancet i:1129–1132
Cuss FM, Dixon CM, Barnes PJ (1986) Effects of platelet activating factor on pulmonary function and bronchial responsiveness in man. Lancet ii:189–192
De Monchy JG, Kauffman HF, Venge P, Koeter GH, Jansen HM, Sleuter HJ, DeVries K (1985a) Bronchoalveolar eosinophilia during allergen-induced late asthmatic reactions. Am Rev Respir Dis 131:373–376
De Monchy JG, Keyzer JJ, Kauffman HF, Beaumont F, DeVries K (1985b) Histamine in late asthmatic reactions following house dust mite inhalation. Agents Actions 16:252–255

Diaz P, Galleguillos FR, Gonzalez MC, Pantin C, Kay AB (1984) Bronchoalveolar lavage in asthma: the effect of DSCG on leucocyte counts, immunoglobulins and complement. J Allergy Clin Immunol 74:41–48
Diaz P, Gonzalez MC, Galleguillos FR, Ancic P, Cromwell O, Shepherd D, Durham SR, Gleich GJ, Kay AB (1989) Leucocytes and mediators in bronchoalveolar lavage during allergen-induced late-phase asthmatic reactions. Am Rev Respir Dis 139:1383–1389
Dunnill MS (1960) The pathology of asthma with special reference to changes in the bronchial mucosa. J Clin Pathol 13:27–33
Dunnill MS (1978) The pathology of asthma. In: Middleton E Jr, Reed CE, Ellis EF (eds) Allergy, principles and practices. Mosby, St Louis, 678–686
Dunnill MS, Massarella GR, Anderson JA (1969) A comparison of the quantitative anatomy of the bronchi in normal subjects, in status asthmaticus, in chronic bronchitis and in emphysema. Thorax 24:176–179
Durham SR, Kay AB (1985) Eosinophils, bronchial hyperreactivity and late-phase asthmatic reactions. Clin Allergy 15:411–418
Durham SR, Carroll M, Walsh GM, Kay AB (1984) Leucocyte activation in allergen-induced late-phase asthmatic reactions. N Engl J Med 311:1398–1402
Durham SR, Dawes J, Kay AB (1985) Platelets in asthma. Lancet ii:36
Evans TW, Chung KF, Rogers DF, Barnes PJ (1987) Effect of platelet-activating factor on airway vascular permeability: possible mechanisms. J Appl Physiol 63:479–484
Fick RB Jr, Metzger WJ, Richerson HB, Zavala DC, Moseley PI, Schoderbek WE, Hunninghake GW (1987) Increased broncho-vascular permeability after allergen exposure in sensitive asthmatics. J Appl Physiol 63:1147–1155
Filley WV, Holley KE, Kephart GM, Gleich GJ (1982) Identification by immunofluorescence of eosinophil granule major basic protein in lung tissues of patients with bronchial asthma. Lancet ii:11–16
Flavahan NA, Slifman NR, Gleich GJ, Vanhoutte PM (1988) Human eosinophil major basic protein causes hyperreactivity of respiratory smooth muscle. Am Rev Respir Dis 138:685–688
Frew AJ, Kay AB (1988) The relationship between infiltrating CD4+ lymphocytes, activated eosinophils and the magnitude of the allergen-induced late phase cutaneous reaction. J Immunol 141:4158–4164
Frew AJ, Moqbel R, Varley J, Azzawi M, Hartnell A, Barkans J, Scheper RJ, Church MK, Holgate ST, Kay AB (1990) T lymphocytes and eosinophils in allergen-induced late phase asthmatic reactions in the guinea pig. Am Rev Respir Dis 141:407–413
Frick WE, Sedgwick JB, Busse WW (1989) The appearance of hypodense eosinophils in antigen-dependent late phase asthma. Am Rev Respir Dis 139:1401–1406
Gerblich AA, Campbell AE, Schuyler MR (1984) Changes in T lymphocyte subpopulations after antigenic bronchial provocation in asthmatics. N Engl J Med 310:1349–1352
Gleich GJ, Frigas E, Loegering DA, Wassom DL, Steinmuller D (1979) Cytotoxic properties of the eosinophil major basic protein. J Immunol 123:2925–2927
Gonzalez MC, Diaz P, Galleguillos FR, Ancic P, Cromwell O, Kay AB (1987) Allergen-induced recruitment of bronchoalveolar (OKT4) and suppressor (OKT8) cells in asthma. Relative increases in OKT8 cells in single early responders compared with those in late-phase responders. Am Rev Respir Dis 136:600–604
Grant IWB, Wyllie AH, Poznansky MC, Gordon ACH, Douglas JG (1984) Corticosteroid resistance in chronic asthma. In: Kay AB, Austen KF, Lichtenstein LM (eds) Asthma. Physiology, immunopharmacology, and treatment. Academic, Press London, pp 359–374
Gregory H, Young J, Schröder J-M, Mrowietz U, Christophers E (1988) Structure determination of a human lymphocyte derived neutrophil activating peptide (LYNAP). Biochem Biophys Res Commun 151:883–890

Harper TB, Gaumer HR, Waring W, Brannon RB, Salvaggio JE (1980) A comparison of cell-mediated immunity and suppressor T cell function in asthmatic and normal children. Clin Allergy 10:555–563

Haslett C, Henson PM (1988) Resolution of inflammation. In: Clark RAF, Henson PM (eds) The molecular and cellular biology of wound repair, Chap 7. Plenum, New York, pp 185–211

Heard BE, Hossain S (1973) Hyperplasia of bronchial muscle in asthma. J Pathol 110:319

Henson PM, Johnston RB Jr (1987) Tissue injury in inflammation: oxidants, proteinases and cationic proteins. J Clin Invest 79:669–674

Hinson JM Jr, Hutchinson AA, Brigham KL, Meyrick BO, Snapper JR (1984) Effect of granulocyte depletion on pulmonary responsiveness to aerosol histamine. J Appl Physiol 56:411–417

Holgate ST, Church MK, Cushley MJ, Robinson C, Mann JS, Howarth PH (1984) Pharmacological modulation of airway calibre and mediator release in human models of bronchial asthma. In: Kay AB, Austen KF, Lichtenstein LM (eds) Asthma. Physiology, immunopharmacology and treatment. Academic, London, pp 391–415

Holtzman MJ, Fabbri LM, O'Byrne PH, Gold BD, Aizawa J, Walters EH, Alpert SE, Nadel JA (1983) Importance of airway inflammation for hyperresponsiveness induced by ozone in dogs. Am Rev Respir Dis 127:686–690

Howarth PH, Durham SR, Lee TH, Kay AB, Church MK, Holgate ST (1985) Influence of albuterol, cromolyn sodium and ipratropium bromide on the airway and circulating mediator responses to antigen bronchial provocation in asthma. Am Rev Respir Dis 132:986–992

Hua X-Y, Dahlen S-E, Lundberg JM, Hammarstrom S, Hedqvist P (1985) Leukotrienes C_4 and E_4 cause widespread and extensive plasma extravasation in the guinea pig. Naunyn-Schmiedeberg's Arch Pharmacol 330:136–141

Hutchinson AA, Hinson JM Jr, Brigham KL, Snapper JR (1983) Effect of endotoxin on airway responsiveness to aerosol histamine in sheep. J Appl Physiol 54:1463–1468

Hutson PA, Church MK, Clay TP, Miller P, Holgate ST (1988) Early and late phase bronchoconstriction after allergen challenge of nonanesthetised guinea pigs. 1. The association of disordered airway physiology to leucocyte infiltration. Am Rev Respir Dis 137:548–557

Hwang KC, Fikrig SM, Friedman HM, Gupta S (1985) Deficient concanavalin A-induced suppressor-cell activity in patients with bronchial asthma, allergic rhinitis, and atopic dermatitis. Clin Allergy 15:67–72

Ilfeld D, Kivity S, Feierman E, Topilsky M, Kuperman O (1985) Effects of in vitro colchicine and oral theophylline on suppressor cell function of asthmatic patients. Clin Exp Immunol 61:360–367

Irvin CG, Baltopoulos G, Henson P (1985) Airways hyperreactivity produced by products from phagocytosing neutrophils. Am Rev Respir Dis 131:A278 (abstr)

Irvin CG, Berend N, Henson PM (1986) Airways hyperreactivity and inflammation produced by aerosolization of human C5a des arg. Am Rev Respir Dis 134: 777–783

Jeffery PK, Nelson FC, Wardlaw AJ, Kay AB (1987) Quantitative analysis of bronchial biopsies in asthma. Am Rev Respir Dis 135:A316

Jeffery PK, Wardlaw AJ, Nelson FC, Collins JV, Kay AB (1990) Bronchial biopsies in asthma: an ultrastructural quantitative study and correlation with hyperreactivity. Am Rev Respir Dis 140:1745–1753

Jorg A, Henderson WR, Murphy RC, Klebanoff JJ (1982) Leukotriene generation by eosinophils. J Exp Med 155:390–402

Joseph M, Tonnel AB, Torpier G, Capron A, Arnoux B, Benveniste J (1983a) Involvement of IgE in the secretory processes of alveolar macrophages from asthmatic patients. J Clin Invest 71:221–230

Joseph M, Auriault C, Capron A, Vorng H, Viens P (1983b) A new function for platelets: IgE-dependent killing of schistosomes. Nature 303:310–312
Josephs LK, Gregg I, Mullee MA, Holgate ST (1989) Nonspecific bronchial reactivity and its relationship to the clinical expression of asthma. Am Rev Respir Dis 140:350–357
Kay AB, Diaz P, Carmichael J, Grant IWB (1981) Corticosteroid-resistant chronic asthma and monocyte complement receptors. Clin Exp Immunol 44:576–580
Kimani G, Tonnesen MG, Henson PM (1988) Stimulation of eosinophil adherence to human vascular endothelial cells in vitro by platelet activating factor. J Immunol 140:3161–3166
King SJ, Miller HRP, Newlands GFJ, Woodbury RG (1985) Depletion of mucosal mast cell protease by corticosteroids: effect on intestinal anaphylaxis in the rat. Proc Natl Acad Sci USA 82:1214–1218
Knauer KA, Lichtenstein LM, Adkinson NF Jr, Fish JE (1981) Platelet activation during antigen-induced airway reactions in asthmatic subjects. N Engl J Med 304:1404–1407
Laitinen LA, Heino M, Laitinen A, Kava T, Haahtela T (1985) Damage to the airway epithelium and bronchial reactivity in patients with asthma. Am Rev Respir Dis (1985) 131:599–606
Lam S, LeRiche JC, Kijek K, Phillips RT (1985) Effect of bronchial lavage volume on cellular and protein recovery. Chest 88:856–859
Lam S, LeRiche J, Phillips D, Chan-Yeung M (1987) Cellular and protein changes in bronchial lavage fluid after late asthmatic reaction in patients with red cedar asthma. J Allergy Clin Immunol 80:44–50
Lee L-Y, Bleecker ER, Nadel JA (1977) Effect of ozone on bronchomotor response to inhaled histamine aerosol in dogs. J Appl Physiol 43:626–631
Lee TC, Lenihan DJ, Malone B, Ruddy LL, Wasserman SI (1984) Increased biosynthesis of platelet-activating factor in activated human eosinophils. J Biol Chem 259:5520–5530
Lee TH, Nagy L, Nagakura T, Walport MJ, Kay AB (1982a) The identification and partial characterisation of an exercise-induced neutrophil chemotactic factor in bronchial asthma. J Clin Invest 69:889–899
Lee TH, Brown MJ, Nagy L, Causon R, Walport MJ, Kay AB (1982b) Exercise-induced release of histamine and neutrophil chemotactic factors in atopic asthmatics. J Allergy Clin Immunol 70:73–81
Lee TH, Nagakura T, Papageorgiou N, Iikura Y, Kay AB (1983a) Exercise-induced late asthmatic reactions with neutrophil chemotactic activity. N Engl J Med 308:1502–1505
Lee TH, Assoufi BK, Kay AB (1983b) The link between exercise, respiratory heat exchange, and the mast cell in bronchial asthma. Lancet i:520–522
Lefort J, Rotilio D, Vargaftig BB (1984) The platelet-independent release of thromboxane A2 by PAF-acether from guinea-pig lungs involves mechanisms distinct from those for leukotriene. Br J Pharmacol 82:565–575
Lellouch-Tubiana A, Lefort J, Pirotzky E, Vargaftig BB, Pfister A (1985) Ultrastructural evidence for extravascular platelet recruitment in the lung upon intravenous injection of platelet-activating factor (PAF-acether) to guinea pigs. Br J Exp Pathol 66:345–355
Leung DYM, Geha RS (1987) Regulation of the human IgE antibody response. Intern Rev Immunol 2:75–91
Lopez AF, Sanderson CJ, Gamble JR, Campbell HD, Young IG, Vadas MA (1988) Recombinant human interleukin-5 is a selective activator of human eosinophil function. J Exp Med 167:219–224
Lundberg JM, Saria A, Lundblad L, Angaard A, Martling C-R, Theodorsson-Norheim E, Stjarne P, Hokfelt T (1987) Bioactive peptides in capsaicin-sensitive C-fiber afferents of the airways: functional and pathophysiological implications. In: Kaliner MA, Barnes PJ (eds) Neural control in health and disease. Dekker, New York, p 417

Maestrelli P, O'Hehir RE, Lamb JR, Tsai J-J, Cromwell O, Kay AB (1988a) Antigen-induced neutrophil chemotactic factor from cloned human T lymphocytes. Immunology 65:605–609

Maestrelli P, Tsai J-J, Cromwell O, Kay AB (1988b) The identification and partial characterization of a human mononuclear cell-derived neutrophil chemotactic factor apparently distinct from IL-1, IL-2, GM-CSF, TNF and IFN-γ. Immunology 64:219–225

Marsh WR, Irvin CG, Murphy KR, Behrens BL, Larsen GL (1985) Increases in airway reactivity to histamine and inflammatory cells in bronchoalveolar lavage after the late asthmatic response in an animal model. Am Rev Respir Dis 131: 875–879

Martin RJ, Cicutto LC, Ballard RD, Szefler SJ (1988) Airway inflammation in nocturnal asthma. Am Rev Respir Dis 137:284

Mazzoni L, Morley J, Page CP, Sanjar S (1985) Induction of airway hyperreactivity by platelet activating factor in the guinea-pig. J Physiol (Lond) 365:107P

McFadden ER Jr (1988) Corticosteroids and cromolyn sodium as modulators of airway inflammation. Chest 94:181–184

Melewicz FM, Kline NE, Cohen AB, Spiegelberg HL (1982) Characterization of Fc receptor for IgE on human alveolar macrophages. Clin Exp Immunol 49:364–370

Metzger WJ, Hunninghake GW, Richerson HB (1985) Late asthmatic reactions: inquiry into mechanisms and significance. Clin Rev Allergy 3:145–165

Metzger WJ, Richerson HB, Warden K, Monick M, Hunninghake GW (1986) Bronchoalveolar lavage of allergic asthmatic patients following allergen provocation. Chest 89:477–483

Metzger WJ, Zavala D, Richerson HB, Moseley P, Iwamota P, Monick M, Sjoerdsma K, Hunninghake GW (1987) Local allergen challenge and bronchoalveolar lavage of allergic asthmatic lungs. Description of the model and local airway inflammation. Am Rev Respir Dis 135:433–440

Moqbel R, Durham SR, Shaw RJ, Walsh GM, MacDonald AJ, Mackay JA, Carroll MP, Kay AB (1986) Enhancement of leukocyte cytotoxicity after exercise-induced asthma. Am Rev Respir Dis 133:609–613

Moqbel R, MacDonald AJ, Kay AB (1990) Release of leukotriene C_4 (LTC_4) from human eosinophils following adherence to IgE- and IgG-coated schistosomula of *Schistosoma mansoni*. Immunology 69:435–442

Murlas C, Roum JH (1985) Bronchial hyperreactivity occurs in steroid-treated guinea pigs depleted of leukocytes by cyclophosphamide. J Appl Physiol 58:1630–1637

Murphy KR, Wilson MC, Irvin CG, Glezen LS, Marsh WR, Haslett C, Henson PM, Larsen GL (1986) The requirement for polymorphonuclear leukocytes on the late asthmatic response and heightened airways reactivity in an animal model. Am Rev Respir Dis 134:62–68

Nagy L, Lee TH, Kay AB (1982) Neutrophil chemotactic activity in antigen-induced late asthmatic reactions. N Engl J Med 306:497–501

Nagy L, Corrigan CJ, Tsai J-J, Kay AB (1989) Neutrophil chemotactic activity (NCA) detectable in the serum of patients with acute severe asthma is elaborated spontaneously by cultured peripheral blood mononuclear cells. Thorax 44:319P (abstr)

O'Byrne PH, Walters EH, Gold BD, Aizawa HA, Fabbri LM, Alpert SE, Nadel JA, Holtzman MJ (1984) Neutrophil depletion inhibits airway hyperresponsiveness induced by ozone exposure. Am Rev Respir Dis 130:214–219

Otsuka H, Denburg JA, Befus AD, Hitch D, Lapp P, Rajan RS, Bienenstock J, Dolovich J (1986) Effect of beclomethasone dipropionate on nasal metachromatic cell subpopulations. Clin Allergy 16:589–595

Papageorgiou N, Carroll M, Durham SR, Lee TH, Walsh GM, Kay AB (1983) Complement receptor enhancement as evidence of neutrophil activation after exercise-induced asthma. Lancet ii:1220–1223

Persson CGA (1987) Leakage of macromolecules from the tracheo-bronchial circulation. Am Rev Respir Dis 135:S71

Poznansky MC, Gordon ACH, Douglas JG, Krajewski AS, Wyllie AH, Grant IWB (1984) Resistance to methylprednisolone in cultures of blood mononuclear cells from glucocorticoid-resistant asthmatic patients. Clin Sci 67:639–645

Poznansky MC, Gordon ACH, Grant IWB, Wyllie AH (1985) A cellular abnormality in glucocorticoid resistant asthma. Clin Exp Immunol 61:135–142

Rivlin J, Kuperman O, Freier S, Godfrey S (1981) Suppressor T-lymphocyte activity in wheezy children with and without treatment by hyposensitization. Clin Allergy 11:353–356

Roche WR, Beasley R, Williams JH, Holgate ST (1989) Subepithelial fibrosis in the bronchi of asthmatics. Lancet i:520–523

Rocklin RE, Sheffer AL, Greineder DR, Melmon KL (1980) Generation of antigen-specific suppressor cells during allergy desensitization. N Engl J Med 302:1213–1219

Rola-Pleszczynski M, Blanchard R (1981) Suppressor cell function in respiratory allergy. Int Arch Allergy Appl Immunol 64:361–370

Ryley HC, Brogan TD (1968) Variation in the composition of sputum in chronic chest diseases. Br J Exp Pathol 49:25

Salvato G (1968) Some histological changes in chronic bronchitis and asthma. Thorax 23:168–172

Saria A, Lundberg JM, Skofitsch G, Lembeck F (1983) Vascular protein leakage in various tissues induced by substance P, capsaicin, bradykinin, serotonin, histamine and by antigen challenge. Naunyn-Schmiedeberg's Arch Pharmacol 324:212–218

Shaw RJ, Cromwell O, Kay AB (1984) Preferential generation of leukotriene C_4 by human eosinophils. Clin Exp Immunol 56:716–722

Shaw RJ, Walsh GM, Cromwell O, Moqbel R, Spry CJF, Kay AB (1985) Activated human eosinophils generate SRS-A leukotrienes following physiological (IgG-dependent) stimulation. Nature 316:150–152

Stimler NP, O'Flaherty JT (1983) Spasmogenic properties of platelet-activating factor: evidence for a direct mechanism in the contractile response of pulmonary tissue. Am J Pathol 113:75–84

Timonen T, Stenius-Aarnala B (1985) Natural killer cell activity in asthma. Clin Exp Immunol 59:85–90

Tonnel AB, Gosset P, Joseph M, Fournier E, Capron A (1983) Stimulation of alveolar macrophages in asthmatic patients after local provocation test. Lancet i:1406–1408

Van Damme J, Van Beeumen J, Opdenakker G, Billiau A (1988) A novel, NH_2-terminal sequence-characterized human monokine possessing neutrophil chemotactic, skin-reactive, and granulocytosis-promoting activity. J Exp Med 167:1364–1376

Vargaftig BB, Lefort J, Chignard M, Benveniste J (1980) Platelet activating factor induces a platelet-dependent bronchoconstriction unrelated to the formation of prostaglandin derivatives. Eur J Pharmacol 65:185–192

Wardlaw AJ, Moqbel R, Cromwell O, Kay AB (1986) Platelet activating factor: a potent chemotactic and chemokinetic factor for human eosinophils. J Clin Invest 78:1701–1706

Wardlaw AJ, Dunnette S, Gleich GJ, Collins JV, Kay AB (1988) Eosinophils and mast cells in bronchoalveolar lavage in mild asthma: relationship to bronchial hyperreactivity. Am Rev Respir Dis 137:62–69

Wardlaw AJ, Hay H, Cromwell O, Collins JV, Kay AB (1989) Leukotrienes, LTC_4 and LTB_4, in bronchoalveolar lavage fluid in bronchial asthma and other respiratory diseases. J Allergy Clin Immunol 84:19–26

Weller PF, Lee CW, Foster DW, Corey EJ, Austen KF, Lewis RA (1983) Generation and metabolism of 5-lipoxygenase pathway leukotrienes by human eosinophils: predominant production of leukotriene C_4. Proc Natl Acad Sci USA 80:7626–7630

Wodnar-Filipowicz A, Heusser CH, Moroni C (1989) Production of the haemopoietic growth factors GM-CSF and interleukin-3 by mast cells in response to IgE receptor-mediated activation. Nature 339:150–152

Yamaguchi Y, Hayashi Y, Sugama Y, Miura Y, Kasahara T, Kitamura S, Torisu M, Mita S, Tominaga A, Takatsu K, Suda T (1988) Highly purified murine interleukin-5 (IL-5) stimulates eosinophil function and prolongs in vitro survival. IL-5 is an eosinophil chemotactic factor. J Exp Med 167:1737–1742

Yoshimura T, Matsushima K, Tanaka S, Robinson EA, Appella E, Oppenheim JJ, Leonard EJ (1987) Purification of a human monocyte-derived neutrophil chemotactic factor that has peptide sequence similarity to other host defense cytokines. Proc Natl Acad Sci USA 84:9233–9237

CHAPTER 3

Inflammatory Mediators

P.J. BARNES, K.F. CHUNG, and C.P. PAGE

A. Introduction

The pathological changes in asthma are likely to be produced by the release of several mediators from inflammatory cells in the airways and the purpose of this chapter is to review some of the inflammatory mediators which have been implicated. There is a vast and rapidly increasing literature dealing with these mediators and knowledge is advancing very rapidly, made possible by improved assays for mediators, by synthetic chemistry which provides pure forms of the mediators and, perhaps most importantly, by the development of potent and specific antagonists, so that the contribution of each mediator to asthma can be evaluated (BARNES et al. 1988a). We have emphasised, where possible, studies in human airways, since it is now increasingly apparent that inflammatory cells, the generation of mediators, and airway responses are markedly different between species, and it is difficult to extrapolate from animal experiments to human airway disease. Although various animal models share some of the features of asthma, there is no entirely satisfactory model. It is, therefore, important that more research should concentrate on human asthma, despite the difficulties involved in such studies.

I. Cellular Origin of Mediators

Many different inflammatory cells may release mediators, which interact in a complex way to produce inflammatory changes in airways (Fig. 1). For many years mast cells have been assumed to play a central role in the pathogenesis of asthma; mast cell mediators, such as histamine, prostaglandin PG D_2 and sulphidopeptide leukotrienes, may explain several of the features of asthma (WASSERMAN 1983; ROBINSON and HOLGATE 1985). It is likely that IgE-dependent release of mediators from mast cells may account for the immediate bronchial response to allergen and mast cells may also be involved in the bronchoconstrictor response to exercise, cold air and fog. Recent evidence, however, questions their central involvment in bronchial hyperresponsiveness and chronic inflammation, since drugs which "stabilise" mast cells, such as β2-adrenoceptor agonists, do not prevent the late-phase response to allergen, nor the subsequent bronchial hyperresponsiveness. On the other hand, corticosteroids which do not have effects on mast cell mediator release are effective (COCKROFT and MURDOCK 1987). This suggests that other inflammatory

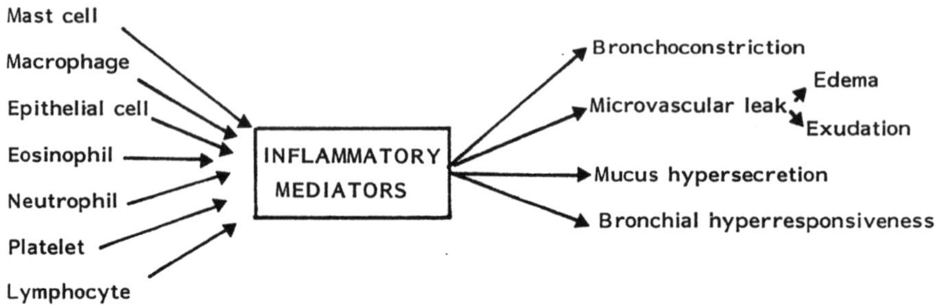

Fig. 1. Several different cells may be involved in the inflammatory process in asthma, leading to the production of many inflammatory mediators which, in combination, lead to the pathophysiological features of asthma

cells may be the source of mediators in asthma. Macrophages are abundant throughout the respiratory tract, and recent evidence that they may be activated by IgE-dependent mechanisms has suggested their involvement in allergic inflammation (JOSEPH et al. 1983b). Macrophages from asthmatic patients release greater amounts of mediators, such as thromboxane, PGs and platelet-activating factor (PAF) than those derived from normal subjects. Eosinophil infiltration is a prominent feature of asthma and differentiates asthma from other inflammatory conditions of the airway. Eosinophils may release a variety of mediators, including leukotrienes (BRUIJNZEEL et al. 1986) and PAF (LEE et al. 1984a), and also release basic proteins, such as major basic protein and eosinophil cationic protein, which are toxic to airway epithelium. Neutrophils are also found in asthmatic airways and may release a number of mediators, including leukotriene B_4, PGs, PAF and adenosine, although their role in asthmatic airways remains less certain than that of eosinophils. Airway epithelial damage is a common feature of even mild asthma (LAITINEN et al. 1985) and may underlie bronchial hyperresponsiveness, since many of the conditions known to increase bronchial responsiveness (ozone exposure, upper respiratory tract viral infection, allergen exposure) are assoicated with epithelial damage. Epithelial cells may release inflammatory mediators, such as leukotriene B_4 (HOLTZMAN et al. 1983) and 15-HETE (HUNTER et al. 1985), which are chemotactic for inflammatory cells. Abnormalities in platelet function have been found in asthma and animal studies suggest that platelets are involved in bronchial hyperresponsiveness (MORLEY et al. 1984). Platelets may release a variety of mediators, such as serotonin, thromboxane, 5- and 12-lipoxygenease products, PAF, and oxygen-free radicals, and may be activated by IgE-dependent mechanisms (JOSEPH et al. 1983a).

II. Mediator Effects

Inflammatory mediators may have a variety of effects on target cells in the airways, which may be relevant to asthma (Fig 2., Table 1). They may lead to

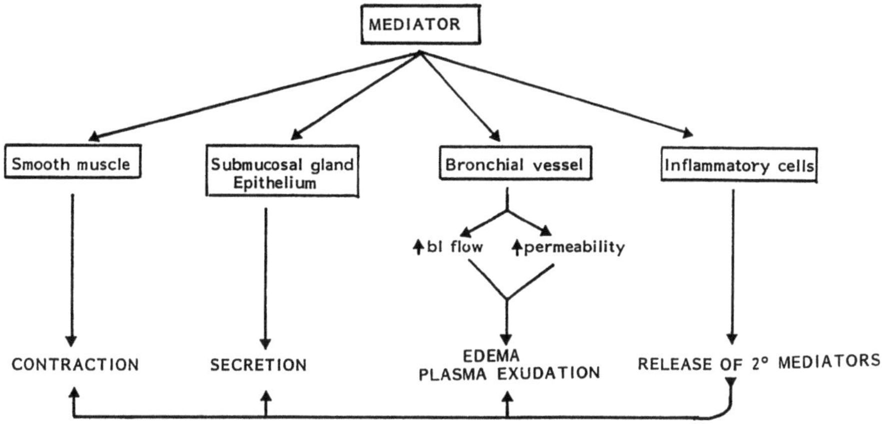

Fig. 2. Inflammatory mediators may have several effects on target cells of the airways, all mediated by activation of specific receptors

contraction of airway smooth muscle, either directly or indirectly, via release of secondary mediators or via neural mechanisms. They may also lead to increased secretion from submucosal glands, to increased fluid transport across airway epithelium and to increased microvascular leakage, which results in oedema of the airway and exudation of plasma into the airway lumen, which may result in the formation of new mediators (PERSSON 1987). Inflammatory mediators may attract and activate inflammatory cells which themselves release a whole array of mediators that serve to perpetuate and amplify the inflammatory response.

III. Mediator Receptors

Mediators produce their effects on target cells by the activation of specific cell surface receptors. More is now known about receptors for mediators in lung and the biochemical pathways involved in receptor activation (BARNES 1987). Using radioligand binding methods it has been possible to characterise and localise these receptors in lung, and it is hoped that further understanding of these receptors may lead to the development of more selective antagonists which will provide evidence for the role of a mediator in such a complex inflammatory disease as asthma. More is now understood about the biochemical pathways involved in pharmacological coupling in airway smooth muscle. Activation of some receptors leads to inhibition of adenylate cyclase via an inhibitory guanine nucleotide regulatory protein (Gi), whereas for other receptors breakdown of membrane phosphoinositides (PI) leads to the generation of inosital trisphosphate (IP_3) which releases intracellular calcium ions. Both mechanisms may be operative and may be interdependent.

Table 1. Effects of inflammatory mediators implicated in asthma

Mediator	Bronchoconstriction	Airway secretion	Microvascular leakage	Chemotaxis	Bronchial hyperresponsiveness
Histamine	+	+	+	+	−
Prostaglandins D2, F2a	++	+	?	?	+
Prostaglandin E2	−	+	−	+	−
Thromboxane	++	?	−	±	+
Leukotriene B4	−	−	±	++	±
Leukotrienes C4, D4, E4	++	++	++	?	±
Platelet activating factor	++	+	++	++	++
Bradykinin	+	+	++	−	−
Adenosine	+	?	?	?	−
Substance P	+	++	++	±	−
Neurokinin A	++	?	+	−	−
Complement fragments	+	+	+	++	−
Serotonin	±	?	+	−	−
Oxygen radicals	+	?	+	?	−

++ pronounced effect; + moderate effect; ± uncertain effect; ? information not available.

IV. Mediator Interactions

The role of each mediator itself is probably complex, but it seems likely that there is an even more complex interaction between different mediators and that this may contribute to airway hyperresponsiveness. Thus, inhalation of PGs E_2 and D_2 may lead to increased responsiveness to inhaled spasmogens (WALTERS et al. 1981; FULLER et al. 1986), although such an increase is only transient, whereas PAF leads to a more sustained increase in bronchial responsiveness (CUSS et al. 1986).

Mediator interaction in the skin is well described and mediators which lead to increased blood flow (such as PGE_2 and PGI_2) potentiate the plasma extravasation caused by other mediators (such as bradykinin) (BASRAN et al. 1982). Whether such interactions occur in the asthmatic airway is not yet certain.

Mediator interaction may also occur by "priming" of inflammatory cells. Thus, exposure of eosinophils to certain cytokines leads to augmented release of mediators (SILBERSTEIN and DAVID 1987). The number of possible interactions is almost limitless and further research in this area should prove fruitful.

V. Mediator Antagonists

The most convincing way to elucidate the role of an individual mediator in a complex inflammatory process, such as asthma, is to study the effect of a selective mediator antagonist or an inhibitor of synthesis. Extensive efforts by the pharmaceutical industry have led to the synthesis of several such agents which are currently being tested in asthma. Of course, any conclusions drawn from such studies must depend on the degree of selectivity of the antagonist. The dose of antagonist must also be adequate to block endogenously generated mediators and, at the very least, should be shown to block the effects of exogenously delivered mediator. Examples of the specific antagonists currently available are given for each mediator. The search for even more potent and specific antagonists may be beneficial to unravelling the components of asthma, but may not necessarily have a major clinical impact, since blocking a single mediator is unlikely to have a major effect if many different mediators are involved.

B. Histamine

Histamine was implicated in the pathogenesis of asthma shortly after its discovery, when it was shown to mimic anaphylactic bronchoconstriction in guinea-pigs (DALE and LAIDLAW 1919). Intravenous histmaine caused bronchoconstriction in asthmatic subjects, and inhaled histamine caused bronchoconstriction in asthmatic but not in normal subjects (CURRY 1946). Histamine is probably the best characterised of all mediators of asthma; there is now a wealth of information about its effects on human airways and the recent introduction of specific and non-sedating antihistamines has made it possible to evaluate the role of histamine in asthma pathophysiology.

I. Synthesis and Metabolism

Histamine is formed by decarboxylation of histidine and stored in preformed cytoplasmic granules of mast cells and basophils in close association with proteoglycans which are predominantly heparin in mast cells and chrondroitin 4-sulphates in human basophils. Histamine forms 5%–10% of the content of human mast cell granules. It is released from lung mast cells or blood basophils by an active secretory process which is calcium-dependent and several triggers to histamine release are recognised (FRIEDMAN and KALINER 1987).

Histamine is metabolized by two major enzymatic pathways and less than 3% is excreted unchanged in the urine. Some 50%–70% of histamine is metabolised to *N* methyl-histamine by *N*-methyl transferase which is found in small intestine, liver, kidney and leukocytes, and the remainder by diamine oxidase (histaminase) to imidazole acetic acid in small intestine, liver, kidney, neutrophils and eosinophils.

II. Histamine Receptors

Histamine produces its effects by interacting with specific receptors on target cells. The existence of more than one receptor subtype was suggested when Ash and Schild found that the classical antihistamine pyrilamine (mepyramine) was able to block some responses, such as contraction of guinea-pig trachea, but not others, such as gastric acid secretion (ASH and SCHILD 1966). The existence of a second histamine receptor subtype (H2) was confirmed with the development of selective antagonists such as cimetidine and ranitidine. There is also a third subtype of receptor (H3) for which selective agonists and antagonists have recently been developed (ARRANG et al. 1987).

H_1-receptors have been identified in animal and human lung homogenates by receptor binding techniques (CARSWELL and NAHORSKI 1982). Using an immunohistochemical technique to study the distribution of cyclic guanosine monophosphate, H_1-receptors have been localised to airway epithelial cells, macrophages and alveolar cells in guinea-pig lung, with surprisingly little localisation to airway or vascular smooth muscle (SERTL et al. 1987). In bovine tracheal smooth muscle H_1-receptors have been determined by direct receptor binding using [^{3}H]pyrilamine. Using phenoxybenzamine to fractionally inactivate receptors, it has been possible to demonstrate that there are few "spare" histamine receptors (GRANDORDY and BARNES 1987).

In this preparation there is a close relationship between H_1-receptor occupancy and PI response. H_2-receptors, which stimulate adenylate cyclase, have been identified in lung using [^{3}H]tiotidine (FOREMAN et al. 1985), but their localisation is not yet certain. H_3-receptors have been differentiated using the selective agonist (*R*)-α-methyl histamine and the antagonist thioperamide. It is possible that they may be involved in feedback inhibition of histamine release from mast cells or basophils. H_3-receptors inhibit neurotransmission in parasympathetic ganglia and the release of acetylcholine from postganglionic cholinergic nerves in guinea-pig and human airways (ICHINOSE and BARNES 1989a, ICHINOSE et al. 1989).

III. Airway Effects

Bronchoconstriction was one of the first properties of histamine which was recognised and histamine was found to contract human bronchi in vitro many years ago. Histamine contracts both large and small human airways in vitro (FINNEY et al. 1985) and this may be modulated by the presence of intact airway epithelium (CUSS and BARNES 1987). In vivo infused histamine causes marked systemic vasodilation but no bronchoconstriction, whereas infused histamine causes bronchoconstriction in asthmatic patients (WHITE and EISER 1983). Similarly, inhaled histamine causes bronchoconstriction in asthmatic patients more readily than in normal subjects, as a manifestation of bronchial hyperreactivity. By contrast, airways from asthmatics do not appear to be more responsive to histamine in vitro (THOMSON 1987), although there is one report of an increased maximal response to histamine (SCHELLENBERG and FOSTER 1984). In animals tachyphylaxis to the bronchoconstrictor effect of histamine may be demonstrated in vitro and may be due to the generation of PGs, since indomethacin prevents its development (HAYE-LEGRAND et al. 1986). Tolerance to histamine challenge may also be found in mild asthmatic subjects, with a reduced bronchoconstrictor response to a second histamine challenge, which is prevented by prior treatment with indomethacin (MANNING et al. 1987).

There is some debate as to whether H_2-receptors are present in human airways. In several animal species H_2-receptors which mediate bronchodilation have been demonstrated (WHITE and EISER 1983). Human peripheral lung strips may show a relaxant response to histamine, via H2-receptors (VINCENC et al. 1984), although this could be an effect on pulmonary vascular smooth muscle rather than on airways. An H_2-selective agonist, impromidine, has no effect on normal or asthmatic airways in vivo (WHITE et al. 1987), and H_2-selective blockers, such as cimetidine and ranitidine, have not been associated with bronchoconstriction or increased sensitivity to bronchoconstrictors in normal or asthmatic subjects. Histamine increases bronchial blood flow in sheep and dogs, an effect which is mediated via H_1- and H_2-receptors (LONG et al. 1985; LAITINEN et al. 1987).

Histamine also causes microvascular leakage in the bronchial microvasculature, which is presumed to be due to contraction of endothelial cells in postcapillary venules (PERSSON 1987). This effect is mediated via H_1-receptors and appears to be greater in larger rather than smaller airways (EVANS et al. 1988b, 1989). Although it has not been possible to study the effect of histamine on human bronchial microvasculature, it is likely that similar effects to those seen in animals occur. Intradermal injection of histamine in humans causes immediate wheal formation, which is completely inhibited by an H_1 antagonist. Histamine increases secretion of mucous glycoproteins from human airways in vitro, and this effect is mediated by H_2-receptors since the effect is blocked by cimetidine and stimulated by dimaprit (SHELHAMER et al. 1980). The effect of histamine is rather weak when compared with other secretagogues, however. In many species the bronchoconstrictor effect of histamine is partially mediated by a vagal reflex, and histamine has been shown to increase action potentials in intrapulmonary vagal afferent nerves, an effect which is mediated by H_1-receptors (SAMPSON and VIDRUK 1979). The role of cholinergic reflexes in the bronchial response to histamine in humans is less certain, since some groups have reported a significant reduction in the bronchoconstrictor response to inhaled histamine following anticholinergic treatment, whereas others have not (WHITE and EISER 1983). It is likely that the vagal component of bronchoconstriction may be greater in normal subjects but becomes relatively less important in hyperresponsive airways. Histamine has an inhibitory effect on airway cholinergic and non-adrenergic, non-cholinergic nerves via H_3-receptors, which are activated by low concentrations of histamine (ICHINOSE and Barnes 1989 a,b, ICHINOSE et al. 1989).

Histamine is also chemotactic to inflammatory cells, such as eosinophils (CLARK et al. 1975) and may, therefore, amplify the inflammatory reaction, although the effects are small when compared to other mediators. Histamine stimulates T lymphocyte suppressor cell function via H_2-receptors and this function may be depressed in atopic individuals (BEER et al. 1982). IgE-mediated release of histamine from human basophils is inhibited by histamine itself, acting on H_2- or possibly H_3-receptors, so that H_2 antagonists could theoretically enhance histamine release, but H_2-receptors have not been demonstrated on mast cells in human airways (KALINER 1978).

IV. Role in Asthma

There is a wealth of evidence which implicates histamine in asthma and the recent introduction of non-sedative antihistamines has made it possible to determine more precisely its contribution to asthma pathophysiology. Measurements of histamine have been made in asthma since the first assays were developed in the 1940s. Fluorimetric assays, which lacked specificity and sensitivity, gave conflicting results but refinement of radio enzymatic assays has made it possible to detect low concentrations of histamine in plasma. Several early studies reported an elevated baseline concentration of plasma histamine in severe asthma, but the concentrations reported were very high and it seems unlikely that they reflected release from mast cells in lung, since such elevations should have caused marked cardiovascular effects. With improved sensitivity of the assay it was shown that even mild asthmatic subjects had elevated values of plasma histamine (BARNES et al. 1982), which has been interpreted as mast cell "leakiness". Several conflicting results of plasma histamine measurements have been reported with various bronchoconstrictor challenges in asthma. Elevated plasma histamine has been reported in exercise-induced bronchoconstriction, but not in matched bronchoconstriction produced by hyperventilation (BARNES and BROWN 1981), and it seems likely that the increase with exercise might be due to the increase in basophil counts which occurs during exercise. Plasma histamine accounts for only 0.5% of total blood histamine, the remainder being contained in basophils, so any contamination of plasma is likely to give marked discrepancies (IND et al. 1983). Plasma histamine is also reported to be increased in allergen challenge in asthmatic subjects (HOWARTH et al. 1985), and there may be a secondary rise associated with the late response. Whether these increases in plasma histamine are a reflection of mediator release from airway mast cells is uncertain, however, and sampling blood from a peripheral vein is probably unlikely to closely reflect the relatively small amount of histamine release in airways, particularly in hyperresponsive patients where a comparatively small amount of histamine released locally may have a profound bronchomotor effect.

More recent studies have therefore measured histamine more locally in bronchoalveolar lavage fluid from asthmatics and demonstrated an elevated concentration in comparison with non-asthmatic subjects (CASALE et al. 1987). While the above studies indicate that histamine may be released in asthma, and thus provide indirect evidence of mast cell degranulation, they do not give information about the contribution of histamine to pathophysiology; this information can only be provided by the use of specific antagonists. If histamine is important in asthma then antihistamines should be effective in its clinical management. Previous experience with antihistamines has not been encouraging in asthma (KARLIN 1972), although the H_1 antagonists used have often lacked specificity and sedative effects have limited the dosage. The recent introduction of potent and selective non-sedative antihistamines, such as terfenadine and astemizole, has made it possible to more easily evaluate the role of histamine in asthma, since it is possible to achieve a greater degree of H1-receptor blockade. Terfenadine casues a degree of bronchodilation similar to that achieved with a beta agonist, suggesting the existence of histamine "tone" in asthma (COOKSON 1987), and partially protects against exercise-induced asthma (PATEL 1984). Additionally, terfenadine has a small inhibitory effect against allergen challenge (CHAN et al. 1986; RAFFERTY et al. 1987), in doses which give a marked shift in the bronchoconstrictor dose-response curves to histamine, suggesting that histamine plays a relatively minor role in immediate bronchoconstriction responses to allergen.

C. Cyclooxygenase Products

I. Synthesis and Metabolism

PGs are formed from arachidonic acid (AA) which is oxidised by cyclooxygenase to the cyclic endoperoxides, PGG_2 and PGH_2, from which $PGF_{2\alpha}$ and

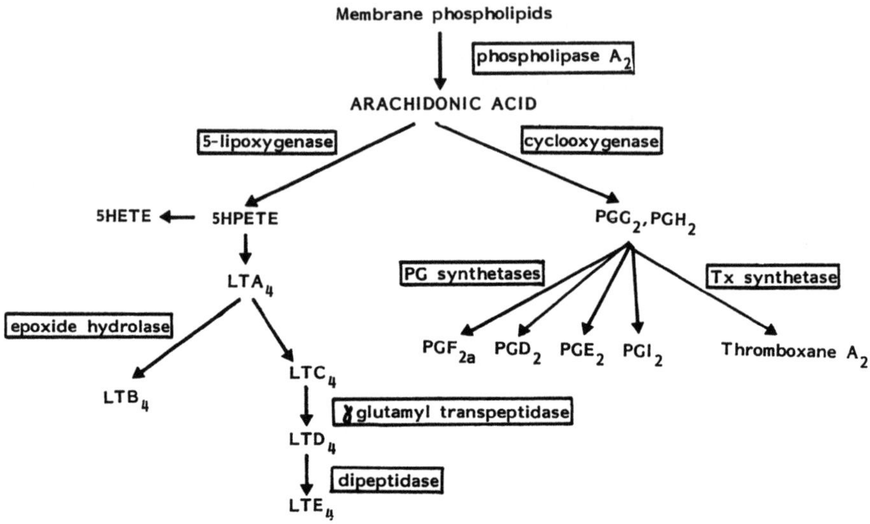

Fig. 3. Pathways involved in the synthesis of eicosanoids. Arachidonic acid is metabolised by 5-lipoxygenase to 5 hydroperoxyeicosatetraenoic acid (5HPETE) and hence to 5 hydroxyeicosatetraenoic acid (5HETE) or leukotriene (LT) A4 which in turn generates the chemotactic leukotriene LTB4, or the sulphidopeptide leukotrienes LTC4, D4 and E4. Metabolism by cyclooxygenase results in the formation of prostaglandins (PG) and thromboxane A2

PGE_2 are formed (Fig 3). Other enzymatic pathways lead to the formation of thromboxane A_2 (TxA_2) and prostacyclin (PGI_2). Inactivation of PGs in the lung is a function of a selective and active uptake by capillary endothelial plasma membranes (BITO 1972). Within one transit in the lung, exogenous PGE_2 and $PGF_{2\alpha}$ are inactivated (MCGIFF et al. 1969; PIPER et al, 1970).

PGD_2 and PGI_2 are released from human lung parenchyma during anaphylaxis in vitro with smaller amounts of TxA_2, PGE_2, and $PGF_{2\alpha}$; by contrast, in the airway, PGI_2, $PGF_{2\alpha}$ and PGE_2 are released in the greatest amount (SCHULMAN et al. 1982). Enriched or purified human lung mast cells undergoing IgE-dependent activation release PGD_2 as the major cyclooxygenase product (SCHLEIMER et al. 1986). Human alveolar macrophages also release PGD_2 (MACDERMOT et al. 1984) in addition to measurable quantities of PGE_2, $PGF_{2\alpha}$ and TxB_2 (GODARD et al. 1982). In vivo, local instillation of antigen in the airways of allergic asthmatics results in the immediate release of PGD_2 in bronchoalveolar lavage fluid (MURRAY et al. 1986). PGE_2 is released from canine airway epithelial cells when stimulated with bradykinin (LEIKAUF et al. 1985) and human pulmonary vascular endothelial cells are an important source of PGI_2 (JOHNSON 1980). Infiltrating cells, such as platelets and neutrophils, may also contribute to the production of cyclooxygenase products, such as TxA_2 and PGE_2 (GOLDSTEIN et al. 1978; NEEDLEMAN et al. 1976).

II. Receptors

Classification of the prostanoid receptors has been based primarily on comparisons of the rank orders of agonist potency (KENNEDY et al. 1982), and receptors for each of the natural prostanoids (i.e., the PGs D_2, E_2, $F_{2\alpha}$, I_2 and TxA_2) has been proposed. Gardiner (GARDINER 1986) recognised only three subtypes of the contractile/stimulant receptor with the following agonists: TxA_2, $PGF_{2\alpha}$, or PGD_2 and PGE_2. In human lung strips, all prostanoid contractile agonists appear to exert their effects via the thromboxane receptor; however, contraction of human bronchioles may be mediated via a novel prostanoid receptor which remains to be identified (McKENNIFF et al. 1989). PGI_2 receptors have been identified in lung homogenates by direct receptor binding (MACDERMOT et al. 1981).

III. Airway Effects

1. Airway Smooth Muscle

PGD_2, $PGF_{2\alpha}$ and TxA_2 contract human airway smooth muscle in vitro (SWEATMAN and COLLIER 1968; GARDINER and COLLIER 1980); PGE_1 (SHEARD 1968) and PGI_2 cause a small degree of relaxation. PGE_2 can either increase or decrease basal tone of isolated human airway muscle preparations (GARDINER 1975). Inhibition of endogenous prostanoids does not modulate the basal tone or the contractile responses to acetylcholine or histamine of isolated asthmatic or non-asthmatic airway smooth muscle (BRINK et al. 1980; HAYE-LEGRAND et al. 1986).

Asthmatics are more responsive to the bronchoconstrictor effect of $PGF_{2\alpha}$ than normal subjects (MATHE and HEDQVIST 1975), and $PGF_{2\alpha}$ is about 100-fold more potent than methacholine. Tachyphylaxis to the bronchoconstrictor effects of $PGF_{2\alpha}$ has been reported in asthmatics (FISH et al. 1983; MATHE and HEDQUIST 1975). Sequential administration of high doses of $PGF_{2\alpha}$ may paradoxically result in bronchodilation, predominantly in the large airways (FISH et al. 1983). PGD_2 is threefold more potent than $PGF_{2\alpha}$ and the duration of its bronchoconstrictor effect is more prolonged (HARDY et al. 1984b). PGD_2 may act via the thromboxane receptor. In dogs, TxB_2 is slightly less potent than $PGF_{2\alpha}$ in causing bronchoconstriction (WASSERMAN and GRIFFIN 1976).

Both PGE_1 and PGE_2 are bronchodilators in normal and asthmatic subjects (CUTHBERT 1971; KAWAKAMI et al. 1973), but bronchoconstrictor responses have also been reported (Mathe and HEDQVIST 1975), possibly by stimulation of airway afferent vagal C-fibres (ROBERTS et al. 1985). Intravenous and aerosolized PGI_2 have no effect on resting pulmonary function in normal or asthmatic subjects, but PGI_2 can prevent the bronchoconstrictor effect of ultrasonic mist, exercise and PGD_2 (BIANCO et al. 1978; HARDY et al. 1984a).

$PGF_{2\alpha}$ and PGD_2 transiently potentiated histamine- and methacholine-induced bronchoconstriction in normal subjects and asthmatic subjects (FULLER et al. 1986; WALTERS et al. 1981). $PGF_{2\alpha}$ also enhanced capsaicin-induced cough in normal subjects (NICHOL et al. 1988). The TxA_2 mimetic (U46619) potentiates cholinergic neurotransmission in canine airways (CHUNG et al. 1985). Because a thromboxane synthetase inhibitor prevents the increased bronchial responsiveness due to PAF allergen and ozone in this species, TxA_2 has been implicated in the pathogenesis of hyperresponsiveness (CHUNG 1986).

2. Secretion

In human airway tissue explants, PGD_2, and $PGF_{2\alpha}$ increased mucous glycoprotein release (MAROM et al. 1981; RICH et al. 1984). $PGF_{2\alpha}$ aerosol increased airway secretions, with the production of mucous glycoproteins probably resulting from submucosal glands rather than from goblet cells.

Using the Ussing short-circuit technique, $PGF_{2\alpha}$ increased net chloride secretion, but PGE_1 decreased both chloride and sodium secretion in canine trachea (AL-BAZZAZ et al. 1981). In bovine trachea, indomethacin reversed net transepithelial flow of sodium and chloride ions (LANGRIDGE-SMITH et al. 1984).

3. Inflammatory Effects

The effect of cyclooxygenase products on airway vascular permeability is poorly documented. In the skin, PGE_1 and PGE_2 are potent vasodilators and can potentiate histamine-, PAF- and bradykinin-induced skin oedema in several species including humans (WILLIAMS and MORLEY 1973; FLOWER et al 1976). PGD_2 induces a wheal and flare response in human skin with a perivascular neutrophil infiltrate (SOTER et al. 1983). In vitro chemokinetic activity of PGD_2 and of TxA_2 for neutrophils, has been reported (SPAGNELLO et al. 1980). By contrast, PGD_2 and PGI_2 are both inhibitors of platelet function (SMITH et al. 1974; MARCUS et al. 1980)

IV. Role in Asthma

Increased levels of PGD_2 in bronchoalveolar lavage fluid and of a circulating metabolite of $PGF_{2\alpha}$ and TxB_2 have been detected during the acute response to inhaled allergen in asthmatic subjects (MURRAY et al. 1986). Cyclooxygenase inhibition with aspirin or indomethacin has no effect on resting pulmonary function of asthmatic and normal subjects (OGILVY et al. 1981; FISH et al. 1981). However, some asthmatics ('aspirin-sensitive') bronchoconstrict after ingesting aspirin or non-steroidal anti-inflammatory drugs (NSAID) by mechanisms that may be related to cyclooxygenase inhibition or platelet activation (AMEISEN et al. 1985). Inhibition of thromboxane synthetase by OKY-

046, an imidazole derivative, improves bronchial hyperresponsiveness in asthmatics (FUJIMURA et al. 1986). Indomethacin or aspirin does not influence the early acute response induced by inhaled antigen (FAIRFAX et al. 1983; SHEPHARD et al. 1985). However, the late phase bronchoconstrictor response to antigen is reduced by indomethacin, aspirin or benoxaprofen (FAIRFAX et al. 1983; SHEPHARD et al. 1985), and an inhibitory effect on indomethacin antigen-induced airway hyperresponsiveness in asthmatics has been observed (KIRBY et al. 1987; KIRBY et al. 1989). Antigen-induced contraction of passively-sensitized human bronchial strips is reduced by indomethacin secondary to an augmented release of histamine, and perhaps of other mediators such as leukotrienes (ADAMS and LICHTENSTEIN 1985). Indomethacin does not influence exercise-induced bronchoconstriction but it prevents the tachyphylactic response to successive bouts of exercise (O'BYRNE and JONES 1986).

Overall, there is reasonable evidence to suggest that cyclooxygenase products play a modulatory role in several aspects of airway function. Whether this role is beneficial or detrimental to the asthmatic airway is likely to depend on the predominance of the cyclooxygenase product because different products may have opposing effects. The availability of more specific antagonists of the prostanoids may help dissect their precise individual contribution to asthma.

D. Lipoxygenase Products

I. Synthesis and Metabolism

AA may be oxygenated at different sites by specific lipoxygenases, initiating the formation of leukotrienes (LTs), lipoxins and several hypdroxyeicosatetraenoic acids (HETEs). The mono-HETEs are the most common derivatives of AA to be detected in the lungs, with 15-HETE being the most predominant (HAMBERG et al. 1980). Both the formation of 5-HPETE from AA and the subsequent conversion of 5-HPETE to LTA_4 are catalysed by 5-lipoxygenase (ROUZER et al. 1986). LTA_4 is hydrolysed enzymatically to the dihydroxyacid LTB_4 and can be conjugated with glutathione to the peptidolipid LTC_4, first identified as a component of slow-reacting substance (SRS). The subsequent conversion of LTC_4 to LTD_4, a cysteinyl glycinyl derivative, is through the action of α-glutamyl transpeptidase (ORNING and HAMMARSTROM 1980). LTD_4 is further metabolized to the cysteinyl derivative, LTE_4, by the action of dipeptidase (PARKER et al. 1980).

Lipoxins are a newly described series of oxygenated derivatives of AA formed from interaction of the 5- and 15-lipoxygenase pathways, which were first isolated from neutrophils incubated with 15-HETE (SERHAN et al. 1987). The profile and quantity of LT generated in vitro are dependent upon the cell type and the stimulus applied. Circulating neutrophils produce approximately five to ten times more LTB_4 than LTC_4 when activated wth the calcium ionophore A23187, but the ratios and quantities are reversed with normal eosinophils (WELLER et al. 1983; SHAW et al. 1984). Eosinophils can also activate the 15-lipoxygenase pathway (HENDERSON et al. 1984). Alveolar macrophages generate 20 times more LTB_4 than LTC_4 (MARTIN et al. 1984). Highly purified human lung mast cells release less than 4 ng of LTB_4 per million mast cells after IgE-mediated activation compared to 10 ng of LTC_4 (MACGLASHAN et al. 1982); only small quantities of LTD_4 and LTE_4 are generated. Human monocytes can also generate substantial amounts of LTB_4 and LTC_4 on stimulation with the calcium ionophore A23187 (PAWLOWSKI et al. 1983). Tracheal epithelial cells from human airways can generate 15-lipoxygenase products (HUNTER et al. 1985).

The transformation of LTC_4 to LTD_4 and LTE_4 represents a bioconversion rather than catabolism of LT. Catabolism of LTE_4 may occur at extrapulmonary sites, but LTE_4 can be partly excreted unchanged from the kidneys (ORNING et al. 1985). Alterations in the peptide portion of LTC_4 do not usually result in a major loss of biological activity (DAHLEN et al. 1983). Sulphidopeptide LT from neutrophils and eosinophils may be rapidly metabolized extracellularly by the concomitantly formed hypochlorous acid (LEE et al. 1982; WELLER et al. 1983). LTB_4 is mostly inactivated intracellularly in neutrophils by omega-oxidation with the involvement of a cytochrome P450-like system (SHAK and GOLDSTEIN 1984). Inactivation by beta-oxidation may occur in vivo.

II. Receptors

Studies of isomers of LTC_4 have demonstrated that differences in binding correspond closely to differences in contractile potency, supporting the concept that the lung binding site is a specific receptor (KUEHL et al. 1984). Discrete receptors for LTC_4 and LTD_4 have been proposed because the molar ratios needed to elicit identical biological effects differ markedly in different tissues (LEWIS et al. 1981) and because the compound FPL 55712 selectively antagonises the effect of LTD_4 (DRAZEN et al. 1980). Two distinct binding sites in guinea-pig lung homogenates, corresponding to the function of LTC_4 and LTD_4 receptors have been demonstrated by radioligand studies (HOGABOOM et al. 1983). The distribution of LTC_4 and LTD_4 binding sites in guinea-pig lung has been mapped by autoradiography, with LTC_4 receptors being more widely distributed and present in higher density than LTD_4 receptors (CARSTAIRS et al. 1988). A proportion of LTC_4 binding may be bound to the enzyme α-glutamyl transpeptidase (SUN et al. 1986). However, pharmacological studies suggest that human bronchi may not contain different receptors for LTC_4 and LTD_4 (BUCKNER et al. 1986). Saturable binding of $[^3H]LTB_4$ to human neutrophils has been demonstrated, but the dissociation constant and number of specific binding sites reported differed significantly between these studies (GOLDMAN and GOETZL 1982; KREISLE and PARKER 1983).

III. Airway Effects

1. Airway Smooth Muscle

LTC_4 and TD_4 are approximately 1000-fold more potent than histamine in contracting human isolated bronchus (DAHLEN et al. 1980), but are less active on human parenchymal strips (SAMHOUN and PIPER 1983). LTE_4 is less potent than LTC_4 and LTD_4, but has a more prolonged effect. LTB_4 also contracts human isolated bronchus but rapid tachyphylaxis develops (SAMHOUN and PIPER 1983). Modest contraction of human bronchial muscle in vitro occurs with 5- and 15-HETE (COPAS et al. 1982). Lipoxin A (LXA) causes long-lasting contraction of guinea-pig lung strip, but is inactive on trachea (DAHLEN et al. 1987).

The sulphidopeptide LT constrict both large and small airways (BARNES et al. 1984b; SMITH et al. 1985) of normal and asthmatic subjects. Inhaled LTC_4 and LTD_4 are 1000–5000 times more potent than histamine, with a longer duration of action (WEISS et al. 1983; BARNES et al. 1984b). LTE_4 is approximately 1/10 as potent as LTD_4 (DAVIDSON et al. 1987), with a longer duration of action, in agreement with its in vitro effect. Asthmatic subjects are hyperresponsive to inhaled LTs (SMITH et al. 1985; ADELROTH et al. 1986; DAVIDSON et al. 1987).

2. Secretion

Both LTC_4 and LTD_4 are potent stimulants of mucous glycoprotein from human airways in vitro, being tenfold more potent than methacholine (MAROM et al. 1982; COLES et al. 1983). The mono-HETEs are less effective in causing mucus secretion in human airways (MAROM et al. 1981). In vivo, LTC_4 and LTD_4 enhance mucus secretion (JOHNSON and MCNEE 1983) and stimulate increased chloride secretion across the epithelium in dog trachea (LEIKAUF et al 1986). LTC_4 stimulates ciliary beat frequency of sheep airways in vitro through the release of cyclooxygenase products (WANNER et al. 1986).

3. Vascular Effects

LTC_4, LTD_4 and LTE_4 cause vasoconstriction and increase microvascular permeability in the airways of guinea-pigs (WOODWARD et al. 1983; HUA et al. 1985), being at least 100–1000 times more active than histamine (WOODWARD et al. 1983), probably through a direct action at the postcapillary venular endothelial cell (DAHLEN et al. 1981; JORIS et al. 1987). In human skin, LTC_4 and LTD_4 are potent vasodilators, producing wheal and flare responses at low concentrations (BISGAARD et al. 1982; CAMP et al. 1983). The increased microvascular permeability induced by LTB_4 probably depends on the emigration and interaction of neutrophils through the endotheleial microvasculature (BRAY et al. 1981; BJORK et al. 1982). LXA causes arteriolar dilation, but has no effect on microvascular permeability (DAHLEN et al. 1987).

4. Cellular Activation

LTB_4 is the most potent chemotactic and chemokinetic lipoxygenase product for neutrophils in vitro (FORD-HUTCHINSON et al. 1986), but is less effective for eosinophils (WARDLAW et al. 1986). Intradermal injection of LTB_4 results in neutrophil accumulation into human skin, associated with a slow-onset tenderness and induration (SOTER et al. 1983). LTB_4 also stimulates the release of lysosomal enzymes (FEINMARK et al. 1981) and enhances the release of oxygen radicals from human neutrophils (SERHAN et al. 1982). The expression of surface complement (C3b) receptors on human neutrophils and eosinophils (NAGY et al. 1982) can also stimulate the chemotaxis of human eosinophils and neutrophils, with maximal responses, similar to those evoked by C5a and formyl-methionyl peptides, (GOETZL et al. 1977; GOETZL and PICKETT 1980) Degranulation of human neutrophils is induced by 5- and 12-HETEs and by LXA (STENSON and PARKER 1980; SERHAN et al. 1984).

5. Effect on Bronchial Responsiveness

LTC_4, D_4 and E_4 and 5-HETE can increase the responsiveness of guinea-pig tracheal muscle to histamine in vitro (CREESE and BACH 1983; LEE et al. 1984b). LTB_4 augments bronchial responsiveness to acetylcholine in dogs but not in humans (O'BYRNE et al. 1985). However, in humans prior inhalation of LTD_4 failed to increase the bronchoconstrictor effect of histamine (BARNES et al. 1984a), although LTE_4 increased histamine airway responsiveness transiently (ARM et al. 1988b).

IV. Role in Asthma

Sulphidopeptide LT can be detected in nasal secretions after allergen challenge in vivo (CRETICOS et al. 1984) and in pooled plasma from subjects with acute asthma (ZAKRZEWSKI et al. 1985). LTE_4, and to a lesser extent LTC_4, have been recovered from bronchoalveolar lavage fluid of asthmatic subjects (LAM et al. 1987). Inhibition of LTC_4 and LTD_4 synthesis through an effect on glutathioine S-transferase by U-60, 257 (Piriprost) (BACH et al. 1982) blocks the non-histamine component of airway smooth muscle contraction induced by allergen in the bronchi of atopic asthmatic subjects in vitro but had no effect on allergen challenge in asthmatic subjects in vivo (MANN et al. 1986b). In vivo, L-649, 923, an LTD_4 receptor antagonist, had no effect on the early and late bronchoconstrictor response after antigen challenge (BRITTON et al. 1987), but at the dose used, L-649, 923 is only weakly effective against LTD_4-induced bronchoconstriction in vivo (BARNES et al. 1987). Short-term treatment of asthmatic subjects with the 5-lipoxygenase inhibitor AA-361 had no effect on the airways hyperresponsiveness of asthma (FUJIMURA et al. 1986). Colchicine, an anti-inflammatory agent, has been shown to inhibit LTB_4 production from human neutrophils in vivo, yet has no effect on either early- or late-phase response to allergen (PETERS et al. 1987).

Dietary supplementation with eicosapentaenoic acid (fish oil) to decrease the formation of lipoxygenase products by diversion to less active eicosapentaenoic derivatives, such as LTB_5, LTC_5 and LTD_5, has no effect on bronchial responsiveness or on the control of asthma in asthmatic subjects (ARM et al. 1988a).

The role of lipoxygenase products, including LT, in asthma, therefore, still remains unclear, perhaps largely because of the difficulty in obtaining convincing pharmacological inhibition of their effects. Apart from LTB_4, the other LTs do not possess significant chemotactic activity for eosinophils.

E. PAF

PAF was first shown to be released by IgE-stimulated basophils (BENVENISTE et al. 1972). It has been chemically characterised as an either-linked phospholipid, 1-*o*-alkyl-2-acetyl-*sn*-gylceryl-3-phosphorylcholine. PAF has many biological properties in addition to platelet activation, and is particularly interesting as a putative mediator of asthma, since it can induce several characteristic features of asthma (BARNES and CHUNG 1987; BARNES et al. 1988b; PAGE 1988).

I. Origin

1. Synthesis

The synthesis of PAF is not secondary to cell damage or physical disruption (TENCE et al. 1980), suggesting that PAF is neither preformed nor stored but rather synthesized denovo. Two distinct synthetic pathways have been described for PAF (SNYDER 1985, 1987). The first is a two-step pathway which has been demonstrated in a number of inflammatory cell types in vitro, including macrophages (NINIO et al. 1982), neutrophils (LOTNER et al. 1980), eosinophils (LEE et al. 1984a) and platelets (CHAP et al. 1981), and which involves the production of the biologically inactive intermediate lyso-PAF from ether-linked phospholipids by the action of phospholipase A2 (PLA2) (Fig. 3). This step is in common with the liberation of AA for the subsequent formation of cyclooxygenase and lipoxygenase metabolites (Fig. 3). The second enzyme has to be concomitantly activated with PLA2, namely an acetly CoA-dependent acetyltransferase, which has been described in a number of inflammatory cell types and is the rate limiting step for PAF production by this pathway (SNYDER 1987).

A second synthetic pathway for PAF involves the enzyme, cholinephosphotransferase, which can synthesise PAF directly from ether-linked phospholipids (SNYDER 1985, 1987). The cholinephosphotransferase pathway may be required to maintain physiological levels of PAF for normal cell function, particularly in the regulation of blood pressure, whereas the rate limiting acetyltransferase pathway is only activated in response in inflammatory signals, such as phagocytosis or chemotaxis (SNYDER 1987).

2. Cellular Origin

Although PAF was originally described as a product of rabbit basophils, it can also be produced by a number of other inflammatory cells. Interestingly, in humans PAF does not appear to be an extracellular product of basophils or mast cells and, although pulmonary mast cells have the capacity to synthesise PAF, it appears to be retained intracellularly (LICHTENSTEIN et al. 1984). This phenomenon has also been observed in human neutrophils, but the precise role of intracellular PAF is uncertain (LYNCH and HENSON 1986). Human platelets produce lesser amounts of PAF than neutrophils but approximately 50 times more lyso-PAF, presumably associated with the production of AA metabolites following PLA2 activation. Eosinophils isolated from patients with eosinophilia release PAF following stimulation with various chemotactic factors, including eosinophilic chemotatic factor of anaphylaxis (ECF-A) and f-Met-Leu-Phe (FMLP), suggesting that PAF release may play a central role in the chemotaxis of human eosinophils (LEE et al. 1984a; SIGAL et al. 1987). Human alveolar macrophages obtained by bronchoalveolar lavage of allergic asthmatics also release PAF following stimulation with the appropriate antigen in vitro (ARNOUX et al. 1987).

Cultured human vascular endothelial cells also release PAF following stimulation with thrombin, calcium ionophore, LT, histamine, bradykinin, ATP or monocyte-derived interleukin-1 (PRESCOTT et al. 1984). As with other cell types, some of the PAF formed by endothelial cell monolayers remains cell-associated rather than being released extracellularly in situations where PGI_2 production can be detected.

3. Metabolism

PAF is very rapidly metabolized by the action of the enzyme phosphatidyl-2-acetylhydrolase, which removes acetate and leads to the formation of lyso-PAF (FARR et al. 1980). Thus, the primary metabolite of PAF is also its precursor in some situations, and in some cell types there is a constant cycle of PAF synthesis and metabolism (TOUQUI et al. 1985). The acetylhydrolase enzyme responsible for the initial metabolism of PAF has been identified in the plasma of a number of mammalian species, including humans, and is extremely active. From studies in the rabbit, 70% of the PAF is metabolized to lyso-PAF 1 min after i.v. injection (LATRIGUE-MATTEI et al. 1984). An acetylhydrolase enzyme (which is capable of metabolizing PAF) has also been reported to be present on the surfact of platelets and is released following activation with PAF (SUZUKI et al. 1988). Acetylhydrolase activity in plasma from asthmatic children is significantly reduced compared with healthy controls, suggesting that PAF may have protracted biological activity in these subjects (MIWA et al. 1988).

II. Receptors

There have been a number of studies using [^{3}H]PAF as a radioligand, which have demonstrated high affinity, saturable binding sites for PAF on human platelets (VALONE et al. 1982), neutrophils (VALONE and GOETZL 1983) and lung membranes (HWANG et al. 1985). Such specific binding is inhibited by a number of PAF antagonists, but there is a high degree of non-specific binding which makes these experiments difficult to interpret. Recently, labelled PAF antagonists, such as [^{3}H] kadsuenone and [^{3}H]WEB 2086, have proved more useful as radioligands (HWANG et al. 1986; UKENA et al. 1988, 1989; DENT et al. 1989). Although none of the above evidence is definitive proof of the existence of a PAF receptor, a protein has been isolated containing the PAF receptor from human platelets (VALONE 1984).

Experiments showing different affinities of the PAF antagonist kadsurenone in peritoneal macrophages and blood leukocytes, have suggested the existence of PAF receptor subtypes (LAMBRECHT and PARNHAM 1986; VOELKEL et al. 1986; STEWART and DUSTING 1988).

III. Airway Effects

1. Airway Smooth Muscle

PAF is a potent inducer of bronchoconstriction in experimental animals and in humans (CUSS et al. 1986; RUBIN et al. 1987). However, PAF does not exert direct contractile effects on human airway smooth muscle preparations in vitro, yet may elicit contraction of airway smooth muscle preparations provided platelets are present in the organ bath (SCHELLENBERG et al. 1983; CERRINA et al. 1983). In experimental animals PAF-induced bronchoconstriction is secondary to platelet activation, since bronchoconstriction is abrogated in animals previously rendered thrombocytopenic by the use of a selective cytotoxic antibody (VARGAFTIG et al. 1980; HALONEN et al. 1980). The nature of the platelet-derived spasmogen is not known, and there is controversy about the effects of antihistamines, serotonin antagonists and inhibitors of AA metabolism (both cyclooxygenase and lipoxygenase) (VARGAFTIG et al. 1980; BONNET et al. 1983; LEWIS et al. 1984; L.J. SMITH et al. 1988). Neutrophils have also been implicated in PAF-induced bronchoconstriction (KLIMEK et al. 1983) and there is a close anatomical relationship between platelets and neutrophils observed throughout the pulmonary vasculature of experimental animals following systemic administration of PAF (DEWAR et al. 1984). As neutrophils and platelets have been observed to cooperate in the formation of novel biologically active materials (MACLOUF et al. 1982; MARCUS et al. 1984), it is conceivable that such products contribute to PAF-induced bronchoconstriction.

In humans, PAF administered by inhalation is a potent bronchoconstrictor, having a rapid onset of action and recovery over 2 h, whereas lyso-PAF has no significant effect (CUSS et al, 1986). The bronchoconstriction induced by PAF in humans is tachyphylactic, preventing cumulative dose-response studies, and PAF-induced bronchoconstriction cannot be inhibited by H1 antagonists or NSAIDs (CHUNG et al. 1988). Surprisingly, there is no relationship between the airway responsiveness to PAF and that to a cholinergic agonist in normal subjects (CUSS et al. 1986). This is in contrast to all other bronchoconstrictor stimuli, such as histamine, LTs and PGs, in which there is a good correlation with the sensitivity to methacholine (BOUSHEY et al. 1980). Even in asthmatic patients showing hyperresponsiveness to methacholine, the airway responsiveness to inhaled PAF is similar to that observed in normal subjects (CHUNG and BARNES 1989; RUBIN et al. 1987).

2. Airway Secretions

There are few reported studies investigating the effect of PAF on airway secretion. PAF increases mucus secretion in the trachea of ferrets both in vitro and in vivo (LANG et al. 1987) and also weakly stimulates mucous glycoprotein from explants of human airways in vitro (GOSWAMI et al. 1987). PAF induces an increase in the protein content of airway secretions, although no alteration in mucus output (ROGERS et al. 1987; PERSSON et al. 1987). It is likely that the increased protein content is secondary to plasma protein extravasation into the airways, as PAF is known to have marked effects on airway microvascular permeability (see below). In isolated porcine trachea PAF stimulates mucus secretion, which is unaffected by antagonists of histamine, acetylcholine and LTD_4 or by inhibitors of PG and LT synthesis (STEIGER et al. 1987). Also, PAF has been observed to stimulate secretion of mucus in explants of rodent airways in organ culture (ADLER et al. 1987). PAF, administered intratracheally or intravenously, slows mucociliary transport, which may result from an effect on ciliated respiratory epithelial cells or from exudation of plasma into the airway lumen (AURSUDKIJ et al. 1987).

3. Vascular Effects

PAF induces microvascular leakage in several tissues including skin (ARCHER et al. 1984b; CHUNG et al. 1987; ROBERTS et al. 1988b) and airways (EVANS et al. 1987; O'DONNELL and BARNETT 1987) at doses over 100 times lower than that of histamine. In humans, the effect of PAF on microvascular permeability has been studied in the skin, where PAF elicits a classic acute wheal and flare response (BASRAN et al. 1984; ARCHER et al. 1985b). The wheal response is unaffected by prior treatment with H1 antagonists (although the flare response is) or cyclooxygenase inhibitors, suggesting that this effect is not secondary to liberated histamine or cyclooxygenase metabolites (ARCHER et al. 1985a). Increased vascular permeability is partly dependent on local blood flow and the addition of local vasodilators, such as PGE1 or PGE2, potentiates PAF-induced vascular permeability in the skin (ARCHER et al. 1984a).

PAF-induced vascular permeability appears to be independent of platelet or neutrophil activation (PIROTZKY et al. 1984), but since endothelial cells have PAF receptors (BUSSOLINO et al. 1987), it is likely that PAF has a direct contractile effect on endothelial cells. PAF-induced vascular permeability can be inhibited in both experimental animals and humans by PAF antagonists, suggesting that PAF is acting via specific PAF receptors (HELLEWELL and WILLIAMS 1986; CHUNG et al. 1987; EVANS et al. 1988a). PAF has potent effects on airway microvascular permeability; as little as 1 ng/kg administered i.v. to guinea-pigs induces a rapid extravasation of Evans blue dye (as a marker of plasma albumin) in central and peripheral airways (EVANS et al. 1987; O'DONNELL and BARNETT 1987). As in the skin, this is a direct effect of PAF, since it is not reduced by platelet depletion, cyclooxygenase inhibition or antagonists of histamine or LTs (EVANS et al. 1987), but is inhibited by PAF antagonists (EVANS et al. 1987, 1988a). PAF induces delayed leakage of plasma proteins into the airways, which may be inhibited by anti-asthma drugs, such as cromoglycate sodium and theophylline (PERSSON 1987).

4. Inflammatory Cell Activation

PAF activates a wide range of inflammatory cells, both in vitro and in vivo. In vitro, PAF induces aggregation of platelets (BENVENISTE et al. 1972), neutrophils (FORD-HUTCHINSON 1983; O'FLAHERTY et al. 1986) and monocytes (YASAKA et al. 1982), with subsequent release of secondary inflammatory mediators, including lipoxygenase and cyclooxygenase products, oxygen radicals and lysosomal enzymes. PAF also induces the chemotaxis of neutrophils (O'FLAHERTY et al. 1981; WARDLAW et al. 1986) and eosinophils (TAMURA et al. 1987; WARDLAW et al. 1986; SIGAL et al. 1987). The response of eosinophils to PAF is of particular interest in the context of asthma, as PAF represents the most potent chemotactic stimuli for human eosinophils so far described (LEE et al. 1984a; WARDLAW et al 1986). Other eosinophil chemotactic stimuli, such as ECF-A and FMLP may act via the release of PAF, suggesting that PAF may play a central role in the chemotactic response of eosinophils. In addition, PAF induces the release of LTC_4 (BRUIJNZEEL et al. 1986) and peroxidase and other granule contents from eosinophils (KROEGEL et al. 1988, 1989). PAF antagonists inhibit IgE-dependent activation and release of oxygen radicals from eosinophils, suggesting that PAF may be involved in allergic stimulation of this cell type (CAPRON et al. 1988). PAF causes much greater activation of eosinophils from asthmatic patients than from other atopic patients (CHANEZ et al. 1988), and eosinophils have a high density of PAF receptors (UKENA et al. 1989). PAF also activates macrophages, with release of oxygen radicals (HARTUNG et al. 1983).

In vivo, PAF results in the recruitment of various inflammatory cells into tissues following either systemic or local administration. After intradermal administration of PAF in normal volunteers, there is a substantial inflammatory cell infiltrate characterised at 4 h by neutrophils, and at 24 h by a mixed cellular infiltrate comprising both neutrophils and mononuclear cells (ARCHER et al. 1985b), whereas in atopic subjects the cellular infiltration is characterised by activated eosinophils and is reminiscent of antigen-induced eosinophil infiltration in the same subjects (HENOCQ and VARGAFTING 1988). This suggests that allergic subjects respond differently to PAF in comparison with healthy individuals. Since the rate limiting acetyltransferase enzyme involved in PAF production is switched on in eosinophils obtained from individuals with eosinophilia compared with healthy subjects (LEE et al. 1984a), these observations indicate that PAF should be considered as a primary mediator involved in the induction and maintenance of the eosinophilic infiltration observed in allergic patients. It is interesting, therefore, that PAF antagonists inhibit antigen-induced eosinophil infiltration into the lungs of sensitized animals (COYLE et al. 1988; LELLOUCH-TUBIANA et al. 1988).

Preliminary studies indicate that inhalation of PAF by normal volunteers results in an increased recovery of neutrophils in bronchoalveolar lavage fluid at 6 h accompanied by activation of neutrophils in the circulation (WARDLAW et al. 1990). In animals, PAF, administered both locally and systemically, induces an eosinophil-rich infiltrate in the lungs (ARNOUX et al. 1988; COYLE et al. 1988; LELLOUCH-TUBIANA et al. 1988).

PAF also induces an extravascular recruitment of platelets into pulmonary tissue, where they are observed to be in close apposition to both airway smooth muscle and infiltrating eosinophils (LELLOUCH-TUBIANA et al. 1985). Such pathological changes have also been reported in allergic animals and have been identified in bronchoalveolar lavage fluid obtained from allergic asthmatics (METZGER et al. 1987). The contribution of extravascular platelets to the pathology of asthma has yet to be fully elucidated, but platelet depletion inhibits both PAF- and antigen-induced eosinophil infiltration in the lungs of animals (LELLOUCH-TUBIANA et al. 1988) and reduces PAF-induced bronchial hyperreactivity in the guinea-pig and rabbit (MAZZONI et al. 1985; COYLE et al. 1989). As platelets are a good source of smooth muscle mitogens, such as platelet-derived growth factor (ROSS et al. 1986), they may contribute to the hyperplasia of bronchial smooth muscle observed both in animals chronically treated with PAF (TOUVAY et al. 1989) and in asthmatic patients at autopsy (HEARD and HOSSAIN 1973).

5. Bronchial Hyperresponsiveness

One of the most important properties of PAF is its ability to induce a non-selective and long-lasting increase in bronchial hyperresponsiveness in both experimental animals and humans. PAF has been shown to elicit increased bronchial responsiveness in guinea-pigs (MAZZONI et al. 1985; BARNES et al. 1987; ROBERTSON et al. 1988), dog (CHUNG et al. 1986), sheep (CHRISTMAN et al. 1987), rabbits (COYLE et al. 1989) and normal human subjects (CUSS et al. 1986; RUBIN et al. 1987). In guinea-pig and rabbit, the increased responsiveness is dependent upon the presence of circulating platelets, since platelet depletion with a specific cytotoxic antibody abrogates

PAF-induced bronchial hyperresponsiveness (Mazzoni et al. 1985), whereas selective depletion of neutrophils is without effect (Morley et al. 1985).

In humans, the maximal increase in bronchial responsiveness to methacholine occurs 3 days after a single exposure to PAF and may persist in some individuals for up to 4 weeks (Cuss et al. 1986; Kim and Kim 1988). Because PAF is rapidly inactivated in the airways, such long-lasting changes must result from secondary mechanisms which are currently under investigation. Although PAF elicits airway hyperresponsiveness to a wide range of spasmogens, including histamine, acetylcholine, serotonin and substance P. the increased responsiveness is not secondary to alterations in receptor number, affinities or post-receptor transduction mechanims (at least for acetylcholine and histamine in the guinea-pig) (Robertson et al. 1988; Coyle et al. 1989). PAF has, however, been observed to elicit a down-regulation of β-adrenoceptors in human lung in vitro (Agrawal and Townley 1987), a phenomenon which may contribute to bronchial hyperresponsiveness and which is a feature of asthmatic airways in vitro. However, in guinea-pigs made hyperresponsive following treatment with i.v. PAF, there is a reduced bronchodilator response to isoproterenol in vivo, but the in vitro responsiveness of tracheal smooth muscle to isoproterenol and tracheal and lung β-receptor density remain unchanged (Barnes et al. 1987). This suggests that the impaired β-adrenoceptor function and is more likely to be due to airway oedema, which would not be reversible by a beta-agonist.

PAF-induced bronchial hyperresponsiveness may be a consequence of eosinophil infiltration (Fig. 4), and the degree of blood eosinophilia is closely related to the degree of bronchial hyperresponsiveness (Frigas and Gleich 1986; Taylor and Luksza 1987; Wardlaw et al. 1988). Eosinophils release cytotoxic materials, such as major basic protein (MBP), eosinophil cationic protein (ECP) and eosinophil peroxidase (EPO), which may lead to damage of airway epithelium (Frigas and Gleich 1986; Motojima et al. 1989). Epithelial disruption is a common feature of asthma, and loss of epithelium may contribute to airway hyperresponsiveness by loss of an epithelial-derived relaxant factor (Cuss and Barnes 1987; Vanhoutte 1988), by exposure of sensory nerve endings (Barnes 1986), or by the loss of enzymes which metabolize sensory neuropeptides (Frossard et al. 1989). Loss of epithelium could also explain the impaired bronchodilator response to beta agonists in vivo following administration of PAF in the guinea-pig, since beta agonists have a reduced effect on airway smooth muscle preparations denuded of airway epithelium (Vanhoutte 1988).

IV. Role in Asthma

1. Release of PAF in Asthma

The precise role of PAF in asthma remains unknown, although PAF may reproduce many features of asthma. The detection of PAF in biological fluids

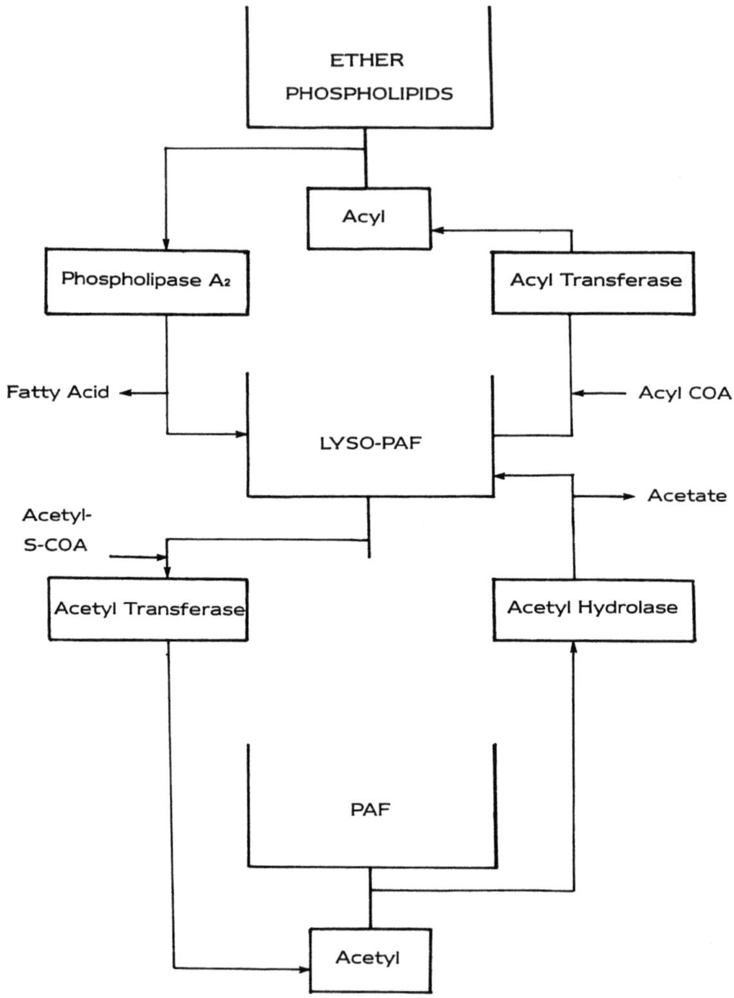

Fig. 4. Synthesis and degradation of platelet activating factor (PAF)

has been hampered by the lack of a simple assay system. Most attempts to measure PAF in biological fluids have relied upon the bioassay developed by HENSON and PINCKARD (1977), which is based upon the fact that PAF is able to selectively desensitize platelets to itself in vitro. A number of investigators have used this technique to show that a PAF-like material is released into the circulation concomitantly with antigen-induced bronchoconstriction (THOMPSON et al. 1984). PAF has also been detected in bronchoalveolar lavage fluid of asthmatics using this bioassay technique (COURT et al. 1987) and lyso-PAF has been detected in blood of allergic asthmatics undergoing allergen-induced late-onset responses (NAKAMURA et al. 1987).

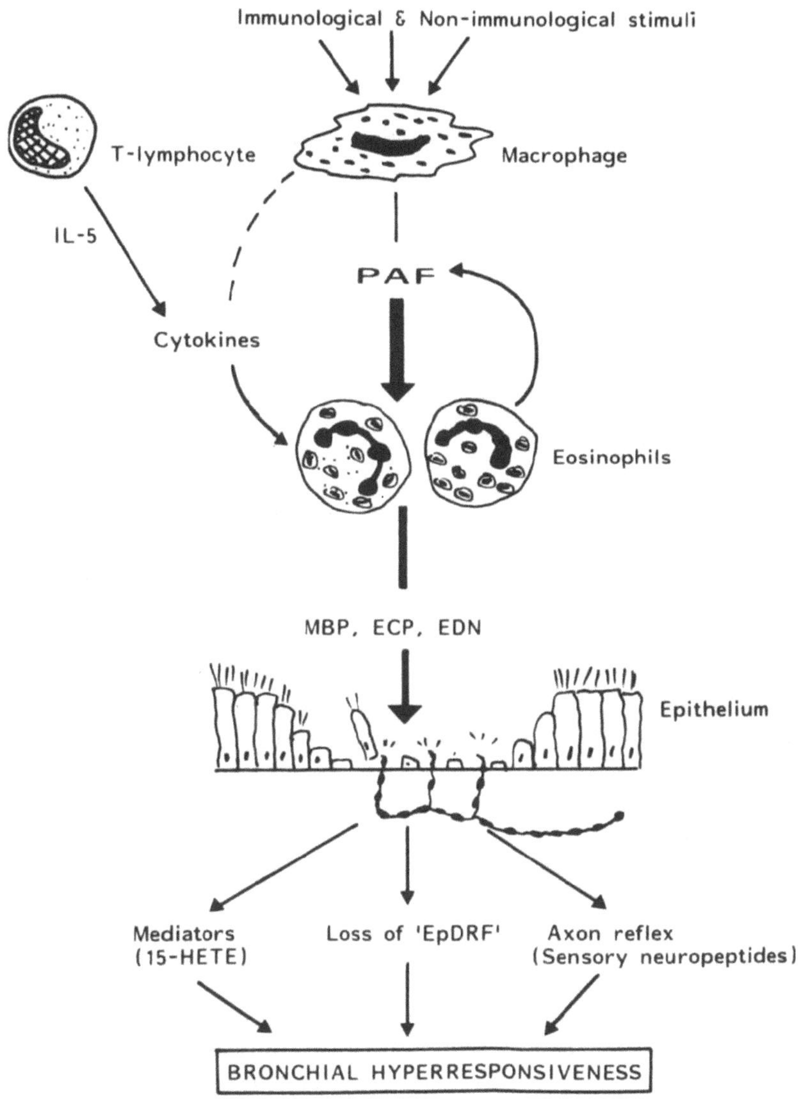

Fig. 5. Possible mechanism of bronchial hyperresponsiveness induced by platelet-activating factor (PAF).PAF may attract and activate eosinophils, which release basic proteins such as major basic protein (MBP), eosinophil cationic protein (ECP) and eosinophil derived neurotoxin (EDN) which are toxic to airway epithelium

2. PAF Antagonists

A more feasible approach to elucidate the role of PAF in asthma is the use of specific PAF antagonists, several of which are now available and some are undergoing clinical trials (CHUNG and BARNES 1988). PAF antagonists inhibit several of the effects of PAF which are relevant to asthma, including eosinophil activation (WARDLAW et al. 1986), bronchoconstriction (DESQUAND et al. 1986; TOUVAY et al. 1985), bronchial hyperresponsiveness (COYLE et al. 1988) and airway oedema (EVANS et al. 1988b). PAF-antagonists also inhibit certain aspects of allergic responses in both experimental animals and humans. BN 52021 and WEB 2086 inhibit allergen-induced bronchoconstriction in sensitized guinea-pigs (BRAQUET et al 1985; CASALS-STENZAL et al. 1987), and BN 52021 reduces the eosinophil activation and bronchial hyperresponsiveness resulting from allergen challenge (COYLE et al. 1988). Furthermore, in ragweed-sensitized rabbits, BN 52021 (COYLE et al. 1990) and L-652–839 (SMITH et al. 1988) inhibit late-onset airways obstruction and the increased bronchial hyperresponsiveness following allergen challenge. However, PAF antagonists do not inhibit propranolol- or indomethacin-induced bronchial hyperresponsiveness in guinea-pigs (DIXON et al. 1989).

Clinical studies with PAF antagonists are still in their infancy, but the ginkgolide mixture, BN 52063, CV 3988 and WEB 2086, appear to function as selective PAF antagonists in humans (CHUNG et al. 1987; ARNOUX et al. 1988; ADAMUS et al. 1988). Additionally, BN 52063 inhibits the late-phase cutaneous response to allergen in atopic subjects, which is assoicated with eosinophil infiltration and has some similarity with the pathology of the late-onset airways obstruction observed in allergic subjects (ROBERTS et al. 1988b). This suggests that PAF may be involved in the late-onset allergic response in humans. However, BN 52063, when taken orally, has recently been shown to be only a modest antagonist of PAF-induced bronchoconstriction in human airways (ROBERTS et al. 1988a), and therefore more potent antagonists may be necessary to evaluate the role of PAF in airway disease.

F. Kinins

Bradykinin and related kinins are formed from plasma precursors as part of the inflammatory response, yet there is relatively little information about the involvement of these potent vasoactive peptides in asthma (FULLER and BARNES 1988). Bradykinin itself was first isolated in 1949 from enzymatic treatment of blood, and later shown to be a nine amino acid peptide. Lysine-bradykinin (kallidin) has also been identified and has similar pharmacological properties (REGOLI and BARABÉ 1980).

I. Formation and Metabolism

Bradykinin in generated from α_2 globulins in plasma (kininogens) by the action of enzymes (kininogenases) which are produced by the liver (plasma kallikrein) and by other tissues (tissue kallikrein). In addition, human lung mast cells and basophils release a kininogenase, which is distinct from kallikrein (PROUD et al. 1983) and which may be identical to tryptase. Both a high molecular weight kininogen (HMWK) and a low molecular weight kininogen are recognised, the former probably acting as a substitute for plasma kallikrein and the latter for tissue kallikrein, since it is formed extravascularly.

Bradykinin and lys-bradykinin are inactivated by various proteolytic enzymes, but the major pathways involve kininase I (carboxypeptidase N) and kininase II (angiotensin converting enzyme, ACE), the latter enzyme being localised to endothelial cells. ACE inhibitors, such as captopril and enalapril, by preventing the action of kininase II may enhance the effects of endogenous bradykinin. Thus, enalapril increases the vascular effects of bradykinin in human skin (FULLER et al. 1987b). However, another potent ACE inhibitor (ramipril) has no effect on the bronchoconstrictor effect of bradykinin (DIXON et al. 1987), suggesting that ACE is not an important mechanism for degradation of bradykinin in human airways. Nor is the effect of bradykinin on human bronchi enhanced by captopril in vitro (FULLER et al. 1987a). Furthermore, there is no evidence that ACE inhibitors cause any deterioration in asthma, although they do produce a dry cough in some patients, which is unrelated to asthma (WEBB et al. 1986). These kinases are less active against lys-bradykinin than against bradykinin, so it may have more prolonged effects. In tissues neutral endopeptidase (enkephalinase) may be more important in degrading bradykinin, and inhibition of this enzyme by phosphoramidon leads to enhanced bronchoconstrictor effects (FROSSARD et al. 1990).

II. Receptors

Kinins activate specific receptors, and, using a series of bradykinin fragments and analogues, it has been possible to distinguish two types of receptor; B_1-receptors are selectively activated by lys-bradykinin and des-Arg-bradykinin, whereas B_2-receptors are more potently stimulated by bradykinin itself. While most responses to kinins appear to be mediated via B_2-receptors, there is some evidence that B_1-receptors might increase in experimental inflammation (REGOLI and BARABE 1980) and so may be relevant to asthma.

III. Airway Effects

Bradykinin is a potent bronchoconstrictor in several animal species. In humans, both intravenous and inhaled bradykinin causes bronchoconstriction in asthmatic but not in normal subjects (SIMONSSON et al. 1973; NEWBALL et al. 1975; FULLER et al. 1987a). In vitro bradykinin has almost no effect on human airways (FULLER et al. 1987a), suggesting that its bronchoconstrictor action is indirect. In contrast to the guinea-pig, aspirin does not reduce its bronchoconstrictor effect in human subjects, but cholinergic antagonists partially inhibit the response, suggesting that a vagal reflex mechanism is involved (FULLER et al. 1987). In dogs, bradykinin is a potent stimulant of bronchial C-fibres (KAUFMAN et al. 1980), and in other tissues produces its effects by releasing sensory neuropeptides from capsaicin-sensitive nerves (UEDA et al. 1984). It is possible that bradykinin causes bronchoconstriction in asthmatic patients by a similar action and activates axon reflex mechanisms. There is no evidence that bradykinin causes bronchial hyperresponsiveness in man, since inhalation of bradykinin does not increase responsiveness to other bronchoconstrictor mediators and even reduces the response to histamine, possibly because of release of bronchodilator PGs (FULLER et al. 1987a). Bradykinin is a potent vasodilator of canine bronchial vessels and also increases airway mucosal thickness (LAITINEN et al. 1987). Bradykinin also causes microvascular leakage in guinea-pig airways (PERSSON 1987) and produces a wheal and flare response in human skin. Injection of kallikrein into human skin causes a late reaction, suggesting that bradykinin may contribute to this inflammatory response (DOR et al. 1983) and raising the possibility that it plays a similar role in the late response to allergens in the airways. Bradykinin stimulates sensory nerve endings in airways and in human subjects produces pronounced dyspnoea (FULLER et al. 1987a), which may be reminiscent of the pain produced by bradykinin application to blister burns in human skin. It is possible that bradykinin may therefore contribute to the symptoms of asthma.

IV. Role in Asthma

Despite the evidence that kinins are released in experimental inflammation (REGOLI and BARABÉ 1980), there is little direct evidence that they are involved in asthma. This is because of difficulties of measurement in biological fluids and because of the lack of specific antagonists suitable for clinical use. Other studies have demonstrated that allergen challenge leads to production of bradykinin and lys-bradykinin in nasal washings of atopic individuals (PROUD et al. 1983). Furthermore, HMWKs could also be detected together with albumin (BAUMGARTEN et al. 1985), suggesting that increased vascular permeability allows entry of HMWK from which kinins are formed by local tissue kallikreins. Such measurements have not been made in the lower respiratory tract, but recently tissue kallikrein has been detected in bronchoalveolar lavage fluid of asthmatic subjects (CHRISTIANSEN et al. 1987).

There are currently no bradykinin antagonists in clinical use, but several potent competitive antagonists, which are peptide analogues of bradykinin, have been developed which may prove suitable in the future (VAVREK and STEWART 1985).

G. Adenosine

Recently there has been increasing interest in the possible involvement of the purine nucleoside adenosine in asthma, since it may be released by allergen challenge and may cause bronchoconstriction (MANN et al. 1986a).

I. Origin

Adenosine is generated extracellularly by dephosphorylation of AMP by the membrane-associated enzyme 5′-nucleotidase. Adenosine is therefore formed under conditions in which AMP is generated within the cell, such as excessive stimulation or under hypoxic conditions. Adenosine may then be taken up into the cells by facilitated transport (which is specifically blocked by dipyridamole), where it is converted back to AMP or broken down by adenosine deaminase to the inactive inosine. Extracellular adenosine is also rapidly inactivated by adenosine deaminase, and therefore adenosine has a very short duration of action, suggesting that it functions as a local hormone. Adenosine may be released from a variety of cells, including leukocytes and mast cells (MANN et al. 1986a).

II. Receptors

Adenosine interacts with specific cell surface receptors, which either inhibit (A_1) or stimulate (A_2) adenylate cyclase and which may be distinguished by selective agonists (LONDOS et al. 1980). Thus, for A_1 receptors N^6-phenylisopropyl adenosine (PIA) is more potent than *N*-ethylcarboxamide adenosine (NECA), whereas for A_2 receptors the order of potency is reversed. While adenosine receptors have been identified in lung by direct receptor binding studies (UKENA et al. 1985), the cellular localisation of the receptors is not known, although adenosine is likely to be active on a wide range of cells.

III. Actions

When administered by aerosol, adenosine induces rapid bronchoconstriction in asthmatic subjects, but has no effect on normal subjects (CUSHLEY et al. 1983); however, the mechanism of bronchoconstriction is not yet certain. Adenosine has little effect on human bronchi in vitro (FINNEY et al. 1985), suggesting that the bronchoconstrictor effect is indirect. Pre-treatment with an anticholinergic drug does not inhibit the bronchoconstriction, suggesting that it is not due to an irritant effect of the inhalation, but it is inhibited by cromoglycate sodium and an antihistamne, raising the possibility that mediator release from mast cells may be important (CUSHLEY and HOLGATE 1985; RAFFERTY et al. 1987). Adenosine is known to enhance the release of histamine from human lung mast cells under certain conditions (HUGHES et al.

1984) and may selectively enhance the secretion of histamine rather than that of newly formed mediators. The adenosine receptor mediating bronchoconstriction paradoxically appears to be the A_2-receptor, since NECA is more potent than PIA in causing bronchoconstriction in rats in vivo (PAUWELS and VAN DER STRAETEN 1986), and presumably this receptor is localised to airway mast cells. The effect of adenosine on airway secretions and other target cells of the airway has not been investigated.

IV. Role in Asthma

After allergen challenge, plasma concentrations of adenosine increase in asthmatics in parallel with bronchoconstriction, but no increase is seen after similar bronchoconstriction induced by methacholine (MANN et al. 1986a). The increase in plasma concentrations is unexpected, since the half-life of adenosine is so short, and may suggest that adenosine is generated secondarily from other cells.

Theophylline, at concentrations which are within the therapeutic range (55–110 μM), is an antagonist of adenosine receptors and its anti-asthma effects have, therefore, been ascribed to adenosine antagonism. Theophylline selectively inhibits the bronchoconstrictor action of inhaled adenosine (CUSHLEY et al. 1984), but it is unlikely that its anti-asthma effects are due to adenosine antagonism, since a related methylxanthine, enprofylline, has even more potent bronchodilator effects but is not an effective antagonist of adenosine receptors (PERSSON 1986). Thus, theophylline cannot be used as a probe to examine the role of endogenous adenosine in asthma, since it has many other effects. Until specific adenosine antagonists that do not have other actions can be used clinically, the role of adenosine in asthma remains difficult to evaluate.

H. Complement

Over 100 years ago it was recognised that serum contained soluble and heat-labile proteins which could lyse bacterial cells. It is now apparent that the complement cascade represents a complex system consisting of a range of plasma proteins that play a role in host defence and in a number of pathological disorders of both immunological and non-immunological origin. The activation sequence and generation of the various components is complex (BROWN et al. 1984) and we have concentrated on the involvement of two complement components, C3a and C5a (anaphylatoxins) which have documented airway effects (BEREND 1988).

I. Origin and Metabolism

The anaphylatoxins are fragments of the complement cascade that play little part in the further activation of the cascade itself, although they may regulate the further production of the C2 component, but may have inflammatory effects. C3a and C5a are generated following activation of the complement pathway by both the classical and the alternative pathways. The complete amino acid sequence of the anaphylatoxins has now been elucidated in several species, including humans, and there is considerable homology (HUGLI 1986). C5a has 74 amino acids and contains an oligosaccharide attached at position 64 with the active site being the carboxy terminal pentapeptide Met-Glu-Leu-Gly-Arg. The remainder of the molecule is required for functional binding to the C5a receptor, which is not so with C3a, although the carboxy terminal of this molecule is again the active site. C3a has 77 amino acids and the carboxy terminal pentapeptide Leu-Gly-Leu-Ala-Arg.

The anaphylatoxins are rapidly inactivated in plasma by the so-called anaphylatoxin inactivator (AI) which expresses a carboxypeptidase B function, removing the carboxy terminal arginine and leaving C3a des Arg and C5a des Arg, products devoid of much of the biological activity of the anaphylatoxins C3a and C5a but still retaining chemotactic activity. Until the recent development of carboxy peptidase inhibitors, the in vivo levels of anaphylatoxins have been difficult to measure.

II. Receptors

Specific membrane receptors have been identified which bind C3 and its various components. C3a receptors have been identified on leukocytes and mast cells, whilst C5a receptors have been identified on mast cells, monocytes, platelets and leukocytes. To date, specific receptors for anaphylatoxins have not been demonstrated in airway preparations, but C5a is able to contract bronchial smooth muscle preparations in vitro (REGAL et al. 1980; REGAL 1989).

III. Airway Effects

1. Smooth Muscle Contraction

Intravenous injection of guinea-pigs with C5a causes bronchoconstriction (REGAL 1989) but the mechanisms involved are unknown, although C5a and C5a des Arg induce the release of histamine (SACKEYFIO 1971), PGs (SACKEYFIO 1971) and LTs (STIMLER et al. 1982, STIMLER 1984) from guinea-pig lung. C5a elicits airway smooth muscle contraction in perfused guinea-pig lungs (REGAL et al. 1980), independently of histamine release. The precise contribution of AA metabolites to C5a-induced bronchoconstriction is not clear, although both cyclooxygenase and lipoxygenase inhibitors inhibit C5a-induced contraction of airway smooth muscle preparations. C3a is a less potent inducer of airway smooth muscle contraction than C5a in the guinea-pig (STIMLER et al. 1983). This effect appears to be mediated predominatly via a cyclooxygenase product, despite the release of histamine.

Both C3a and C5a induce marked tachyphylaxis in airway smooth muscle preparations, although there is no cross-desensitization between them, indicating that they are likely to activate discrete receptors (REGAL et al. 1980).

2. Vascular Effects

Complement activation has long been recognised as a trigger of increased vascular permeability in skin, which was believed to be secondary to histamine release (DIAS DA SILVA and LEPOW 1967). C5a and C3a induce vascular permeability through neutrophil activation, although the role of the neutrophil has not been fully elucidated (WEDMORE and WILLIAMS 1981d). Although C5a releases PAF from neutrophils (WEDMORE and WILLIAMS 1981), it is unlikely that PAF is the mediator responsible for the neutrophil-dependent vascular permeability induced by C5a as PAF antagonists do not inhibit C5a-induced oedema formation (HELLEWELL and WILLIAMS 1986). In humans, C5a produces immediate wheal and flare reactions in skin; an H1 anti-histamine reduced the flare response but not the wheal (YANCEY et al. 1985); in addition, biopsies of skin showed neutrophil infiltration, endothelial cell oedema and mast cell degranulation. There is little work on the role of C5a in airways, although preliminary studies have shown that C5a is associated with neutrophil recruitment in airways (IRVIN et al. 1986).

3. Mucus Secretion

Little is known about the effects of the anaphylatoxins on airway secretion of mucociliary clearance. C3a stimulates mucous glycoprotein secretion from human airways in vitro (MAROM et al. 1985), probably via a direct effect on secretory cells.

4. Chemotaxis and Cell Activation

One of the most widely studied effects of anaphylatoxins in their ability to induce activation of inflammatory cells. C5a and C5a des Arg are chemotactic for neutrophils, with a potency even greater than that of LTB_4 (MOVAT et al. 1984). C5a is also chemotactic for macrophages (RICHARDS et al. 1984), basophils (KAY and AUSTIN 1972) and eosinophils (KAY 1970). In contrast, human C3a is devoid of chemotactic activity (FERNANDEZ et al. 1978). Both C5a and C5a des Arg stimulate the adhesion of inflammatory cells and elicit the release of other mediators, including lysosomal enzymes, free oxygen radicals (BENDER and VAN EPPS 1985), lipoxygenase and cyclooxygenase products of AA metabolism (CLANCY et al. 1983) and PAF from both neutrophils (CAMUSSI et al. 1981) and eosinophils (et al. 1984a).

5. Bronchial Hyperresponsiveness

Inhalation of C5a des Arg causes increase airway responsiveness to histamine 4 h later (IRVIN et al. 1986), a time when neutrophil infiltration occurs in the airways. The increased airway responsiveness is reduced in animals rendered neutropenic, suggesting that neutrophils contribute to the induction of bronchial hyperresponsiveness by C5a.

IV. Role in Asthma

Little is known of the role of anaphylatoxins in human asthma, since studies with inhibitors of complement activation have not yet been reported in humans. Measurement of C5a and C3a have proved to be difficult in plasma and their release has not yet been demonstrated in asthma. The potent effects of these mediators on microvascular leakage and bronchial smooth muscle have not been confirmed in human subjects and no specific antagonists are available. Several clinical investigations have reported the activation of the complement cascade during asthma. Plasma C4 concentrations have been found to be elevated in childhood asthma and depressed in non-atopic adult asthmatics (KAY et al. 1974), although other investigators have not confirmed this observation (GODFREY and HAWKSLEY 1975; TOPILSKY et al. 1976; DIAZ et al. 1984). Furthermore, no changes in complement are detected in allergic asthmatics following either early or late reactions to allergen provocation (HUTCHCROFT and GUZ 1978; KAUFFMAN et al. 1983; ALAM et al. 1983). A few patients develop reduced hemolytic complement activity following allergen provocation (STALENHEIM and MACHEDO 1985), whereas others have reported an increase (BAUER et al. 1980). The role of complement in aspirin-sensitive asthma is equally controversial, since some investigators have reported decreased complement levels after oral challenge with aspirin (ARROYAVE et al. 1977), whereas others report changes in hemolytic complement activity or C4 in arterial or venous blood following aspirin provocation (PLESKOW et al. 1983). These studies do not exclude the possibility that there may be local complement activation within the airways in asthma. The use of specific inhibitors of the complement cascade, such as *N*-acetyl-aspartyl-glutanic acid (NAAGA), may be useful in asthma, and preliminary results have already indicated a beneficial effect in allergic rhinitis (GHAEM 1987).

I. Serotonin

Serotonin (5-hydroyxtryptamine) was once considered to be an important mediator of asthma, since it caused bronchoconstriction in several animal species, but its relevance to human asthma now seems doubtful.

I. Origin

Serotonin is formed by decarboxylation of tryptophan in the diet and is stored in secretory granules. In rodents serotonin is present in mast cell granules, but this is not the case in humans. Serotonin in humans is localised to neuroendocrine cells of the gastrointestinal and respiratory tract, to certain nerves and to secretory granules in platelets. The possible involvement in platelets in asthma (MORLEY et al. 1984) has, therefore, re-awakened interest in serotonin.

II. Receptors

The development of specific antagonists has made it possible to recognise at least three types of serotonin receptor. $5HT_1$-receptors are usually inhibitory, $5HT_2$-receptors are excitatory and mediate smooth muscle contraction and $5HT_3$-receptors are present on nerves and stimulate neurotransmitter release from certain peripheral nerves (RICHARDSON and ENGEL 1986). The development of more and more selective antagonists has provided evidence for heterogeneity within $5HT_1$-and $5HT_3$-receptor subtypes.

III. Airway Effects

Serotonin causes bronchoconstriction in several species, including guinea-pig, cat, rat, dog and monkey, but there is considerable doubt about its effect in human airways. Serotonin, paradoxically, relaxes human airways in vitro (RAFFESTIN et al. 1985). In vivo inhaled serotonin was reported to cause bronchoconstriction in some asthmatic patients (PANZANI 1962), but other studies have found no consistent bronchoconstrictor response in either normal or asthmatic subjects (TONNESEN 1985).

Serotonin also causes microvascular leakage in guinea-pig airways (SARIA et al. 1983) and could have a similar effect in humans since it causes a wheal response in human skin and stimulates an axon reflex (RICHARDSON and ENGEL 1986).

Few studies have been performed with antagonists of serotonin in asthma. Ketanserin, a $5HT_2$ antagonist, has no protective action against exercise-induced asthma (So et al. 1985).

J. Chemotactic Factors

Many of the mediators discussed above, and particularly LTB_4, 15–HETE, PAF and C5a, have potent chemotactic activity. In addition, a number of poorly defined large molecules have been identified as chemoattractants and investigated for their potential contribution to allergic inflammation. However, almost all the work in this area has relied upon in vitro observations of chemoattractant activity and no conclusive proof of the involvement of these materials in vivo has been obtained. Materials displaying chemotactic activity by neutrophils in vitro have been identified as products released from human leukaemic basophils, rat mast cells and extracts of lung tissue (O'DRISCOLL et al. 1982). Neutrophil chemotactic activity (NCA) has also been detected in the serum of patients undergoing experimentally-induced physical and temperature-induced urticaria, allergic and non-allergic bronchoconstriction (KAY and LEE 1982). However, many of the defined low molecular weight chemotactic factors (e.g. LTB_4 and PAF) avidly bind to plasma proteins and it still remains plausible that such chemotactic activity could be secondary to these low molecular weight materials bound to plasma proteins. The specificity of NCA as a marker of allergic responses is dubious because similar NCA chemotactic activity has been reported in patients with active bronchitis and pneumonia (CUNDELL et al. 1984). It seems likely that NCA may represent an indication of an acute inflammatory process.

There is another group of chemotactic factors which are selective for eosinophils. ECF-A has been identified in supernatants from IgE-challenged tissue extracts of human lung and isolated cell preparations, such as human leukaemic basophils and human mast cells. Furthermore, ECF-A activity has been identified in the serum of patients undergoing antigen-induced bronchoconstriction (METZGER et al. 1986). ECF-A was originally described as two tetrapeptides, having the sequence Val-Gly-Ser-Glu and Ala-Gly-Ser-Glu (GOETZL and AUSTEN 1975), but it is now clear that such tetrapeptides only form a very small component of the original ECF-A (WASSERMAN 1983). ECF-A is far less potent than PAF as a chemotactic agent from human eosinophils, however (WARDLAW et al. 1986).

K. Oxygen Radicals

Oxygen radicals are generated as part of the inflammatory reaction and are therefore likely to be involved in the pathophysiology of asthma. Activation of various inflammatory cells, including macrophages, neutrophils, eosinophils and mast cells generates the oxygen intermediates superoxide anion (O_2^-) and hydrogen peroxide (H_2O_2); the hydroxyl radical (OH^-) is formed non-enzymatically as a secondary reaction (BABIOR 1984). Oxygen radicals may have various toxic effects on cellular function, including inhibition of certain enzymes (especially those dependent on SH groups), damage to DNA and the formation of lipid peroxides from the polyunsaturated fatty acids present in the cell membrane. It is surprising that so little is known about the effects of oxygen radicals on airway function.

Oxygen radicals have effects on airway smooth muscle in vitro. H_2O_2 is the oxygen radical which appears to have the major effect on airway tone and causes contraction both in bovine (STEWART et al. 1981) and guinea-pig airways (RHODEN and BARNES 1989). In guinea-pig, the contractile effect of H_2O_2 is greatly enhanced by removal of epithelium, suggesting that oxygen radicals release a relaxant factor. The bronchoconstriction is also reduced by indomethacin, suggesting that H_2O_2 also releases constrictor cyclooxygenase products. Oxygen radicals may also affect airway smooth muscle by an action on β-adrenoceptor function. Thus, alveolar macrophages incubated with guinea-pig trachea lead to reduced relaxation responses to isoprenaline, an effect which is prevented by free radical scavengers (ENGELS et al. 1985). However, direct incubation of oxygen radicals with airways fails to alter β-receptor function (RHODEN and BARNES 1989).

Oxygen radicals also cause increased vascular permeability. possibly via a direct toxic effect on vascular endothelial cells (DEL MAESTRO et al. 1981) and might therefore contribute to oedema formation in asthmatic airways.

The role of oxygen-derived free radicals in asthma is still not certain, but perhaps studies using antioxidants or free radical scavengers might show some benefit. Ascorbic acid is an effective antioxidant and reduces methacholine-induced bronchoconstriction in asthmatic subjects (MOHSENIN et al. 1983), although this could be mediated through an alternative mechanism.

L. Conclusions

Many different mediators have now been implicated in asthma and we have discussed the evidence for their involvement in asthma. In most cases the evidence is circumstantial, and it will be necessary to develop more potent and selective antagonists before the role of each mediator in a complex inflammatory disease, such as asthma, can be elucidated. There is increasing evidence that there are complex interactions between mediators with amplification or modification of their effects, which may make it even more difficult to determine the contribution of a single mediator. The therapeutic implication is that an antagonist of a single mediator is unlikely to have a major clinical effect. Thus, even potent antihistamines have not proved to be effective in the management of clinical asthma. Perhaps PAF might prove to be the exception, since this mediator most closely mimics the pathological features of asthma and the imminent availability of specific PAF antagonists for clinical studies should shortly answer this question. It seems likely that other mediators will be described in the future and may contribute to the inflammatory reaction of asthma.

We have emphasised human studies where possible, since there appear to be marked differences between species in production of and response to inflammatory mediators. Furthermore, there is no animal model which duplicates all the features of human asthma, although animal studies have provided important information about the processes involved in asthma, such as microvascular leakage, which cannot yet be measured in human airways. In the future, there should be greater emphasis on human studies, particularly studies in asthmatic patients, in order to unravel the complexities of the inflammatory response and the contribution of different mediators.

Acknowledgement. We are very grateful to Madeleine Wray for her careful and detailed preparation of the manuscript.

References

Adams GK, Lichtenstein LM (1985) Indomethacin enhances response of human bronchus to antigen. Am Rev Respir Dis 131:8–10

Adamus WS, Heuer H, Meade CJ, Frey G, Brecht HM (1988) Inhibitory effect of oral WEB 2086, a novel selective PAF-acether antagonist on ex vivo platelet aggregation. Eur J Clin Pharmacol 35:237–240

Adelroth E, Morris MM, Hargreave FE, O'Byrne PM (1986) Airway responsiveness to leukotrienes C_4 and D_4 and to methacholine in patients with asthma and normal controls. N Engl J Med 315:480–484

Adler KB, Schwartz JE, Anderson WH, Welton AF (1987) Platelet activating factor stimulates secretion of mucin by explants of rodent airways in organ culture. Exp Lung Res 13:25–43

Agrawal DK, Townley RD (1987) Effects of platelet activating factor on beta-adrenoceptors in human lung. Biochem Biophys Res Commun 143:1–6

Alam R, Roznniecki J, Swatko A, Kuzminska B (1983) Complement in allergen-induced bronchospasm in house dust RAST negative asthmatic patients. Allergol Immunopathol (Madr) 11:431–433

Al-Bazzaz FJ, Yadava VP, Westenfelder C (1981) Modification of Na and Cl transport in canine tracheal mucosa by prostaglandins. Am J Physiol 240: F101–105

Ameisen JC, Capron A, Joseph M, Macloof J, Voring H, Pancre V, Fournier E, Wallaert B, Tonnel AB (1985) Aspirin-sensitive asthma:abnormal platelet response to drugs including asthmatic attacks; diagnostic and physiopathological implications. Int Arch Allergy Appl Immunol 77:107–114

Archer CB, Froehlich W, Page CP, Paul W, Morley J, MacDonald DM (1984a) Synergistic interaction between prostaglandins and Paf-acether in experimental animals and man. Prostaglandins 27:495–501

Archer CB, Page CP, Paul W, Morley J, MacDonald DM (1984b) Inflammatory characteristics of platelet activating factor (PAF-acether) in human skin. Br J Dermatol 110:45–50

Archer CB, MacDonald DM, Morley J, Page CP, Paul W, Sanjar S (1985a) Effects of serum albumin, indomethacin and histamine H1-antagonists on paf-acether-induced inflammatory responses in the skin of experimental animals and man. Br J Pharmacol 85:109–113

Archer CB, Page CP, Morley J, MacDonald DM (1985b) Accumulation of inflammatory cells in response to intracutaneous platelet-activating factor (Paf-acether) in man. Br J Dermatol 112:285–290

Arm JP, Horton CE, Mencia-Huerta JM, House F, Eiser NM, Clark TJH, Spur BW, Lee TH (1988a) Effect of dietary supplementation with fish oil lipids on mild asthma. Thorax 43:84–92

Arm JP, Spur BW, Lee TH (1988b) The effects of inhaled leukotriene E_4 on the airway responsiveness to histamine in subjects with asthma and normal subjects. Allergy Clin Immunol 82:654–660

Arnout J, van Hecken A, de Lepleleire I, Miyamoto Y, Holmes I, de Schepper P, Vemylen J (1988) Effectiveness and tolerability of CV-3988, a selective PAF antagonist, after intravenous administration to man. Br J Clin Pharmacol 25:445–451

Arnoux B, Joseph M, Simoes MH, Tonnel AB, Duroux P, Capron A, Benveniste J (1987) Antigenic release of Paf-acether and beta-glucuronidase from alveolar macrophages of asthmatics. Bull Eur Physiopathol Respir 23:119–124

Arnoux B, Denjean A, Page CP, Nolibe D, Morley J, Benveniste J (1988) Accumulation of platelets and eosinophils in baboon lung after PAF-acether challenge: inhibition by ketotifen. Am Rev Respir Dis 137:855–860

Arrang JM, Garbang M, Lancelot JC et al. (1987) Highly potent and selective ligands for histamine H_3-receptors. Nature 327:117–123

Arroyave CM, Stevenson DD, Vaughan JH, Tan EM (1977) Plasma complement changes during bronchospasm produced in asthmatic patients. Clin Allergy 7:173–182
Ash ASP, Schild HO (1966) Receptors mediating some actions of histamine. Br J Pharmacol 27:427–439
Aursudkij B, Rogers DF, Evans TW, Alton EWFW, Chung KF, Barnes PJ (1987) Reduced tracheal mucus velocity in guinea-pig in vivo by platelet activating factor. Am Rev Respir Dir 35:A160
Babior BM (1984) The respiratory burst of phagocytes. J Clin Invest 73:599–601
Bach MK, Brashler JR, Smith HW, Fitzpatrick FA, Sun FF, McGuire JC (1982) 6, 9-deexpoxy-6, 9-(phenylimino)-delta6, 8-prostaglandin I1, (U-60, 257), a new inhibitor of leukotriene C and D synthesis: in vitro studies. Prostaglandins 23:759–770
Barnes NC, Piper PJ, Costello JF (1984a) Actions of inhaled leukotrienes and their interactions with other allergic mediators. Prostaglandins 28:629–631
Barnes NC, Piper PJ, Costello JF (1984b) Comparative effects of inhaled leukotriene C_4, leukotriene D_4, and histamine in normal human subjects. Thorax 39: 500–504
Barnes NC, Piper PJ, Costello J (1987) The effect of an oral leukotriene antagonist L-649, 923 on histamine and leukotriene D_4-induced bronchoconstriction in normal man. J Allergy Clin Immunol 79:816–821
Barnes PJ (1986) Asthma as an axon reflex. Lancet 1:242–245
Barnes PJ (1987) Inflammatory mediator receptors and asthma. Am Rev Respir Dis 135:S26–31
Barnes PJ, Brown MJ (1981) Venous plasma histamine in exercise and hyperventilation induced asthma in man. Clin Sci 61: 159–162
Barnes PJ, Chung KF (1987) PAF closely mimics pathology of asthma. Trends Pharmacol Sci 8:285–287
Barnes PJ, Ind PW, Brown MJ (1982) Plasma histamine and catecholamines in stable asthmatic subjects. Clin Sci 62:661–665
Barnes PJ, Grandordy BM, Page CP, Rhoden KJ, Robertson DN (1987) The effect of platelet activating factor on pulmonary beta-adrenoceptors. Br J Pharmacol 90:709–715
Barnes PJ, Chung KF, Page CP (1988a) Inflammatory mediators in asthma. Pharmacol Rev 40:49–84
Barnes PJ, Chung KF, Page CP (1988b) Platelet-activating factor as a mediator of allergic disease. J Allergy Clin Immunol 81:919–934
Basran GS, Morley J, Paul W, Turner-Warwick M (1982) Evidence in man for synergistic interactions between putative mediators of acute inflammation and asthma. Lancet 1:935–937
Basran GS, Page CP, Paul W, Morley J (1984) Platelet activating factor: a possible mediator of the dual response to allergen. Clin Allergy 14:75–79
Bauer X, Dorsch W, Becker T (1980) Levels of complement factors in human serum during immediate and late asthmatic reactions and during acute hypersensitivity pneumonitis. Allergy 35:383–390
Baumgarten CR, Togias AG, Naclerio RM, Lichtenstein IM, Norman PS, Proud D (1985) Influx of kininogens into nasal secretions after antigen challenge of allergic individuals. J Clin Invest 76:191–197
Beer DJ, Osband ME, McCaffrey RP, Soter NA, Rocklin RE (1982) Abnormal histamine-induced suppressor-cell function in atopic subjects. N Engl J Med 306:454–458
Bender JG, Van Epps DE (1985) Stimulus interactions in release of superoxide anions (O_2^-) from human neutrophils. Further evidence for multiple pathways of activation. Inflammation 9:67–79
Benveniste J, Henson PM, Cochrane CG (1972) Leucocyte dependent histamine release from rabbit platelets: the role of IgE, basophils and a platelet activating factor. J Exp Med 136:1356–1377

Berend N (1988) Complement. In: Barnes PJ, Rodger IW, Thomson NC (eds) Asthma: basic mechanisms and clinical management. Academic, London, pp 325–339

Bianco S, Robuschi M, Cesarani C, Ganfoldi C, Kamburoff P (1978) Prevention of aspecifically induced bronchoconstriction by PGI_2 and 20-methyl-PGI_2 in asthmatic patients. Pharmacol Res Commun 10:657–675

Bisgaard H, Kristensen J, Sondergaard J (1982) The effect of leukotriene C_4 and D_4 on cutaneous blood flow in humans. Prostaglandins 23:797–801

Bito LZ (1972) Accumalation and apparent active transport of prostaglandins by some rabbit tissues in vitro. J Physiol (Lond) 221:371–387

Bjork J, Hedqvist P, Afors K-E (1982) Increase of vascular permeability induced by leukotriene B_4 and the role of polymorphonuclear leukocytes. Inflammation 6:189–200

Bonnet J, Thibaudean D, Bessin P (1983) Dependency of PAF-acether induced bronchospasm on the lipoxygenase pathway in the guinea-pig. Prostaglandins 26:457–466

Boushey HA, Holtzman MJ, Sheller JR, Nadel JA (1980) Bronchial hyperreactivity. Am Rev Respir Dis 121:389–413

Braquet P, Etienne A, Touvay C, Bourgain R, Lefort J, Vargaftig BB (1985) Letter – Involvement of platelet activating factor in respiratory anaphylaxis, demonstrated by Paf-acether inhibitor BN 52021. Lancet 1:1501

Bray MA, Cunningham FM, Ford-Hutchinson AW, Smith MJH (1981) Leukotriene B_4: a mediator of vascular permeability. Br J Pharmacol 72:483–486

Brink C, Grimaud C, Guillot C, Orehek J (1980) The interaction between indomethacin and contractile agents on human isolated airway muscle. Br J Pharmacol 69:383–388

Britton JR, Hanley SP, Tattersfield AE (1987) The effect of an oral leukotriene D_4 antagonist L-649, 923 on the response to inhaled antigen in asthma. J Allergy Clin Immunol 79:811–816

Brown EJ, Joiner KA, Frank MM (1984) Complement. In: Paul WE (ed) Fundamental immunology. Raven, New York, pp 645 – 668

Bruijnzeel PLB, Koenderman L, Kok PTM, Hamelink ML, Verhagen JL (1986) Platelet activating factor (PAF-acether) induced leukotriene C_4 formation and luminol dependent chemi-luminescence of human eosiniphils. Pharmacol Res Commun 18:61–69

Buckner CK, Krell RD, Laravuso RB, Coursin DB, Bernistein PR, Will JA (1986) Pharmacological evidence that human intralobar airways do not contain different receptors that mediate contractions to leukotrienes C_4 and D_4. J Pharmacol Exp Ther 237:558–562

Bussolino F, Cammussi G, Aglietta M, Braquet P, Bosia A, Pescarmona G, Sanavio F, D'Urso N, Marchisio PC (1987) Human endothelial cells are targets for platelet activating factor (PAF) I. PAF induces changes in cytoskeleton structures. J Immunol 139:2439–2446

Camp RDR, Coutts AA, Greaves MW, Kay AB, Walport MJ (1983) Responses of human skin to intradermal injection of leukotrienes C_4, D_4 and B_4. Br J Pharmacol 80:497–502

Camussi G. Tetta C, Segoloni G, Deregibus MC, Bussolino F (1981) Neutropenia induced by platelet-activating factor (PAF-acether) released from neutrophils: The inhibitory effect of prostacyclin (PGI_2). Agents Actions 11:550–552

Capron M, Benveniste J, Braquet P, Capron A (1988) Role of PAF-acether in IgE-dependent activation of eosinophils. In:Braquet P (ed) New trends in lipid mediator research. Karger, Basel, pp 10–17

Carstairs JR, Norman P, Abram T, Barnes PJ (1988) Autoradiographic localization of leukotriene C_4 and D_4 binding sites in guinea-pig lung. Prostaglandins 35:503–514

Carswell H, Nahorski SR (1982) Distribution and characteristics of histamine H1-receptors in guinea-pig airways identified by [^{3}H] mepyramine. Eur J Pharmacol 81:301–307

Casale TB, Wood D, Richerson HB, Trapp S, Metzger WJ, Zavala D, Hunninghake GW (1987) Elevated bronchoalveolar lavage fluid histamine levels in allergic asthmatics are associated with methacholine bronchial hyperresponsiveness. J Clin Invest 79:1197–1203
Casals-Stenzel J, Muacevic G, Weber K-H (1987) Pharmacological actions of WEB 2086, a new specific antagonist of platelet activating factor. J Pharmacol Exp Ther 241:974–981
Cerrina J, Raffestin B, Labat C, Boullet C, Bayol A, Gateau O, Brink C (1983) Effects of PAF-acether on isolated muscle preparations from the rat, guinea-pig and human lung. In: Benveniste J, Arnoux B (eds) Platelet-activating factor INSERM symposium no 23. Elsevier, Amstardam, pp 205–212
Chan TB, Shelton DM, Eiser NM (1986) Effect of an oral H1-receptor antagonist, terfenadine, on antigen-induced asthma. Br J Dis Chest 80:375–384
Chanez P, Dent G, Yukawa T, Chung KF, Barnes PJ (1988) Increased eosinophil responsiveness to platelet-activating factor in asthma. Clin Sci 74:5P
Chap H, Mauco G, Simon MF, Benveniste J, Douste-Blazy L (1981) Biosynthetic labelling of platelet activating factor (PAF-acether) from radioactive acetate by stimulated platelets. Nature 289:312–314
Christiansen SC, Proud D, Cochrane CG (1987) Detection of tissue kallikrein in the bronchoalveolar lavage fluid of asthmatic patients. J Clin Invest 79:188–197
Christman BW, Lefferts PL, Snapper JR (1987) Effect of platelet activating factor on aerosol histamine responses in awake sheep. Am Rev Respir Dis 135:1267–1270
Chung KF (1986) Role of inflammation in the hyperreactivity of the airways in asthma. Thorax 41:657–662
Chung KF, Barnes PJ (1988) PAF antagonists: their therapeutic potential in asthma. Drugs 35:93–103
Chung KF, Barnes PJ (1989) Effects of platelet activating factor on airway calibre, airway responsiveness and circulating cells in asthmatic subjects. Thorax 44: 108–115
Chung KF, Evans TW, Graf PD, Nadel J (1985) Modulation of cholinergic neurotransmission in canine airways by thromboxane-mimetic, U 46619. Eur J Pharmacol 117:373–375
Chung KF, Aizawa H, Leikauf GD, Ueki IF, Evans TW, Nadel JA (1986) Airway hyperresponsiveness induced by platelet-activating factor: role of thromboxane generation. J Pharmacol Exp Ther 236:580–584
Chung KF, Dent G, McCusker M, Guinot Ph, Page CP., Barnes PJ (1987) Effect of a ginkgolide mixture (BN 52063) in antagonising skin and platelet responses to platelet activating factor in man. Lancet 1:248–251
Chung KF, Minette P, McCusker M, Barnes PJ (1988) Ketotifen inhibits the cutaneous but not the airway responses to platelet-activating factor in man. J Allergy Clin Immunol 81:1192–1197
Clancy RM, Dahinden CA, Hugli TE (1983) Arachidonate metabolism of human polymorphonuclear leukocytes stimulated by *N*-formyl-Met-Leu-Phe or complement component C5a is independent of phospholipase activation. Proc Natl Acad Sci USA 80:7200–7204
Clark RA, Gallin JI, Kaplan AP (1975) The selective eosinophil chemotactic activity of histamine. J Exp Med 142:1462–1476
Cockroft DW, Murdock KY (1987) Comparative effects of inhaled salbutamol, sodium cromoglycate and beclomethasone diproprionate on allergen-induced early asthmatic responses, late asthmatic responses and increased bronchial responsiveness to histamine. J Allergy Clin Immunol 79:734–740
Coles SJ, Neill KH, Reid LM, Austen KF, Nii Y, Corey EJ, Lewis RA (1983) Effects of leukotrienes C_4 and D_4 an glycoprotein and lysozyme secretion by human bronchial mucosa. Prostaglandins 25:155–170
Cookson WOCM (1987) Bronchodilator action of the antihistamine terfenadine. Br J Clin Pharmacol 24:120–121
Copas JL, Borgeat P, Gardiner PJ (1982) The actions of 5-, 12-, and 15-HETE on tracheobronchial smooth muscle. Prostaglandins Leukotrienes Med 8:105–114

Court EN, Goadby P, Hendrick DJ, Kelly CA, Kingston W, Stenton SC, Walters EH (1987) Platelet-activating factor in bronchoalveolar lavage fluid from asthmatic patients. Br J Clin Pharmacol 24:258P
Coyle A, Spina D, Page CP (1990) PAF-induced bronchial hyperreactivity in the rabbit: contribution of platelets and airway smooth muscle. Br J Pharmacol 101: 31–38
Coyle AJ, Urwin SC, Page CP, Touvay C, Villain B, Braquet P (1988) The effect of the selective antagonist BN 52021 on PAF and antigen-induced bronchial hyperreactivity and eosinophil accumulation. Eur J Pharamcol 148:51–8
Creese BR, Bach MK (1983) Hyperreactivity of airways smooth muscle produced *in vitro* by leukotrienes. Prostaglandins Leukotrienes Med 11:161–169
Creticos PS, Peters SP, Adkinson NF et al. (1984) Peptide leukotriene release after antigen challenge in patients sensitive to ragweed. N Engl J Med 310:1626–1630
Cundell DR, Morgan DJR, Davis RJ (1984) N. C. F. – A mast cell specific chemotactic factor? Clin Sci 66:50P
Curry JJ (1946) The action of histamine on the respiratory tract in normal and asthmatic subjects. J Clin Ivest 25:785–791
Cushley MJ, Holgate ST (1985) Adenosine-induced bronchoconstriction in asthma: role of mast cell-mediator release. J Allergy Clin Immunol 75:272–278
Cushley MJ, Tattersfield AE, Holgate ST (1983) Inhaled adenosine and guanosine on airway resistance in normal and asthmatic subjects. Br J Clin Pharmacol 15:161–165
Cushley MJ, Tattersfield AE, Holgate ST (1984) Adenosine-induced bronchoconstriction in asthma: antagonism by inhaled theophylline. Am Rev Respir Dis 129:380–384
Cushley MJ, Wee LH, Holgate ST (1986) The effect of inhaled 5-hydroxytryptamine (5 HT, serotonin) on airway calibre in man. Br J Clin Pharmacol 22:487–490
Cuss FM, Barnes PJ (1987) Epithelial mediators. Am Rev Respir Dis 136:S32–35
Cuss FM, Dixon CMS, Barnes PJ (1986) Effects of inhaled platelet activating factor on pulmonary function and bronchial responsiveness in man. Lancet 2:189–192
Cuthbert MF (1971) Bronchodilatory activity of aerosols of prostaglandins E_1 and E_2 in asthmatic subjects. Proc R Soc Med 64:15–16
Dahlen SE, Hedqvist P, Hammarstrom B, Samuelsson B (1980) Leukotrienes are potent constrictors of human bronchi. Nature 288: 484–486
Dahlen SE, Bjork J, Hedqvist P, Arfors K-E, Hammarstrom S, Lindgren J-A, Samuelsson B (1981) Leukotrienes promote plasma leakage and leukocyte adhesion in postcapilliary venules: in vivo effects with relevance to the acute inflammatory response. Proc Natl Acad Sci USA 78:3887–3891
Dahlen SE, Hedqvist P, Hammarstrom S (1983) Contractile activities of several cysteine-containing leukotrienes in the guinea-pig lung strip. Eur J Pharmacol 86:207–215
Dahlen SE, Raud J, Serhan CN, Bjork B, Samuelsson B (1987) Biological activities of lipoxin A include lung strip contraction and dilation of arterioles in vivo. Acta Physiol Scand 130:643–648
Dale HH, Laidlaw PP (1919) Histamine shock. J Physiol 52:355
Davidson AB, Lee TH, Scamon PD, Solway J, McFadden ER, Ingram RH, Corey EJ, Austen KF, Drazen JM (1987) Bronchoconstrictor effects of leukotriene E_4 in normal and asthmatic subjects. Am Rev Respir Dis 135:333–337
del Maestro RF, Bjork J, Arfors KE (1981) Increase in microvascular permeability induced by enzymatically generated free radicals. I. In vivo study. Microvasc Res 22:239–254
Dent G, Ukena D, Chanez P, Sybrecht G, Barnes P (1989) Characterization of PAF receptors on human neutrophils using the specific antagonist WEB 2086. Correlation between receptor binding and function. FEBS Lett 244:365–368
Desquand S, Touvay C, Randon J, Lagente V, Vilain B, Maridonneau-Parini I, Etienne A, Lefort J, Braquet P, Vargaftig BB (1986) Interference of BN 52021 (ginkgolide B) with the bronchopulmonary effects of PAF-acether in the guinea pig. Eur J Pharmacol 127:83–95

Dewar A, Archer CB, Paul W, Page CP, MacDonald DM, Morley J (1984) Cutaneous and pulmonary histopathological response to platelet activating factor (Paf-acether) in the guinea-pig. J Pathol 144:25–34
Dias da Silva W, Lepow IH (1967) Complement as a mediator of inflammation. II. Biological properties of anaphylotoxin prepared with purified components of human complement. J Exp Med 125:921–946
Diaz PD, Galleguillos FR, Gonzalez MC, Pantin CFA, Kay AB (1984) Bronchoalveolar lavage in asthma: the effect of disodium cromoglycate on leukocyte counts, immunoglobulins and complement. J Allergy Clin Immunol 74:41–48
Dixon CMS, Fuller RW, Barnes PJ (1987) The effect of an angiotensin converting enzyme inhibitor, ramipril, on bronchial responses to inhaled histamine and bradykinin in asthmatic subjects. Br J Clin Pharmacol 23:91–93
Dixon EJA, Wilsoncroft P, Robertson DN, Page CP (1989) PAF does not contribute to bronchial hyperreactivity induced by indomethacin and propranolol. Br J Pharmacol 97:717–722
Dor PJ, Vervolet D, Sarene M, Andrac L, Bunerandi JJ, Charpin J (1983) Induction of late cutaneous reaction by kallikrein injection: comparison with allergic-like response to compound 48/80. J Allergy Clin Immunol 71:363–370
Drazen JM, Austen KF, Lewis RA, Clark DA, Goto G, Marfat A, Corey EJ (1980) Comparative airway and vascular activities of leukotrienes C_1 and D in vivo and in vitro. Proc Natl Acad Sci USA 77: 4354–4358
Engels F, Oosting RS, Nijkamp FP (1985) Pulmonary macrophages induce deterioration of guinea-pig tracheal beta-adrenergic function through release of oxygen radicals. Eur J Pharmacol 111:143–144
Evans TW, Chung K, Rogers DF, Barnes PJ (1987) Effect of platelet-activating factor on airway vascular permeability: possible mechanisms. J Appl Physiol 63: 479–484
Evans TW, Dent G, Rogers DF, Aursudkij B, Chung KF, Barnes PJ (1988a) Effect of a PAF antagonist, WEB 2086, on airway microvascular leakage in the guinea pig and platelet aggregation in man. Br J Pharmacol 94:164–168
Evans TW, Rogers DF, Aursudkij B, Chung KF, Barnes PJ (1988b) Microvascular leakage associated with antigen-induced bronchial anaphylaxis is inhibited by nedocromil sodium or BW 755C in the guinea-pig. Am Rev Respir Dis 138:395–399
Evans TW, Rogers DF, Aursudkij B, Chung KF, Barnes PJ (1989) Regional and time-dependent effects of inflammatory mediators on airway microvascular permeability in guinea pigs. Clin Sci 76:479–485
Fairfax AJ, Hanson JM, Morley J (1983) The late reaction following bronchial provocation with house dust mite allergen: dependence on arachidonic acid metabolism. Clin Exp Immunol 52:393–398
Farr RS, Cox CP, Wardlow ML, Jorenson R (1980) Preliminary studies of an acid labile factor (ALF) in human sera that inactivates platelet-activating factor (PAF). Clin Immunol Immunopathol 15:318–330
Feinmark SJ, Lindgren JA, Claesson H-E, Malmsten C, Samuelsson B (1981) Stimulation of human leukocyte degranulation by leukotriene B_4 and its omega-oxidized metabolites. FEBS Lett 136:141–144
Fernandez HN, Henson PM, Otani A, Hugli TE (1978) Chemotactic response to human C3a and C5a anaphylotoxins I. Evaluation of C3a and C5a leukotaxis in vitro and under simulated in vivo conditions. J Immunol 120:109–115
Finney MJB, Karlsson J-A, Persson CGA (1985) Effects of bronchoconstrictors and bronchodilators on a novel human small airway preparation. Br J Pharmacol 85:29–36
Fish JE, Ankin MG, Adkinson Jr NF, Peterman VI (1981) Indomethacin modification of immediate-type immunologic airway responses in allergic asthmatic and nonasthmatic subjects. Am Rev Respir Dis 123:609–614
Fish JE, Newball HH, Norman PS, Peterman VE (1983) Novel effects of $PGF_{2\alpha}$ in airway function in asthmatic subjects. J Appl Physiol 54:105–122

Flower RJ, Harvey EA, Kingston WP (1976) Inflammatory effects of prostaglandin D_2 in rat and human skin. Br J Pharmacol 56:229–233
Ford-Hutchinson AW (1983) Neutrophil aggregating properties of PAF-acether and leukotriene B_4. Int J Immunopharmacol 5:17–21
Ford-Hutchinson AW, Bray MA, Doig MV, Shipley ME, Smith MJH (1981) Leukotriene B_4, a potent chemokinetic and aggregating substance released from polymorphonuclear leukocytes. Nature 286:264–265
Foreman JC, Norris DB, Rising TJ, Webber SE (1985) The binding of [^{3}H]-tiotidine to homogenates of guinea-pig lung parenchyma. Br J Pharmacol 86:475–482
Friedman MM, Kaliner MA (1987) Human mast cells and asthma. Am Rev Respir Dis 135:1157–1164
Frigas E, Gleich GJ (1986) The eosinophil and the pathology of asthma. J Allergy Clin Immunol 77:527–537
Frossard N, Stretton CP, Barnes PJ (1990) Modulation of bradykinin responses in airway smooth muscle by epithelial factors Agents Actions (in press)
Frossard N, Rhoden KJ, Barnes PJ (1989) Influence of epithelium on guinea pig airway responses to tachykinins: role of endopeptidase and cyclooxygenase. J Pharmacol Exp Ther 248:292–299
Fujimura M, Sasaki F, Natsumi Y, Takahashi Y, Hipfumi S, Taga K, Mifune J, Tanaka T, Matsuda T (1986) Effects of a thromboxane synthetase inhibitor (OKY-046) and a lipoxygenase inhibitor (AA-861) on bronchial responsiveness to acetylcholine in asthmatic subjects. Thorax 41:955–959
Fuller RW, Barnes PJ (1988) Kinins. In: Barnes PJ, Rodger IW, Thomson NK (eds) Basic mechanisms and clinical management. Academic, London, pp 259–272
Fuller RW, Dixon CMS, Dollery CT, Barnes PJ (1986) Prostaglandin D_2 potentiates airway responses to histamine and methacholine. Am Rev Respir Dis 133:252–254
Fuller RW, Dixon CMS, Cuss FMC, Barnes PJ (1987a) Bradykinin-induced bronchoconstriction in man: mode of action. Am Rev Respir Dis 135:176–180
Fuller RW, Warren JB, McCusker M, Dollery CT (1987b) Effect of enalapril on skin responses to bradykinin in man. Br J Clin Pharmacol 23:88–90
Gardiner PJ (1975) The effects of some natural prostaglandins on isolated human circular bronchial muscle. Prostaglandins 10:607–616
Gardiner PJ (1986) Characterisation of prostanoid relaxant/inhibitory receptors (u) using a highly selective agonist, TR4979. Br J Pharmacol 87:45–56
Gardiner PJ, Collier HOJ (1980) Specific receptors for prostaglandins in airways. Prostaglandins 19:819–841
Ghaem A (1987) A preliminary evaluation of the effect of *N*-acetyl aspartyl glutanate on pollen nasal challenge and measured by rhinomanometry and symptomatology. Allergy 42:626–630
Godard P, Chaintreuil J, Damon M, Coupe M, Flandre O, Crastes de Paulet A, Michel FB (1982) Functional assessment of alveolar macrophages: comparison of cells from asthmatics and normal subjects. J Allergy Clin Immunol 70:88–93
Godfrey RC, Hawkesley MR (1975) C4 levels and the classification of bronchial asthma. Lancet 1:464–465
Goetzl EJ, Austen KF (1975) Purification and synthesis of eosinophilotactic tetrapeptides of human lung tissue: identification as eosinophil chemotactic factor of anaphylaxis. Proc Natl Acad Sci USA 72:4123–4127
Goetzl EJ, Pickett WC (1980) The human PMN leukocyte chemotactic activity of complex hydroxy-eicosatetraenoic acids (HETEs). J Immunol 125:1789–1791
Goetzl EJ, Woods JM, Gorman RR (1977) Stimulation of human eosinophil and neutrophil polymorphonuclear leukocyte chemotaxis and random migration by 12-L-hydroxy-5, 8, 10, 14-eicosatetraenoic acid (HETE). J Clin Invest 59:179–183
Goldman DW, Goetzl EJ (1982) Characterization of a receptor on human neutrophils for the chemotactic mediator 5, 12-dihydroxy-6, 14 *cis*-8, 10-*trans*-eicosatetraenoic acid (leukotriene B4). J Immunol 129:1600–1604

Goldstein IM, Malmsten CL, Kindahl H et al. (1978) Thromboxane generation by human peripheral blood polymorphonuclear leukocytes. J Exp Med 148:787–792
Goswami SK, Ohashi M, Panagiotis S, Marom Z (1987) Platelet activating factor enhances mucous glycoprotein release from human airways in vitro. Am Rev Respir Dis 135:A159
Grandordy B, Barnes PJ (1987) Phophoinositide turnover in airway smooth muscle. Am Rev Respir Dis 136:817–820
Halonen M, Palmer JD, Lohman IC, McManus LM, Pinckard RN (1980) Respiratory and circulatory alterations induced by acetyl glyceryl ether phosphorylcholine, a mediator of IgE anaphylaxis in the rabbit. Am Rev Respir Dis 122:915–924
Hamberg M, Hedqvist P, Radegran K (1980) Identification of 15-hydroxy-5, 8, 11, 13-eicosatetraenoic acid (15-HETE) as the major metabolite of arachidonic acid in human lung. Acta Physiol Scand 110:219–221
Hardy C, Robinson C, Bradding P, Holgate ST (1984a) Prostacyclin: a functional antagonist of prostaglandin D_2 induced bronchoconstriction (abstract). Thorax 39:696
Hardy CC, Robinson C, Tattersfield AE, Holgate ST (1984b) The bronchoconstrictor effect of inhaled prostaglandin D_2 in normal and asthmatic men. N Engl J Med 311:209–213
Hartung HP, Parnham MJ, Winkleman J, Engleberger W, Hadding U (1983) Platelet activating factor (PAF) induces the oxidative burst in macrophages. Int J Immunopharmacol 5:115–121
Haye-Legrand I, Cerrina J, Raffestin B, Labat C, Boullet C, Bayol A, Benveniste J, Brink C (1986) Histamine contraction of isolated human airway muscle preparations: role of prostaglandins. J Pharmacol Exp Ther 239:536–541
Heard BE, Hossain S (1973) Hyperplasia of bronchial muscle in asthma. J Pathol 110:319
Hellewell PG, Williams TJ (1986) A specific antagonist of platelet-activating factor suppresses oedema formation in an Arthus reaction but not oedema induced by leukocyte chemoattractants in rabbit skin. J Immunol 137:302–307
Henderson WR, Harley JB, Favci AS (1984) Arachidonic acid metabolism in normal and hypereosinophilic syndrome eosinophils: generation of leukotrienes B_4, C_4, D_4 and 15-lipoxygenase products. Immunology 51:679–686
Henocq E, Vargaftig BB (1988) Skin eosinophilia in atopic patients. J Allergy Clin Immunol 81:691–695
Henson PM, Pinckard RN (1977) Basophil-derived platelet-activating factor (PAF) as an in vivo mediator of acute allergic reactions: demonstration of specific desensitization of platelets to PAF during IgE-induced anaphylaxis in the rabbit. J Immunol 119:2179–2184
Hogaboom GK, Mong S, Wu H, Crooke ST (1983) Peptidoleukotrienes: distinct receptors for leukotriene C_4 and D_4 in the guinea-pig lung. Biochem Biophys Res Commun 116:1136
Holtzman MJ, Aizawa H, Nadel JA, Goetzl EJ (1983) Selective generation of leukotriene B_4 by tracheal epithelial cells from dogs. Biochem Biophys Res Commun 114:1071–1076
Howarth PH, Durham SR, Lee TH, Kay AB, Church MK, Holgate ST (1985) Influence of albuterol, cromolyn sodium and ipratropium bromide on the airway and circulating mediator responses to allergen bronchial provocation in asthma. Am Rev Respir Dis 132:986–992
Hua X-Y, Dahlen S-E, Lundberg JM, Hammerstrom S, Hedqvist P (1985) Leukotrienes C_4 and E_4 cause widespread and extensive plasma extravasation in the guinea-pig. Naunyn-Schmiedebergs Arch Pharmacol 330:136–141
Hughes PJ, Holgate ST, Church MK (1984) Adenosine inhibits and potentiates IgE-dependent histamine release from human lung mast cells by an A2-purinoceptor mediated mechanism. Biochem Pharmacol 33:3847–3852
Hugli TE (1986) Biochemistry and biology of anaphylatoxins. Complement 3:111–127

Hunter JA, Finkbeiner WE, Nadel JA, Goetzl EJ, Holtzman MJ (1985) Predominant generation of 15-lipoxygenase metabolites of arachidonic acid by epithelial cells from human trachea. Proc Natl Acad Sci USA 82:4633–4637
Hutchcroft BJ, Guz A (1978) Levels of complement components during allergen-induced asthma. Clin Allergy 8:59–64
Hwang S-B, Lam M-H, Shen TY (1985) Specific binding sites for platelet activating factor in human lung tissues. Biochem Biophys Res Commun 128:972–979
Hwang SB, Lam MH, Chang MN (1986) Specific binding of [^{3}H] dihydrokadsurenone to rabbit platelet membranes and its inhibition by the receptor agonists and antagonists of platelet-activating factor. J Biol Chem 261:13720–13726
Ichinose M, Barnes PJ (1989a) Inhibitory histamine H3-receptors on cholinergic nerves in human airways. Eur J Pharmacol 164:139–146
Ichinose M, Barnes PJ (1989b) Histamine H3-receptors mediated inhibition of non-adrenergic, non-cholinergic bronchoconstriction in guinea-pig in vivo. Am Rev Respir Dis 139:236
Ichinose M, Stretton CD, Schwartz J-C, Barnes PJ (1989) Histamine H3-receptors inhibit cholinergic neurotransmission in guinea pig airways. Br J Pharmacol 97:13–15
Ind PW, Barnes PJ, Brown MJ, Causon R, Dollery CT (1983) Measurement of plasma histamine. Clin Allergy 13:61–67
Irvin CG, Berend N, Henson PM (1986) Airways hyperreactivity and inflammation produced by aerosolization of human C5a des arg. Am Rev Respir Dis 134:777–783
Johnson AR (1980) Human pulmonary endothelial cells in culture activities of cells from arteries and cells from veins. J Clin Invest 65:841–850
Johnson HG, McNee ML (1983) Secretagogue responses of leukotriene C_4, D_4: comparison of potency in canine trachea in vivo. Prostaglandins 25:237–243
Joris I, Majno G, Corey EJ, Lewis RA (1987) The mechanism of vascular leakage induced by leukotriene E_4: endothelial contraction. Am J Pathol 126:19–24
Joseph M, Auriault C, Capron A, Vorng H, Viens P (1983) A new function for platelets: IgE-dependent killing of schistosomes. Nature 303:310–312
Joseph M, Tonnel AB, Tarpier G, Capron A (1983b) Involvement of immunoglobulin E in the secretory process of alveolar macrophages from asthmatic patients. J Clin Invest 71:221–230
Kaliner M (1978) Human lung tissue and anaphylaxis. The effects of histamine on the immunologic release of mediators. Am Rev Respir Dis 118:1015–1022
Karlin JM (1972) The use of antihistamines in asthma. Ann Allergy 30:342–347
Kauffman HF, van der Heide S, de Monchy JGR, de Vries K (1983) Plasma histamine concentrations and complement activation during house dust mite-produced bronchial obstructive reactions. Clin Allergy 13:219–228
Kaufman MP, Coleridge HM, Coleridge JCG, Baker DG (1980) Bradykinin stimulates afferent vagal C-fibers in intrapulmonary airways of dogs. J Appl Physiol 48:511–517
Kawakami Y, Uchiyama K, Irie T, Murao M (1973) Evaluation of aerosols of prostaglandins E_1 and E_2 as bronchodilators. Eur J Clin Pharmacol 6:127–132
Kay AB (1970) Studies on eosinophil leucocyte migration I. Factors specifically chemotactic for eosinophils and neutropils generated from guinea-pig serum by antigen-antibody complexes. Clin Exp Immunol 7:723–737
Kay AB, Austen KF (1972) Chemotaxis of human baspohil leukocytes. Clin Exp Immunol 11:557–563
Kay AB, Lee TH (1982) Neutrophil chemotactic factor of anaphylaxis. J Allergy Clin Immunol 70:317–320
Kay AB, Bacon GD, Mercer BA, Simpson H, Grafton JW (1974) Complement components and IgE in bronchial asthma. Lancet 11:916–920
Kennedy I, Coleman RA, Humphrey RPA, Levy GP, Lumley P (1982) Studies on the characterisation of prostanoid receptors: a proposed classification. Prostaglandins 24:667–689

Kim YY, Kim MK (1988) The effect of PAF on non-specific bronchial reactivity in bronchial asthmatics. Allergy 9:149
Kirby JG, Hargreave FE, O'Byrne PM (1987) Indomethacin inhibits allergen-induced airway hyperresponsiveness but not allergen-induced asthmatic responses. Am Rev Respir Dis 135:312
Kirby JG, Hargreave FE, Cockcroft DW, O'Byrne PM (1989) Effect of indomethacin on allergen-induced asthmatic responses. J Appl Physiol 66:578–583
Klimek JJ, Winslow CM, Saunders RN (1983) Platelet and neutrophil interactions in platelet-activating factor (PAF)-induced changes in vascular permeability (VP) and pulmonary inflation pressure in guinea-pigs. Fed Proc 42:693
Kreisle RA, Parker CW (1983) Specific binding of leukotriene B_4 to a receptor on human polymorphonuclear leukocytes. J Exp Med 157:628–641
Kroegel C, Yukawa T, Dent G, Chanez P, Chung KF, Barnes PJ (1988) Platelet activating factor induces eosinophil peroxidase release from purified human eosinophils. Immunology 64:559–562
Kroegel C, Yukawa T, Dent G, Venge P, Chung KF, Barnes PJ (1989) Stimulation of degranulation from human eosinophils by platelet activating factor. J Immunol 142:3518–3526
Kuehl FA, De Haven RN, Pong SS (1984) Lung tissue receptors for sulfidopeptide leukotrienes. J Allergy Clin Immunol 74:378–381
Laitinen LA, Heino M, Laitinen A, Kava T, Haahtela T (1985) Damage of the airway epithelium and bronchial reactivity in patients with asthma. Am Rev Respir Dis 131:599–606
Laitinen LA, Laitinen A, Widdicombe JG (1987) Effects of inflammatory and other mediators on airway vascular beds. Am Rev Respir Dis 135:S567–S570
Lam S, Leriche J, Phillips D, Chan-Yeung M (1987) Cellular and protein changes in bronchial lavage fluid after late asthmatic reaction in patients with red cedar asthma. J Allergy Clin Immunol 80:44–50
Lambrecht G, Parnham MJ (1986) Kadsurenone distinguishes between different platelet activating factor receptor subtypes on macrophages and polymorphonuclear leucocytes. Br J Pharmacol 87:287–299
Lang M, Hansen D, Hahn HL (1987) Effect of the PAF antagonist CV-3988 on PAF-induced changes in mucus secretion and in respiratory and circulatory variables in ferret. Agents Actions [Suppl] 21:245–251
Langridge-Smith JE, Rao MC, Field M (1984) Chloride and sodium transport across bovine tracheal epithelium: effects of secretagogues and indomethacin. Pflugers Arch 402:42–47
Latrigue-Mattei C, Godeneche D, Chabard JL, Petie J, Berger JA (1984) Pharmakokinetic study of ^{3}H-labelled Paf-acether. II Comparison with ^{3}H-labelled lyso-Paf-acether after intravenous administration in the rabbit and protein binding. Agents Actions 15:643–648
Lee CW, Lewis RA, Corey EJ, Barton A, Oh H, Tauber AI, Austen KF (1982) Oxidative inactivation of leukotriene C_4 by stimulated human polymorphonuclear leukocytes. Proc Natl Acad Sci USA 79:4166–4170
Lee TC, Lenihan DJ, Malone B, Roddy LL, Wasserman SI (1984a) Increased biosynthesis of platelet activating factor in activated huma eosinophils J Biol Chem 259:5526–5530
Lee TH, Austen KF, Corey EJ, Drazen JM (1984) LTE_4-induced airway hyperresponsiveness of guinea pig tracheal smooth muscel to histamine and evidence for 3 separate sulfidopeptide leukotriene receptors. Proc Natl Acad Sci USA 81:4922–4925
Leikauf GD, Ueki IF, Nadel JA, Widdicombe JH (1985) Bradykinin stimulates chloride secretion and prostaglandin E_2 release by canine tracheal epithelium. Am J Physiol 248:48–55
Leikauf GD, Ueki IF, Widdicombe JH, Nadel JA (1986) Alteration of chloride secretion across canine tracheal epithelium by lipoxygenase products of arachidonic acid. J Appl Physiol 250:F47–53

Lellouch-Tubiana A, Lefort J, Pirotzky E, Vargaftig BB, Pfister A (1985) Ultrastructural evidence for extravascular platelet recruitment in the lung upon intravenous injection of platelet activating factor (Paf-acether) to guinea-pigs. Br J Exp Pathol 66:345–355
Lellouch-Tubiana A, Lefort J, Simon M-T, Pfister A, Vargaftig BB (1988) Eosinophil recruitment into guinea pig lungs after PAF-acether and allergen administration: modulation by prostacyclin, platelet depletion, and selective antagonists. Am Rev Respir Dis 137:948–954
Lewis AJ, Dervinis A, Chang J (1984) The effects of antiallergic and bronchodilator drugs on platelet-activating factor (PAF-acether) induced bronchospasm and platelet aggregation. Agents Actions 15:636–641
Lewis RA, Drazen JM, Austen KF, Toda M, Brion F, Marfat A, Corey EJ (1981) Contractile activities of structural analogues of leukotrienes C and D: role of the polar substituents. Proc Natl Acad Sci USA 78:4579–4583
Lichtenstein LM, Schleimer RP, MacGlashan DW, Peters SP, Schulman ES, Proud D, Creticos PS, Naclerio RM, Kagey-Sobotka A (1984) In vitro and in vivo studies of mediator release from human mast cells. Asthma: Physiology, immunopharmacology and treatment. Academic, London, pp 1–15
Londos C, Cooper DMF, Wolff J (1980) Subclass of external adenosine receptors. Proc Natl Acad Sci USA 77:2551–2554
Long WM, Sprung CL, Elfawal H (1985) Effects of histamine on bronchial artery blood flow and bronchomotor tone. J Appl Physiol 59:254–261
Lotner GZ, Lynch JM, Betz SJ, Henson PM (1980) Human neutrophil-derived platelet activating factor. J Immunol 124:676–684
Lynch JM, Henson PM (1986) The intracellular retention of newly-synthesized platelet-activating factor. J Immunol 137:2653–2661
MacDermot JM, Barnes PJ, Wadell K, Dollery CT, Blair IA (1981) Prostacyclin binding to guinea pig pulmonary receptors. Eur J Pharmacol 68:127–130
MacDermot J, Kelsey CR, Waddell KA et al. (1984) Synthesis of leukotriene B_4, and prostanoids by human alveolar macrophages: analysis by gas chromatography/mass spectrometry. Prostaglandins 27:163–179
MacGlashan DW, Schleimer RP, Peters SP, Schulman ES, Adams GK, Newball HH, Lichtenstein LM (1982) Generation of leukotrienes by purified human lung mast cells. J Clin Invest 70:747–751
Maclouf J, de Laclos BF, Borgeat P (1982) Stimulation of leukotriene biosynthesis in human blood leukocytes by platelet-derived 12-hydroxy-icosatetraenoic acid. Proc Natl Acad Sci USA 79:6042–6046
Mann JS, Holgate ST, Renwick AG, Cushley MJ (1986a) Airway effects of purine nucleosides and nucleotides and release with bronchial provocation in asthma. J Appl Physiol 61:1667–1676
Mann JS, Robinson C, Sheridan AQ, Clement P, Bach MK, Holgate ST (1986b) Effect of inhaled Piriprost (U-60257), a novel leukotriene inhibitor, on allergen and exercise induced bronchoconstriction in asthma. Thorax 41:746–752
Manning PJ, Jones GL, O'Byrne PM (1987) Indomethacin prevents histamine tachyphylaxis in asthmatics. Am Rev Respir Dis 135:A313
Marcus AJ, Weksler BB, Jaffe EA, Broekman MJ (1980) Synthesis of prostacyclin from platelet derived endoperoxides by cultured human endothelial cells. J Clin Invest 66:979–986
Marcus AJ, Safier LB, Ullman HL, Broekman MJ, Islam N, Oglesby TD, Gorman RR (1984) 12*S*, 20-Dihydroxyicosatetraenoic acid: a new eicosanoid synthesized by neutrophils from 12*S*-hydroxyicosatetraenoic acid produced by thrombin- or collagen-stimulated platelets. Proc Natl Acad Sci USA 81:903–907
Marom Z, Shelhamer JH, Kaliner M (1981) Effects of arachidonic acid, monohydroxyeicosatetraenoic aicd and prostaglandins on the release of mucous glycoproteins from human airways in vitro. J Clin Invest 67:1695–1702
Marom Z, Schelhamer JH, Bach MK, Morton DR, Kaliner M (1982) Slow reacting

substances LTC_4 and D_4 increases the release of mucus from human airways in vitro. Am Rev Respir Dis 126:449
Marom Z, Shelhamer J, Berger M, Frank M, Karliner M (1985) Anaphylatoxin C3a enhances mucous glycoprotein release from human ariways in vitro. J Exp Med 161:657–668
Martin TR, Altman LC, Albert RK, Henderson WR (1984) Leukotriene B_4 production by the human alveolar macrophage: a potential mechanism for lung amplification. Am Rev Respir Dis 129:106–111
Mathe AA, Hedqvist P (1975) Effect of prostaglandin $F2_\alpha$ and E_2 on airway conductance in healthy subjects and asthmatic patients. Am Rev Respir Dis 11:313–320
Mazzoni L, Morley J, Page CP, Sanjar S (1985) Induction of airway hyper-reactivity by platelet activating factor in the guinea-pig. J Physiol (Lond) 365:107
McGiff JC, Terragno NA, Strand JC, Lee JB, Lonigro AJ, Ng KKF (1969) Selective passage of prostaglandins by the lung. Nature 223:742–745
McKenniff M, Rodger IW, Norman P, Gardiner PJ (1989) Characterisation of receptors mediating the contractile effects of prostanoids in guinea pig and human airways. Eur J Pharmacol 153:149–159
Metzger WJ, Richerson HB, Wasserman SI (1986) Generation and partial characterization of eosinophil chemotactic activity and neutrophil chemotactic activity during early and late phase airway responses. J Allergy Clin Immunol 78:282–290
Metzger WJ, Zavala D, Richerson HB, Mosely P, Iwamota P, Monick M, Sjoersdma K, Hunninghake GW (1987) Local allergen challenge and bronchoalveolar lavage of allergic asthmatic lungs. Am Rev Respir Dis 135:433–440
Miwa M, Miyake T, Yamanaka T, Sugatani J, Suzuki Y, Sakata S, Araki Y, Matsumoto M (1988) Characterization of serum platelet-activating factor (PAF) acetylhydrolase: correlation between deficiency of serum PAF acetylhydrolase and respiratory symptoms in asthmatic children. J Clin Invest 82:1983–1991
Mohsenin V, Dubois AB, Douglas JS (1983) Effect of ascorbic acid on responses to methacholine challenge in asthmatic subjects. Am Rev Respir Dis 127:143–147
Morley J, Sanjar S, Page CP (1984) Platelet activation as a basis for asthma exacerbation. Lancet 2:1142–1144
Morley J, Page CP, Sanjar S (1985) Pulmonary responses to platelet-activating factor. Prog Respir Res 19:117–123
Motojima S, Frigas E, Loegering DA, Gleich GJ (1989) Toxicity of eosinophil cationic proteins for guinea pig tracheal epithelium in vitro. Am Rev Respir Dis 139: 301–305
Movat HZ, Rettl C, Burrowes CE, Johnston MG (1984) The in vivo effect of leukotriene B_4 on polymorphonuclear leukocytes and the microcirculation. Comparison with activated complement (C5a des Arg) and enhancement of prostaglandin E_2. Am J Pathol 115:233–244
Murray JJ, Tonnel AB, Brash AR, Roberts LJ, Gosset P, Workman R, Cafron A, Oates JA (1986) Release of prostaglandin D_2 into human airways during acute antigen challenge. N Engl J Med 315:800–804
Nagy L, Lee TH, Goetzl EJ, Pickett W, Kay AB (1982) Complement receptor enhancement and chemotaxis of human neutrophils and eosinophils by leukotrienes and other lipoxygenase products. Clin Exp Immunol 47:541–547
Nakamura T, Morita Y, Kuriyama M, Ishihara K, Ito K, Miyamoto T (1987) Platelet activating factor in late asthmatic responses. Int Arch Allergy Appl Immunol 82:57–61
Needleman P, Moncada S, Bunting S, Vane JR, Hamberg M, Samuelsson B (1976) Identification of an enzyme in platelet microsomes which generates thromboxane A_2 from prostaglandin endoperoxides. Nature 1:558–560
Newball HM, Keiser HR, Pisano JJ (1975) Bradykinin and human airways. Respir Physiol 24:139–146
Nichol GM, Nix A, Barnes PJ, Chung KF (1988) Enhancement of capsaicin-induced cough by inhaled prostaglandin $F_{2\alpha}$ ($PGF_{1\alpha}$). Thorax 43:836–837

Ninio EW, Mencia-Heurata JM, Heymans F, Benveniste J (1982) Biosynthesis of platelet activating factor: evidence for an acetyl-transferase activity in murine macrophages. Biochim Biophys Acta 710:23–31
O'Byrne PM, Jones GL (1986) The effect of indomethacin on exercise-induced broncho-constriction and refractiveness after exercise. Am Rev Respir Dis 134:69–72
O'Byrne PM, Leikauf GD, Aizawa H, Bethel RA, Ueki IF, Holtzman MJ, Nadel JA (1985) Leukotriene B_4 induces airway hyperresponsiveness in dogs. J Appl Physiol 59:1941–1946
O'Driscoll BR, Lee TH, Kay AB (1982) Immunological release of neutrophil chemotactic activity from isolated lung fragments. J Allergy Clin Immunol 72:695–701
O'Donnell SR, Barnett CJK (1987) Microvascular leakage to platelet activating factor in guinea-pig trachea and bronchi. Eur J Pharmacol 138:385–396
O'Flaherty JT, Surles JR, Redman J, Jacobson D, Piantadosi C, Wykle RL (1986) Binding and metabolism of platelet-activating factor by human neutrophils. J Clin Invest 78:381–388
Ogilvy CS, Dubois AB, Douglas JS (1981) Effects of ascorbic acid and indomethacin on the airways of healthy male subjects with and without induced bronchoconstriction. J Allergy Clin Immunol 67:363–369
Orning L, Hammarstrom S (1980) Inhibition of leukotriene C and leukotriene D biosynthesis. J Biol Chem 255:8023–8026
Orning L, Kaïjser L, Hammarstrom S (1985) In vivo metabolism of leukotriene C_4 in man: urinary excretion of leukotriene E_4. Biochem Biophys Res Commun 130:214–217
Panzani R (1962) 5-hydroxytryptamine (serotonin) in human bronchial asthma. Ann Allergy 20:721–732
Page CP (1988) Platelet activating factor and asthma. J Allergy Clin Immunol 81:142–151
Parker CW, Falkenhein SF, Huber MM (1980) Sequential conversion of the glutathionyl side chain of slow reacting substance (SRS) to cysteinyl-glycine and cysteine in rat basophilic leukaemia cells stimulated with A23187. Prostaglandins 20:863–866
Patel KR (1984) Terfenadine in exercise-induced asthma. Br Med J 288:1496–1497
Pauwels R, van der Straeten M (1986) The bronchial effect of adenosine in the rat. Arch Int Pharmacodyn Ther 280:229–239
Pawlowski NA, Kaplan G, Hamill AL et al. (1983) Arachidonic acid metabolism by human monocytes studies with platelet-depleted cultures. J Exp Med 158:593
Persson CGA (1986) Development of safer xanthine drugs for the treatment of obstructive airways disease. J Allergy Clin Immunol 78:817–824
Persson CGA (1987) Leakage of macromolecules from the tracheobronchial circulation. Am Rev Respir Dis 135:71–75
Persson CGA, Erjefalt I, Sundler F (1987) Airway microvascular and epithelial leakage of plasma induced by PAF-acether (PAF) and capsaicin (CAP). Am Rev Respir Dis 135:A401
Peters SP, Freeland HS, Kelly SJ, Pipkorn U, Naclerio RM, Proud D, Schleimer RP, Lichtenstein LM, Fish JE (1987) Is Leukotriene B_4 an important mediator in IgE-mediated allergic reactions? Am Rev Respir Dis 135:42–45
Piper PJ, Vane JR, Wyllie JH (1970) Inactivation of prostaglandins by the lungs. Nature 225:600–605
Pirotzky E, Page CP, Roubin R, Pfister A, Paul W, Bonnet J, Benveniste J (1984) Paf-acether-induced plasma exudation in rat skin is independent of platelets and neutrophils. Microcirc Endothelium Lymphatics 1:107–122
Pleskow WW, Chenoweth DE, Simon RA, Stevenson DD, Curn JG (1983) The absence of detectable complement activation in aspirin-sensitive asthmatic patients during aspirin challenge. J Allergy Clin Immunol 72:462–468
Prescott SM, Zimmerman GA, McIntyre TM (1984) Human endothelial cells

in culture produce platelet-activating factor (1-alkyl-2-acetyl-sn-glycero-3-phosphocholine) when stimulated with thrombin. Proc Natl Acad Sci USA 81:3534–3538

Proud D, Togias A, Naclerio RM, Crush SA, Norman PS, Lichtenstein LM (1983) Kinins are generated in vivo following nasal airway challenge of allergic individuals with allergen. J Clin Invest 72:1678–1685

Rafferty P, Beasley R, Holgate ST (1987) The contribution of histamine to immediate bronchoconstriction provoked by inhaled allergen and adenosine 5′ monophosphate in atopic asthma. Am Rev Respir Dis 136:369–373

Raffestin B, Cerrina J, Boullet C, Labat C, Benveniste J, Brink C (1985) Response and sensitivity of isolated human pulmonary muscle preparations to pharmacological agents. J Pharmacol Exp Ther 233:186–194

Regal JF (1989) The role of C5a in hypersensitivity reactions in the lung. Pulm Pharmacol 2:3–12

Regal JF, Eastman AJ, Pickering RJ (1980) C5a-induced tracheal contraction. A histamine independent mechanism. J Immunol 124:2876–2878

Regoli D, Barabe J (1980) Pharmacology of bradykinin and related kinins. Pharmacol Rev 32:1–46

Rhoden KJ, Barnes PJ (1989) Effect of hydrogen peroxide radicals on responses of guinea-pig tracheal smooth muscle in vitro: role of cyclooxygenase and airway epithelium. Br J Pharmacol 98:325–330

Rich B, Peatfield AC, Williams IP, Richardson PS (1984) Effects of prostaglandins E_1, E_2 and F_{2alpha} on mucin secretion from human bronchi in vitor. Thorax 39:420–423

Richards SW, Peterson PK, Verbaugh HA, Nelson RD, Hammerschmidt DE, Hoidal JR (1984) Chemotactic and phagocytic responses of human alveolar macrophages to activated complement components. Infect Immun 43:775–778

Richardson BP, Engel G (1986) The pharmacology and function of 5-HT3 receptors. TINS 3:424–428

Roberts AM, Schultz HO, Green JF, Armstrong DJ, Kaufman MP, Coleridge HM, Coleridge JCG (1985) Reflex tracheal contraction evoked in dogs by bronchodilator prostaglandins E_2 and I_2. J Appl Physiol 58:1823–1831

Roberts NM, McCusker M, Barnes PJ (1988a) Effect of a PAF antagonist, BN 52063, on PAF-induced bronchoconstriction in normal subjects. Br J Clin Pharmacol 26: 65–72.

Roberts NM, Page CP, Chung KF, Barnes PJ (1988b) The effect of a specific PAF antagonist, BN 52063, on antigen-induced cutaneous responses in man. J Allergy Clin Immunol 82:236–241

Robertson DN, Rhoden KJ, Grandordy B, Page CP, Barnes PJ (1988) The effect of platelet activating factor on histamine and muscarinic receptor function in guinea-pig airways. Am Rev Respir Dis 137:1317–1322

Robinson C, Holgate ST (1985) New Perspectives on the putative role of eicosanoids in airway hyperresponsiveness. J Allergy Clin Immunol 76:140–144

Rogers DF, Aursudkij B, Evans TW, Belvisi MG, Chung KF, Barnes PJ (1987) Platelet activating factor increases protein exudation but not mucus secretion in guinea-pig trachea in vivo. Am Rev Respir Dis 135:A160

Ross R, Raines EW, Bowen-Pope DF (1986) The biology of platelet-derived growth factor. Cell 46:155–169

Rouzer CA, Matsumoto T, Samuelsson B (1986) Single protein from human leukocytes possesses 5-lipo- oxygenase and leukotriene A synthase activities. Proc Natl Acad Sci USA 83:857–861

Rubin A-H, Smith LJ, Patterson R (1987) The bronchoconstrictor properties of platelet-activating factor in humans. Am Rev Respir Dis 136:1145–1151

Sackeyfio AC (1971) Definition of the histamine component of the bronchoconstrictor and cardiovascular effects of anaphylotoxins in the guinea-pig. Br J Pharmacol 43:424

Samhoun MN, Piper PJ (1983) Comparative actions of leukotrienes in lung from various species. In: PJ Piper (ed) Leukotrienes and other lipoxygenease products. Wiley, New York, pp 161–177
Sampson SR, Vidruk DH (1979) The nature of the receptor mediating stimulant effects of histamine on rapidly adapting vagal afferents in lung. J Physiol (Lond) 287:509–518
Saria A, Lundberg JM, Skofitsch G, Lembeck F (1983) Vascular protein leakage in various tissues induced by substance P, capsaicin, bradykinin, serotonin, histamine and by antigen challenge. Naunyn Schmiedebergs Arch Pharmacol 324:212–218
Schellenberg RR, Foster A (1984) In vitro responses of human asthmatic airway and pulmonary vascular smooth muscle. Int Arch Allergy Appl Immunol 75:237–241
Schellenberg RR, Walker B, Snyder F (1983) Platelet-dependent contraction of human bronchus by platelet activating factor. J Allergy Clin Immunol 71:145
Schleimer RP, MacGlashan DW, Peters SP, Pinckard RN, Adkinson NF, Lichtenstein LM (1986) Characterization of inflammatory mediator release from purified human lung mast cells. Am Rev Respir Dis 133:614–617
Schulman ES, Adkinson Jr NF, Newball HH (1982) Cyclooxygenase metabolites in human lung anaphylaxis airway vs parenchyma. J Appl Physiol 53:589–595
Serhan CN, Radin A, Smolen JE, Korchak H, Samuelsson B, Weissmann G (1982) Leukotriene B_4 is a complete secretagogue in human neutrophils: a kinetic analysis. Biochem Biophys Res Commun 107:1006–1012
Serhan CN, Hamberg M, Samuelsson B (1984) Lipoxins, a novel series of compounds formed from arachidonic acid in human leukocytes. Proc Natl Acad Sci USA 81:5335–5339
Serhan CN, Hirsch U, Palmblad, J Samuelsson B (1987) Formation of lipoxin A by granulocytes from eosinophilic donors. FEBS Lett 217:242–246
Sertl K, Casale TB, Wescott SL, Kaliner MA (1987) Immunohistochemical localization of histamine-stimulated increases in cyclic GMP in guinea pig lung. Am Rev Respir Dis 135:456–462
Shak S, Goldstein IM (1984) Omega-oxidation is the major pathway for the catabolism of leukotriene B_4 in human polymorphonuclear leukocytes. J Biol Chem 259:10181–10187
Shaw RJ, Cromwell O, Kay AB (1984) Preferential generation of leukotriene C_4 by human eosinophils. Clin Exp Immunol 56:716–722
Sheard P (1986) The effect of prostaglandin E_1 on isolated bronchial muscles from man. J Pharm Pharmacol 20:232–233
Shelhamer J, Marom Z, Kaliner M (1980) Immunologic and neuropharmacologic stimulation of mucous glycoprotein release from human airways in vitro. J Clin Invest 66:1400–1408
Shephard EG, Malon L, Macfarlane CM, Mouton W, Joubert JR (1985) Lung function and plasma levels of thromboxane B2, 6-keto-prostaglandin F_{1alpha}, beta-thromboglobulin in antigen-induced asthma before and after indomethacin pretreatment. Br J Clin Pharmacol 19:459–470
Sigal CE, Valone FH, Holtzman J, Goetzl EJ (1987) Preferential human eosinophil chemotactic activity of the platelet activating factor (PAF): 1-*O*-hexadecyl-2-acetyl-sn-glyceryl-3-phosphocholine (AGEPC). J Clin Immunol 7:179–188
Silberstein DS, David JR (1987) The regulation of human eosinophil function by cytokines. Immunol Today 8:380–385
Simonsson BG, Skoogh BE, Bergh NP Anderson R, Svedmyr N (1973) In vivo and in vitro effect of bradykinin on bronchial motor tone in normal subjects and in patients with airway obstruction. Respiration 30:378–388
Smith HR, Henson PM, Clay KL, Larsen GL (1988) Effect of the PAF antagonist L-659, 989 on the late asthmatic response and increased airway reactivity in the rabbit. Am Rev Respir Dis 137:283

Smith JB, Silver MJ, Ingerman CM, Kocsis JJ (1974) Prostaglandin D_2 inhibits the aggregation of human platelets. Thromb Res 5:291–299
Smith LJ, Greenberger PA, Patterson R, Krell RD, Bernstein PR (1985) The effect of inhaled leukotriene D_4 in humans. Am Rev Respir Dis 131:368–372
Smith LJ, Rubin A-H, Patterson R (1988) Mechanism of platelet activating factor-induced bronchoconstriction in humans. Am Rev Respir Dis 137:1015–1019
Snyder F (1985) Chemical and biochemical aspects of platelet activating factor: a novel class of acetylated either-linked choline phospholipids. Med Res Rev 5:107–140
Snyder F (1987) The significance of dual pathways for the biosynthesis of platelet activating factor: 1-alkyl-2-lyso-sn-glycero-3-phosphate as a branchpoint. In: Winsolw CM, Lee ML (eds) New horizons in platelet activating factor research. Wiley. New York, pp 13–25
So SY, Lam WK, Kvens S (1985) Selective 5-HT2 receptor blockade in exercise-induced asthma. Clin Allergy 15:371–376
Soter NA, Lewis RA, Corey EJ, Anster KF (1983) Local Effects of synthetic leukotrienes (LTC_4, LTD_4, LTE_4 and LTB_4) in human skin, J Invest Dermatol 80:115–119
Spagnello PJ, Ellner JJ, Hassid A, Dunn MJ (1980) Thromboxane A_2 mediators augmented polymorphonuclear leukocyte adhesiveness. J Clin Invest 66:406–414
Stalenheim S, Machado L (1985) Late allergic bronchial reactions and the effect of allergen provocation on the complement system. J Allergy Clin Immunol 75:508–512
Steiger J, Bray MA, Subramanian N (1987) Platelet activating factor (PAF) is a potent stimulator of porcine tracheal fluid secretion in vitro. Eur J Pharmacol 142:367–372
Stenson WF, Parker CW (1980) Monohydroxyeicosatetraenoic acids (HETEs) induce degranulation of human neutrophils. J Immunol 124:2100–2104
Stewart AG, Dusting GJ (1988) Characterization of receptors for platelet activating factor on platelets, polymorphonuclear leukocytes and macrophages. Br J Pharmacol 94:1225–1233
Stewart RM, Weir EK, Mongomery MR, Niewoehner DE (1981) Hydrogen peroxide contracts airway smooth muscle: a possible endogenous mechanism. Respir Physiol 45:333–342
Stimler NP (1984) Spasmogenic activity of C5a des Arg anaphylotoxins on guinea-pig lung parenchymal strips: sensitivity of the leukotriene-mediated component to cyclo-oxygenase inhibitors. Biochem Biophys Res Commun 125:852–858
Stimler NP, Bach MK, Blour CM, Hugli TE (1982) Release of leukotrienes from guinea-pig lung stimulated by C5a des arg anaphylotoxin. J Immunol 128:2247–2252
Stimler NP, Blour CM, Hugli TE (1983) C3a-induced contraction of guinea-pig lung parenchyma. Role of cyclo-oxygenase metabolites. Immunopharmacology 5:251–257
Sun FF, Chau LY, Spur B, Corey EJ, Lewis RA, Austen KF (1986) Identification of a high affinity leukotriene C_4-binding protein in rat liver cytosol as glututhione-S-transferase. J Biol Chem 261:8540–8546
Suzuki Y, Miwa M, Harada M, Matsumoto M (1988) Acetylhydrolase released from platelets on aggregation with platelet activating factor. Clin Exp Pharmacol Physiol 13 [Suppl]:15 (abstract)
Sweatman WJF, Collier HOJ (1968) Effects of prostaglandins on human bronchial muscle. Nature 217:69
Tamura N, Agrawal DK, Townley RG (1987) Effects of platelet activating factor on the chemotaxis of normodense eosinophils from normal subjects. Biochem Biophys Res Commun 142:638–644
Taylor KJ, Luksza AR (1987) Peripheral blood eosinophil counts and bronchial responsiveness. Thorax 42:452–456

Tence M, Polonsky J, LeCouedic JP, Benveniste J (1980) Release, purification and characterisation of platelet activating factor (PAF). Biochimie 62:251–259
Thompson PJ, Hanson JM, Bilani H, Turner-Warwick M, Morley J (1984) Platelets, platelet activating factor and asthma. Am Rev Respir Dis 129:A3
Thomson NC (1987) In vivo versus in vitro human airway responsiveness to different pharmacologic stimuli. Am Rev Respir Dis 136:S58–62
Tonnesen P (1985) Bronchial challenge with serotonin in asthmatics. Allergy 40:136–140
Topilsky M, Spitzer S, Pick AI, Weiss H (1976) Complement in asthma. Lancet 1:813
Touqui L, Jacquemin C, Dumarey C, Vargaftig BB (1985) Alkyl-2-acyl-sn-glycero-3-phosphorylcholine is the precursor of platelet activating factor in stimulated rabbit platelets. Evidence for an alkylacetyl-glycerophosphorylcholine cycle. Biochim Biophys Acta 833:111–118
Touvay C, Etienne A, Braquet P (1985) Inhibition of antigen-induced lung anaphylaxis in the guinea-pig by BN 52021 a new specific PAF-acether receptor antagonist isolated from Ginkgo biloba. Agents Actions 17:371–372
Touvay C, Pfister A, Vilain B, Coyle AJ, Page CP, Lellouch-Tubiana A, Pignol B, Mencia-Huerta JM, Braquet P (1990) Effect of long-term infusion of platelet-activating factor via osmotic minipumps on pulmonary responsiveness and morphology in the guinea-pig. Pulm Pharmacol (in press)
Ueda N, Maramatsu I, Fujiwara M (1984) Capsaicin and bradykin-induced substance P-ergic responses in the iris sphincter muscle of the rabbit. J Pharmacol Exp Ther 230:469–473
Ukena D, Schirren CG, Schwab U (1985) Effect of xanthine derivatives on adenosine-receptors of guinea pig lung. In: Andersson K-E, Persson CGA (eds) Anti-asthma xanthines and adenosine. Excerpta Medica, Amsterdam, pp 390–398
Ukena D, Dent G, Birke FW, Robaut C, Sybrecht GW, Barnes PJ (1988) Radioligand binding of antagonists of platelet-activating factor to intact human platelets. FEBS Lett 2:285–289
Ukena D, Kroegel C, Dent G, Yukawa T, Sybrecht G, Barnes PJ (1989) PAF-receptors in eosinophils: identification with a novel ligand [^{3}H]WEB 2086. Biochem Pharmacol 38:1702–1705
Valone FH (1984) Isolation of a platelet membrane protein which binds the platelet activating factor. Immunology 52:169–174
Valone FH, Goetzl EJ (1983) Specific binding by human polymorphonuclear leucocytes of the immunological mediator 1-*o*-hexadecyl/octadecyl-2-acetyl-sn-glycero-3-phosphorylcholine. Immunology 48:141–149
Valone FH, Coles E, Reinhold VR, Goetzl EJ (1982) Specific binding of phospholipid platelet activating factor by human platelets. J Immunol 129:1637:1641
Vanhoutte PM (1988) Epithelium-derived relaxing factor: myth or reality? Thorax 43:665–668
Vargaftig BB, Lefort J, Chignard M, Benveniste J (1980) Platelet-activating factor induces a platelet-dependant bronchoconstriction unrelated to the formation of prostaglandin derivatives. Eur J Pharmacol 65:185–192
Vavrek RJ, Stewart JM (1985) Competitive antagonists of bradykinin. Peptides 6:161–164
Vincenc K, Block J, Shaw J (1984) Relaxation and contraction responses to histamine in the human lung parenchymal strip. Eur J Pharmacol 84:201–210
Voelkel NF, Chang SW, Pfeffer KD, Worthen SG, McMurty IF, Henson PM (1986) PAF antagonists: different effects of platelets, neutrophils, guinea-pig ileum and PAF-induced vasodilation in isolated rat lung. Prostaglandins 32:359–372
Walters EH, Parrish RW, Bevan C, Smith AP (1982) Induction of bronchial hypersensitivity: evidence for a role of prostaglandins. Thorax 36:571–574
Wanner A, Sielczak M, Mella JF, Abraham WM (1986) Ciliary responsiveness in allergic and nonallergic airways. J Appl Physiol 60:1967–1971

Wardlaw AJ, Moqbel R, Cromwell O, Kay AB (1986) Platelet-activating factor. A potent chemotactic and chemokinetic factor for human eosinophils. J Clin Invest 78:1701–1706
Wardlaw AJ, Dunnette S, Gleich GJ, Collins JV, Kay AB (1988) Eosinophils and mast cells in bronchoalveolar lavage in subjects with mild asthma. Am Rev Respir Dis 88:62–69
Wardlaw AJ, Chung KF, Mobel R, Macdonald AJ, McCusker M, Hartnell A, Collins JV, Barnes PJ, Kay AB (1990) Effects of inhaled PAF in man on circulating and bronchoalveolar lavage neutrophils: relationship to bronchoconstrictor and changes in airway responsiveness. Am Rev Respir Dis 141:386–392
Wasserman MA, Griffin RL (1976) In vivo and in vitro bronchoconstrictor responses to prostaglandin F_{2alpha}, cyclic endoperoxide analogues, and thromboxane B_2. In: A Bouhuys (ed) Lung cells in disease. North-Holland, Amsterdam, pp 309–312
Wasserman SI (1983) Mediators of immediate hypersensitivity. J Allergy Clin Immunol 72:101–115
Webb D, Benjamin N, Collier J, Robinson B (1986) Enalopril-induced cough. Lancet 2:1094
Wedmore CV, Williams TJ (1981) Control of vascular permeability by polymorphonuclear leucocytes in inflammation. Nature 289:646–650
Weiss JW, Drazen JM, McFadden ER et al. (1983) Airway constriction in normal humans produced by inhalation of leukotriene D: potency, time course, and effect of aspirin therapy, JAMA 249:2814–2817
Weller PF, Lee CW, Foster DW, Corey EJ, Ansten KF, Lewis RA (1983) Generation and metabolism of 5-lipoxygenase pathway leukotrienes by human eosinophils: predominant production of leukotriene C_4. Proc Natl Acad Sci USA 80:7626–7630
White J, Eiser NM (1983) The role of histamine and its receptors in the pathogenesis of asthma. Br J Dis Chest 77:215–226
White JP, Mills J, Eiser NM (1987) Comparison of the effects of histamine H1- and H2-receptor agonists on large and small airways in normal and asthmatic subjects. Br J Dis Chest 81:155–169
Williams TJ, Morley J (197) Prostaglandins as potentiators of increased vascular permeability in inflammation. Nature 246:215–217
Woodward DF, Weichman BM, Gill CA, Wasserman MA (1983) The effect of synthetic leukotrienes on tracheal microvascular permeability. Prostaglandins 25:131
Yancey KB, Hammer CH, Harvath L, Renfer L, Frank MM, Lawley TJ (1985) Studies of human C5a as a mediator of inflammation in normal human skin. J Clin Invest 75:486–495
Yasaka T, Boxer LA, Baehner RL (1982) Monocyte aggregation and superoxide anion response to formyl-methionyl-leucyl-phenylalanine (FMLP) and platelet-activating factor (PAF). J Immunol 128:1939–1944
Zakrzewski JT, Barnes NC, Piper PJ, Costello JF (1985) Measurement of leukotriene in arterial and venous blood from normal and asthmatic subjects by radioimmunoassay. Br J Clin Pharmacol 19:574

CHAPTER 4

Pharmacology of Airway Smooth Muscle

I.W. RODGER and R.C. SMALL

A. Introduction

In undertaking to write on the subject of the pharmacology of airway smooth muscle we have deliberately chosen not to produce an exhaustive review of the literature on the subject. Given the space constraints, it would be inappropriate to try to do so. Instead we have chosen to focus upon the excitation-contraction (E/C) coupling and uncoupling mechanisms that exist in airway smooth muscle. Thus, this chapter presents both biochemical and pharmacological information on contraction and relaxation mechanisms. In the context of drug-induced relaxation of airway smooth muscle we have been selective in concentrating on the actions of those drugs (β-adrenoceptor agonists and alkylxanthines) that are currently the principal bronchodilators in the therapy of asthma. In presenting much of the information contained in this chapter we have attempted to cite key references. Where appropriate, reference to relevant review articles has also been made.

B. E/C Coupling Mechanisms

I. The Airway Smooth Muscle Cell

Contraction of smooth muscle is dependent upon the level of free calcium ions (Ca^{2+}) within the cytoplam of the cell (BOLTON 1979; TAKUWA et al. 1987; FELBEL et al. 1988; TAYLOR and STULL 1988; TAYLOR et al. 1989). In the resting (relaxed) state the intracellular concentration of free Ca^{2+} ($[Ca^{2+}]_i$) of airway smooth muscle is thought to lie somewhere between 0.05 and 0.35 μM (KOTLIKOFF et al. 1987; FELBEL et al. 1988; TAYLOR and STULL 1988; TAYLOR et al. 1989; PANETTIERI et al. 1989). This level of $[Ca^{2+}]_i$ is regarded as being insufficient to activate the contractile apparatus. In contrast, the $[Ca^{2+}]$ in the extracellular fluid is of the order of 1–2 mM. Thus, there exists a large, inwardly-directed concentration gradient (a transmembrane concentration difference of approximately 10^4) down which Ca^{2+} will tend to flow. In the resting state, however, few Ca^{2+} ions gain admission to the cell. This is a testament to the functional integrity of the cell membrane which effectively, and efficiently, partitions the intra- and extracellular compartments. This partitioning is, however, not absolute. There is a small quantity of Ca^{2+} that enters the cell, albeit slowly, via a passive Ca^{2+} leak process. The intracellular Ca^{2+} levels achieved by passive leakage, however, do not rise either sufficiently rapidly or to a sufficient magnitude to activate the contractile machinery.

Homeostatic regulation of Ca^{2+} is essential for the survival of smooth muscle cells since elevated $[Ca^{2+}]_i$ is toxic to cellular metabolism. Reliance solely upon the relative impermeability of the cell membrane, as a means of controlling Ca^{2+} levels within cells, would be primitive and wholly inadequate. Thus, cells have developed several additional homeostatic control mechanisms which serve to ensure that any elevation of $[Ca^{2+}]_i$ is as brief as possible. These auxilliary mechanisms comprise: (a) a sodium ion-calcium ion exchange process, (b) a Ca^{2+} efflux pump and (c) a mechanism for the intracellular sequestration of Ca^{2+} (RODGER 1988). The first two processes are concerned with active extrusion of Ca^{2+} from the cell. The Ca^{2+} sequestration process involves intracellular organelles, in particular the sarcoplasmic reticulum. Each of these processes is dealt with in more detail later in this chapter when considering relaxation mechanisms.

II. Activator Calcium Ions and E/C Coupling

1. Coupling Mechanisms

When a contractile agonist interacts with its receptor the $[Ca^{2+}]_i$ of airway smooth muscle cells rises abruptly from its resting level to between 0.3 and 1.0 μM (KOTLIKOFF et al. 1987; FELBEL et al. 1988; TAYLOR and STULL 1988; TAYLOR et al. 1989; PANETTIERI et al. 1989). This so-called activator Ca^{2+} can only be derived from two sources; from the extracellular compartment or from intracellular stores such as the sarcoplasmic reticulum. The relative contribution of activator Ca^{2+} from these two sources is dependent upon the nature and concentration of the agonist and the component (phasic or tonic) of the contractile response in question (BOLTON 1979; CREESE 1983; RODGER 1986).

Essentially two forms of E/C coupling are recognised: electromechanical and pharmacomechanical. Electromechanical coupling depends either upon electrical depolarisation of the plasmalemma which opens calcium channels leading to Ca^{2+} influx and hence an increase in $[Ca^{2+}]_i$ or on a voltage-dependent release of stored intracellular Ca^{2+}. In contrast, the pharmacomechanical coupling mechanism is voltage-independent. It may involve either extracellular Ca^{2+} influx via ligand-gated calcium channels or release of activator Ca^{2+} from intracellular stores. The release mechanism is mediated either via ligand-generated intracellular second messengers or by direct action of a ligand on sarcoplasmic reticular stores.

2. Extracellular Activator Ca^{2+}

Activator Ca^{2+} originating in the extracellular compartment can only gain admission to the cell once the cell membrane has been rendered permeable to them. This is achieved by the opening of specific Ca^{2+} channels in the plasmalemma through which Ca^{2+} flow down their electrochemical and concentration gradient. Two types of calcium channel have been proposed: (a) voltage-operated (VOC) and (b) receptor-operated (ROC) (Bolton 1979). It is pertinent to this review that brief consideration be given to the existence and functional significance of these two types of calcium ion channels in airway smooth muscle. For more comprehensive reading on the subject see Rodger 1986, 1987; Small and Foster 1986, 1988; Giembycz and Rodger 1987.

a) Voltage-Operated Channels

Of the two types of calcium channel that have been proposed the characteristics and properties of the VOC are much the better understood. As the name implies the probability of VOC opening is governed by the potential difference that exists across the plasma membrane (Bolton 1979; Kotlikoff 1988; Marthan et al. 1989). Membrane depolarisation increases both the probability of VOC opening and the duration of the open channel time. Additionally, VOCs display a susceptibility to blockade by certain calcium channel blocking drugs, e.g. verapamil and nifedipine.

There is a substantial body of convincing evidence, from both electrophysiological and ion-flux studies, that supports the view that VOCs are present in airway smooth muscle and that they exhibit selectivity for Ca^{2+} (Rodger 1987; Kotlikoff 1988; Marthan et al. 1989). Similarly, with regard to contractile events, there is overwhelming evidence showing that VOCs are primarily responsible for contractions induced by KCl and tetraethylammonium (TEA). For example, in bovine (Kirkpatrick et al. 1975), canine (Suzuki et al. 1976; Coburn and Yamaguchi 1977; Farley and Miles 1977; Coburn 1979) and guinea-pig (Foster et al. 1983a,) airway smooth muscle, KCl elicits a contraction that is accompanied by, and correlated with, both graded membrane depolarisation and the influx of ^{45}Ca as assessed by the lanthanum technique (Foster et al. 1983a; Raeburn and Rodger 1984). Additionally, calcium antagonists not only suppress slow wave and action potential discharge evoked by both KCl and TEA but also the associated ^{45}Ca uptake and tension changes in airway smooth muscle preparations (Coburn 1979; Kannan et al. 1983; Foster et al. 1984; Raeburn and Rodger 1984; Ahmed et al. 1985; Baba et al. 1985; Raeburn et al 1986; Takuwa et al. 1987). Finally, the dihydropyridine Ca^{2+} agonist BAY k 8644 has been shown to augment both ^{45}Ca uptake and KCl-induced contractions of airway smooth muscle (Allen et al. 1985b; Advenier et al. 1986).

In summary, therefore, there is convincing evidence for the existence of VOCs for Ca^{2+} in airway smooth muscle. Opening of these channels is, in all likelihood, the principal mechanism that underlies the entry of extracellular

Ca^{2+} into airway smooth cells in response to substances such an TEA and KCl. In stark contrast, however, there is little evidence to support the involvement of VOCs in the mechanisms underlying contraction initiated by physiologically relevant agonists, for example cholinomimetics, histamine and eicosanoids (SMALL and FOSTER 1986; 1987; RODGER 1987).

b) Receptor-Operated Channels (ROCs)

ROCs are ion channels opened or operated by a receptor for a stimulant substance (BOLTON 1979). It is envisaged that ROCs are not wholly selective for Ca^{2+}, have an ionic permeability determined by the controlling receptor, can be gated by either voltage-dependent or voltage-independent events and are not readily inhibited by organic calcium antagonists.

Different pharmacological agonists (e.g. acetylcholine, methacholine, histamine, serotonin and leukotrienes C_4 and D_4) have been shown to elicit contraction of airway smooth muscle that is associated with graded depolarisation of the cell membrane. Action potential discharge, however, is never observed (SMALL and FOSTER 1986). Thus, if VOC opening were occurring it would have to be via graded depolarisation. This, however, is not the case since some of these same agonists have been shown to elicit contractions in fully depolarised airway preparations (KROEGER and STEPHENS 1975; KIRKPATRICK et al. 1975; SUZUKI et al. 1976; FARLEY and MILES 1977; CAMERON and KIRKPATRICK 1977; COBURN 1979; McCAIG and SOUHRADA 1980; AHMED et al. 1984; ITO and ITOH 1984a; McCAIG and RODGER 1988; MURLAS and DOUPNIK 1989). Furthermore, acetylcholine, histamine and leukotriene D_4 do not, apparently, induce increases in ^{45}Ca uptake into airway smooth muscle during the period assocaited with tension development (AHMED et al. 1984; RAEBURN and RODGER 1984).

Collectively, therefore, these data argue against both the existence of ROCs in airway smooth muscle and the involvement of VOCs (to any significant extent) in the actions underlying receptor stimulation by pharmacological agonists. This latter view is strengthened by the many observations that calcium antagonists, in concentrations that block the contractile effects of KCl and TEA, fail to inhibit the contractions elicited by a wide range of agonists in airway preparations from several species including man (see RODGER 1987). Additionally, BAY k 8644 does not potentiate either acetylcholine or histamine in contracting guinea-pig or human airway preparations (ALLEN et al. 1985b; ADVENIER et al. 1986).

Thus, at the present time there exists no compelling evidence that supports the notion that ROCs are present in the plasma membrane of airway smooth muscle cells. However, before discarding the idea of ROCs, it is important to point out that the inability to demonstrate their presence (or absence) is severely hampered by the lack of a suitable, selective inhibitor of Ca^{2+} entry through such channels and a suitably refined technique for detecting any low levels of Ca^{2+} entry that might be occurring in response to an agonist (RODGER 1987). The recent report that SKF 96365 is an inhibitor of ROCs in non-excitable cells (MERRITT et al. 1989) signals the possibility that, before long, selective inhibitors of Ca^{2+} influx through ROCs will be available to help address the above question.

3. Intracellular Activator Ca^{2+}

The data presented above highlight the fact that, at best, there is only a minor involvement of extracellular Ca^{2+} in the *initiation* (as opposed to *maintenance*) of contractions produced by pharmacological agonists interacting with their cell surface receptors. This being the case, then such agonists must presumably rely principally upon activator Ca^{2+} released from intracellular stores for induction of contraction. This contention is sustained by the results of in vitro studies undertaken to examine the dependence of different agonists on extracellular Ca^{2+} for contraction. Thus, while contractions elicited by KCl and A23187 (the Ca^{2+} ionophore) are readily inhibited in Ca^{2+}-free medium, those elicited by acetylcholine, carbachol, methacholine, histamine, serotonin and the leukotrienes C_4 and D_4 are relatively unaffected (for references see RODGER 1988; MARTHAN et al. 1985, 1988; NOUAILHETAS et al. 1988).

With the gradual acceptance that the sarcoplasmic reticulum in smooth muscle can act not only as a sink for Ca^{2+} but also as a physiologically important source of activator Ca^{2+} there has developed an acute interest in the signal transduction mechanism responsible for the release of Ca^{2+} from such stores. The question that has been posed is how a contractile agonist, via interaction with its extracellularly-located receptors, is able to communicate with the sarcoplasmic reticulum to induce it to release some of its stored Ca^{2+}. Whilst there is evidence for a Ca^{2+}-induced Ca^{2+} release mechanism (ITO and ITOH 1984a) most attention has focussed upon a possible, pivotal role played by certain metabolites of cell membrane inositol phospholipids, acting as second messengers, in the control of intracellular Ca^{2+} release and the maintenance of the contractile state. In a variety of cell types, including airway smooth muscle, certain agonists induce the diesteratic (phosphoinositidase C) cleavage of phosphatidylinositol-4,5-bisphosphate (PIP_2) to yield two putative intracellular second messengers, *myo*-inositol-1,4,5-trisphosphate (IP_3) and 1,2-diacylglycerol (BERRIDGE and IRVINE 1984, 1989).

a) Inositol-1,4,5-Trisphosphate and Intracellular Ca^{2+} Release

The proposal that IP_3 is an intracellular mediator of Ca^{2+} release in many different cell types is well-sustained by recent findings (see BERRIDGE and IRVINE 1989). With regard to airway smooth muscle, generation of inositol phospholipid metabolites in response to cholinergic agonists, histamine and leukotrienes has been demonstrated by several groups of workers in lung tissue from different species including man (BARON et al. 1984, 1989;

HASHIMOTO et al. 1985; GRANDORDY et al. 1986; TAKUWA et al. 1986; GRANDORDY and BARNES 1987; DUNCAN et al. 1987; MONG et al. 1987, 1988; MILLER-HANCE et al. 1988; MEURS et al. 1988, 1989; HALL and HILL 1988, 1989). In these studies, however, either total inositol phosphates or total (unresolved) IP_3 was measured. Only the 1,4,5,-isomer of IP_3 is active in mobilising Ca^{2+} from non-mitochondrial intracellular stores (see BERRIDGE and IRVINE 1989). Thus the recent reports (CHILVERS et al. 1989; LANGLANDS et al. 1989) that this isomer is rapidly generated in airway tissues after contractile agonist stimulation are of paramount importance. It is also important to the hypothesis that the agonist-induced increases in the levels of IP_3 precede any detectable mechanical response. Additionally, IP_3 has been shown to elicit both Ca^{2+} release from sarcoplasmic reticular stores and contraction of saponin-permeabilised airway smooth muscle (HASHIMOTO et al. 1985; TWORT and VAN BREEMEN 1989). Finally, it has been shown that there is a direct relationship between inositol phosphate accumulation and contraction using different muscarinic cholinoceptor agonists (MEURS et al. 1988, 1989). Taken together the above evidence is strongly supportive of the hypothesis that IP_3 is a second messenger responsible for the initiation of airway smooth muscle contraction via the mobilisation of activator Ca^{2+} from intracellular stores. Precisely how IP_3 induces the liberation of Ca^{2+} from sarcoplasmic reticular stores is not yet known although specific binding sites have been identified (SPAT et al. 1986; ROSS et al. 1989). Recent evidence suggests that IP_3 directly activates a channel in the sarcoplasmic reticular membrane that remains open only for as long as IP_3 is associated with its binding site (GHOSH et al. 1988).

Understanding the mechanisms underlying pharmacomechanical coupling, however, is complicated by the fact that intracellular Ca^{2+} release in smooth muscle is not solely dependent upon IP_3 generation (GHOSH et al. 1988; SAIDA et al. 1988; KOBAYASHI et al. 1988; KITAZAWA et al. 1989). Furthermore, some release mechanisms require activation of regulatory G proteins whilst others do not. To complicate matters still further a guanosine triphosphate (GTP)-activated intracellular Ca^{2+} translocation process has been identified which may allow Ca^{2+} to be moved between compartments without it entering the cytoplasm. It has been proposed that such a mechanism may be responsible for filling/refilling the IP_3-releasable Ca^{2+} pool (MULLANEY et al. 1988; THOMAS 1988; GHOSH et al. 1989). Further contradictory evidence to the IP_3 postulate exists. In the study of LANGLANDS et al. (1989) the cyclic guanosine monophosphate (cGMP) phosphodiesterase inhibitor M&B 22948 (zaprinast) prevented the histamine- and methacholine-induced generation of IP_3 without affecting the contractile response. Taken at face value this result indicates a dissociation between IP_3 generation and tension development and posts a question mark over the signal transduction pathway involving IP_3 as the principal mechanism involved in Ca^{2+} release in airway smooth muscle. Clearly a great deal more work is required to unravel the intricacies of the mechanism(s) underlying pharmacomechanical coupling.

The evidence for major involvement of a Ca^{2+}-induced Ca^{2+} release mechanism (ITO and ITOH 1984a) should not be ignored.

b) Diacylglycerol and Protein Kinase C

Diacylglycerol is not implicated in the intracellular Ca^{2+} release process and, consequently, the initial generation of tension. Instead, it has been suggested that diacylglycerol may be more intimately involved in those events that govern tension maintenance of smooth muscle via an effect on Ca^{2+}-activated, phospholipid-dependent protein kinase C (PARK and RASMUSSEN 1986a, RODGER 1986; RASMUSSEN et al. 1987). Activation of protein kinase C is critically dependent upon the level of free Ca^{2+} within the myoplasm and upon certain phospholipids, notably phosphatidylserine. Under normal physiological conditions when a smooth muscle cell is at rest the basal levels of intracellular Ca^{2+} are insufficient to activate protein kinase C. However, upon stimulation with a contractile agonist that mobilises intracellular Ca^{2+}, two things happen that influence protein kinase C. Initially, the $[Ca^{2+}]_i$ increases abruptly, but transiently, and quicky returns towards baseline values. Diacylglycerol then initiates the translocation of protein kinase C from the cytosol (when it exists in an inactive form) to the endoplasmic face of the plasmalemma. In this location the affinity of protein kinase C for Ca^{2+} is significantly enhanced such that the enzyme is capable of being maximally activated at a $[Ca^{2+}]_i$ that is less than 1 μM (NISHIZUKA 1983, 1988; RASMUSSEN et al. 1984). Thus, in the presence of diacylglycerol, protein kinase C is capable of being activated at the levels of intracellular Ca^{2+} that are thought to exist during the tonic (sustained) phase of a smooth muscle contraction.

Certain phorbol ester tumour promoters, most notably phorbol myristate acetate, 12-*O*-tetradecanoyl phorbol-13-acetate and phorbol dibutyrate, can substitute for diacylglycerol in the activation of protein kinase C. Whilst these agents are not wholly selective in activating protein kinase C they have, nevertheless, proved to be useful pharmacological tools. For example, it has been shown that these phorbol esters can induce slow monotonic contractions in airway smooth muscle,in some instances without increasing cytosolic Ca^{2+} levels (PARK and RASMUSSEN 1986a, b; TAKUWA et al. 1987; DALE and OBIANIME 1985, 1987; OBIANIME et al. 1988). Importantly, the time courses of the late-onset protein phosphorylation changes induced by cholinomimetic agonists and phorbol dibutyrate in airway preparations are very similar (PARK and RASMUSSEN 1986b). Furthermore, phorbol esters cause contraction of detergent-skinned smooth muscle when the free Ca^{2+} concentration is maintained at 0.1 μM, i.e. at a level beneath that required to induce a contraction in itself (CHATTERJEE and TEJADA 1986; MILLER et al. 1986). In not all instances, however, do phorbol esters induce contractile responses. Pretreatment of airway smooth muscle with phorbol esters inhibits agonist-induced contractions (MENKES et al. 1986) and inhibits agonist-induced increases in $[Ca^{2+}]_i$ (KOTLIKOFF et al. 1987).

The precise physiological functions of protein kinase C are only slowly being unravelled. Exactly how the enzyme mediates its effects is also uncertain. What is known, however, is that there is more than one species of protein kinase C molecule; several discrete sub-species have been identified (NISHIZUKA 1988). These proteins are derived both from multiple genes and from alternative splicing of a single mRNA transcript, yet they retain a primary structure containing conserved structural motifs with a high degree of sequence homology. To date seven isoenzymes have been identified. Each shows subtle differences in its mode of activation, sensitivity to Ca^{2+} and catalytic activity towards substrates. It is also known that the enzyme can phosphorylate a variety of different proteins, some of which are involved in the contractile mechanism. In general terms, such phosphotransferase activity is associated with an enhanced sensitivity to Ca^{2+}(see RODGER 1988). In spite of some contradictory results, protein kinase C is an attractive candidate mediator responsible for the enhanced sensitivity of contractile proteins to Ca^{2+}, a step regarded as essential for the generation and maintenance of the tonic phase of contraction.

III. Biochemical Basis of Airway Smooth Muscle Contraction

Contraction of smooth muscle is thought to occur via a mechanism similar to the sliding filament process first proposed for skeletal muscle. The sliding of actin and myosin past each other is achieved by the cyclic attachment of the globular heads of myosin molecules to actin (so-called crossbridge formation), a flexing change in the configuration of the myosin head with respect to actin, detachment of myosin from actin followed by subsequent reattachment at another site further down the actin molecule. It is this rapid cyclical attachment and detachment of crossbridges between actin and myosin (crossbridge cycling) that is responsible for the active force development in smooth muscle. The process is fuelled by the energy derived from the breakdown of adenosine triphosphate (ATP) by actin-activated myosin. ATPase. Furthermore, the rate of force development is directly proportional to the rate of crossbridge cycling (KAMM and STULL 1985a,b).

It is widely accepted that myosin phosphorylation plays a major regulatory role in the contraction of airway smooth muscle. Adoption of this view stems from the results of numerous studies that have clearly shown that during contraction the levels of myosin light chain phosphorylation are significantly increased over resting levels (DELANEROLLE et al. 1982; SILVER and STULL 1982; GERTHOFFER and MURPHY 1983; KAMM and STULL 1985a, b, 1986; GERTHOFFER 1986; PERSECHINI et al. 1986; TAYLOR and STULL 1988; TAYLOR et al. 1989). Recently, however, it has been established that the magnitude of myosin light chain phosphorylation is not directly proportional to the development of tension *and* its maintenance, since myosin phosphorylation can decline whilst developed tension (the tonic phase of a contraction) is maintained (Fig. 1) (SILVER and STULL 1982, 1984; GERTHOFFER and

MURPHY 1983; KAMM and STULL 1985b; GERTHOFFER 1986). The current interpretation of these events is that the level of myosin phosphorylation does not correlate with the absolute level of tension developed (i.e. the *number* of crossbridges formed). There is, however, a good correlation between the level of myosin phosphorylation and the rate of tension development (i.e. the *rate* of crossbridge cycling) (GERTHOFFER and MURPHY 1983; KAMM and STULL 1985a, b). Interestingly, the time courses for myosin phosphorylation and dephosphorylation are closely paralled by agonist-induced changes in the intracellular free Ca^{2+} concentration in smooth muscle cells (the so-called Ca^{2+} transient) (Fig. 1) (KOTLIKOFF et al. 1987; FELBEL et al. 1988; TAYLOR and STULL 1988; TAYLOR et al. 1989; PANETTIERI et al. 1989). Taken together, these two sets of observations have led to the suggestion that two Ca^{2+}-dependent pathways operate to control contraction of smooth muscle (AKSOY et al. 1982; RASMUSSEN and BARRETT 1984; RODGER 1986; HAL and MURPHY 1989).

In simplified terms, one current view of the sequence of biochemical events that underlies contraction of airway smooth muscle is as follows. Contractile agonists, on combining with their specific cell-surface receptors, activate phosphoinositidase C via a GTP-binding regulatory protein (in all likelihood G_i). This cleaves PIP_2 creating the two second messengers IP_3 and diacylglycerol. The IP_3 promptly induces the release of Ca^{2+} from intracellular stores. As soon as the $[Ca^{2+}]_i$ increases, the free Ca^{2+} combine with the Ca^{2+}-binding protein calmodulin. The Ca^{2+}-calmodulin complex in turn activates myosin light chain kinase. This kinase then phosphorylates the 20 kDa light chain of myosin. The rate at which tension is then generated (the phasic component of the contraction) is determined by the rate at which crossbridges cycle, which in itself is determined by the magnitude of myosin light chain phosphorylation. The generation of IP_3 by a contractile agonist, however, is transient (Fig. 1) (CHILVERS et al. 1989; LANGLANDS et al. 1989). Consequently, the second messenger-induced release of Ca^{2+} from the sarcoplasmic reticulum is not continuous. Thus, it would appear that IP_3 is adapted only to provide an initial burst of intracellular Ca^{2+} to induce rapid crossbridge cycling and thus tension development. In the absence of further release of intracellular Ca^{2+}, $[Ca^{2+}]_i$ returns towards (although not precisely to) resting levels (Fig. 1). This fall in $[Ca^{2+}]_i$ results in the dissociation of Ca^{2+} from calmodulin which consequently switches off the activity of myosin light chain kinase. Myosin phosphorylation must, therefore, decline and so too myosin ATPase activity (Fig. 1). Despite these events, which one might expect to uncouple the E/C sequence and produce relaxation, the developed tension is well-sustained (Fig. 1). It has been proposed, therefore, that such steady-state force maintenance does not require continued myosin light chain phosphorylation because of an alteration in the type of attachment formed between actin and myosin. Thus, instead of rapidly cycling crossbridges, so-called latchbridges (dephosphorylated myosin crossbridges), which cycle slowly or not at all, are formed (AKSOY et al. 1982, 1983; HAI and MURPHY

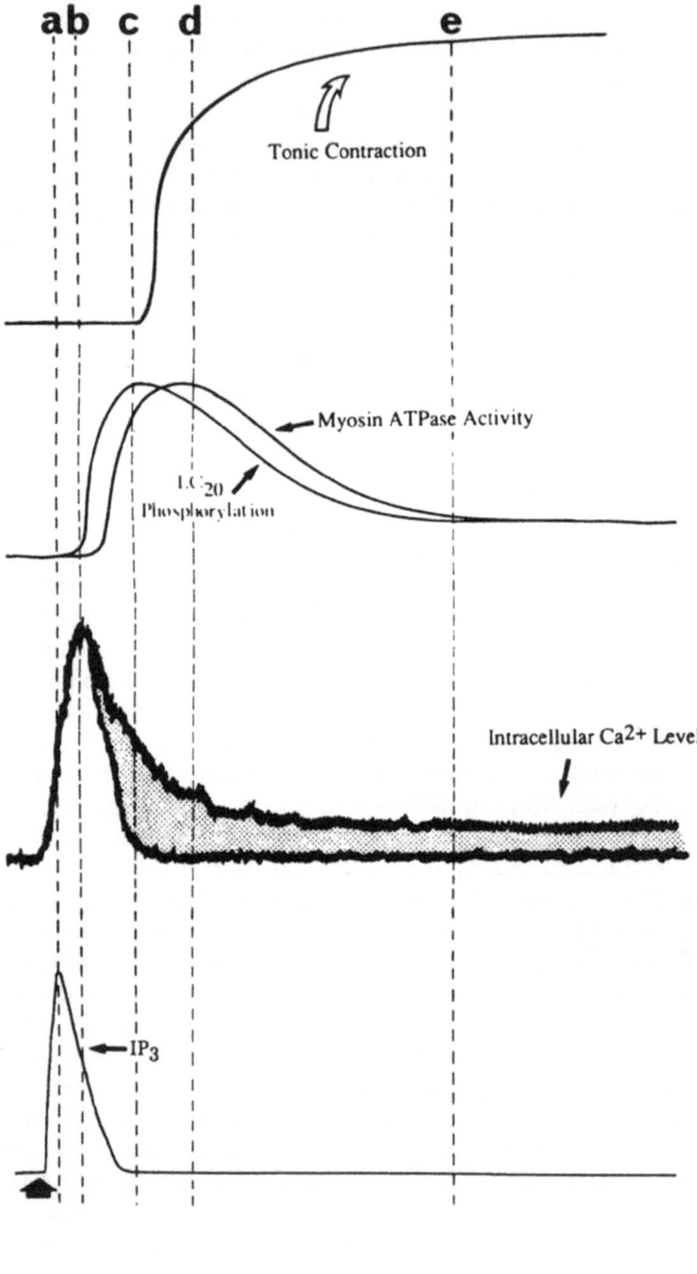
a
b
c
d
e
Tonic Contraction
Myosin ATPase Activity
LC20
Phosphorylation
Intracellular Ca2+ Levels
IP3

Fig. 1.

1988, 1989; MURPHY 1989). The maintenance of these latchbridges is Ca^{2+}-dependent and a smooth muscle in the latch state can be relaxed by removal of Ca^{2+}. This does not necessarily imply that a latchbridge is directly regulated by Ca^{2+}. It does mean, however, that some Ca^{2+}-dependent process must be involved in the formation of latchbridges (HAL and MURPHY 1989; MURPHY 1989). In view of the low $[Ca^{2+}]_i$ that exists during the tonic phase of contraction (Fig. 1) one of the characteristic features of the latchbridge state must be the enhanced sensitivity of the contractile apparatus to Ca^{2+}. Precisely how this is achieved is not yet known although a possible role for protein kinase C has been proposed (RASMUSSEN and BARRETT 1984; RODGER 1986; RASMUSSEN et al. 1987). Thus two separate, although in all likelihood interdependent, pathways exist to control contraction of airway smooth muscle. These are diagrammatically illustrated in Fig. 2. The first, the calmodulin/myosin light chain kinase pathway, is responsible for the initial rapid development of tension. The second pathway adapts the cell to accommodate sustained contraction at minimum cost to itself, in terms of both energy expenditure (ATPase activity) and protection from the toxic consequences of intracellular Ca^{2+} overload.

There is another point that must be considered. The sarcoplasmic reticulum contains only a finite pool of activator Ca^{2+} and is, consequently, unable to maintain a prolonged low level of Ca^{2+} release in response to IP_3. Where, therefore, do the activator Ca^{2+} come from that enables the tonic phase of contraction to be maintained? One must presume that they come from the extracellular compartment. This being the case then the Ca^{2+} must gain entry to the cell via some type of ion channel. Precisely how they do this remains unknown. One postulate is that emptying of the agonist/IP_3-sensitive Ca^{2+} pool can, in some cell types, cause enhanced permeability of the plasmalemma permitting Ca^{2+} entry (PUTNEY 1986). This has been demonstrated, for example, in recent experiments using the non-phorbol ester

Fig. 1. Schematic representation of the time courses of several interlinked biochemical events involved in the generation of airway smooth muscle contraction. One of the first events to occur following receptor activation by a contractile agonist, added at the *arrow* (*bottom left*), is the rapid formation of 1,4,5-IP_3. IP levels rise to a peak (*a*) and then rapidly wane to control values. Closely following this event there is a rapid elevation of $[Ca^{2+}]_i$ which peaks at (*b*). Like IP_3, the increase in $[Ca^{2+}]_i$ is only transient, levels returning to, or close to, basal values very quickly–a consequence of the protective homeostatic Ca^{2+} regulatory mechanisms operated by the smooth muscle cells. The *shaded area* illustrates the different $[Ca^{2+}]_i$ decay profiles measured by different investigators. Following the transient surge in $[Ca^{2+}]_i$ the 20 kDa light chain of myosin is phosphorylated (*LC_{20} phosphorylation* peaking at *c*) and closely thereafter myosin is activated as an enzyme (*myosin ATPase*, peaking at *d*). This allows rapid crossbridge cycling between actin and myosin and hence the development of airway smooth muscle contraction. Note, however, that as tension develops LC_{20} phosphorylation and myosin ATPase activity wane. Despite these events, and the very low $[Ca^{2+}]_i$, the tonic contraction (*e*) is well-sustained. See text for further explanation

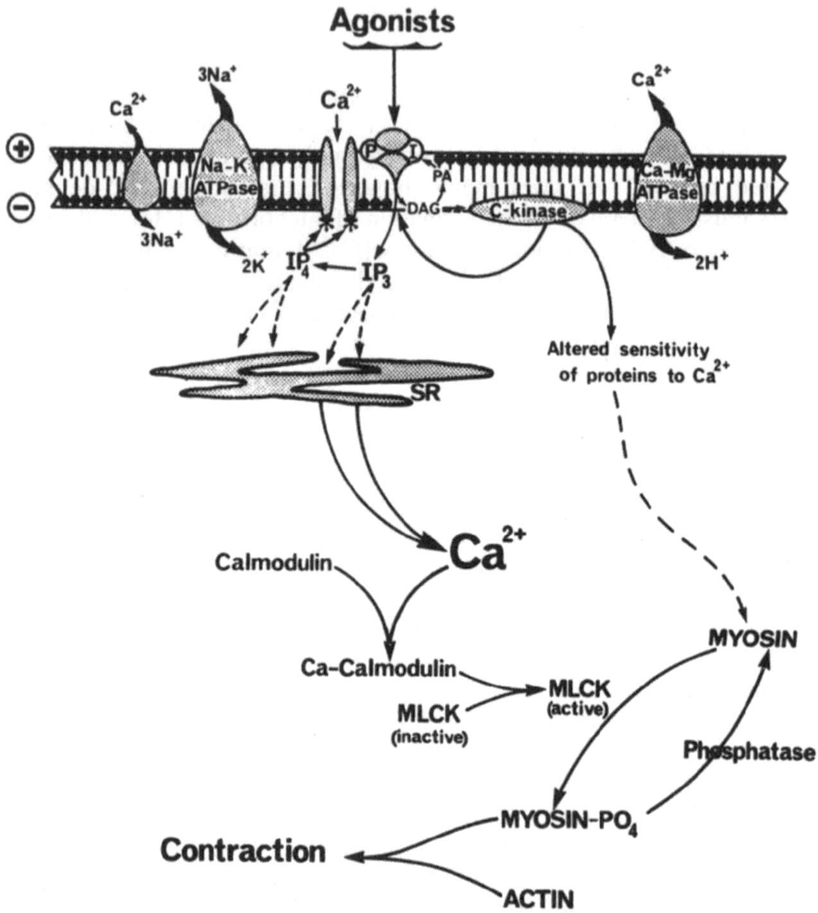

Fig. 2. Diagrammatic summary of the sequence of events thought to be involved in excitation-contraction coupling in airway smooth muscle. The cell membrane is shown containing the sodium-calcium exhange system, the sodium pump (*Na-K ATPase*), the Ca^{2+} efflux pump (*Ca-Mg ATPase*), protein kinase C (*C-kinase*) and a receptor complex coupled to the phosphatidylinositol (*PI*) system. On interacting with their cell surface receptors, agonists activate phosphoinositidase C which results in the formation of the intracellular second messengers inositol 1,4,5-trisphoshate (*IP_3*) and diacylglycerol (*DAG*). The initial tension development (phasic response) is determined by the activity of the calmodulin/myosin light chain kinase (*MLCK*) pathway which is turned on as a consequence of the release of activator Ca^{2+} from the sarcoplasmic reticulum (*SR*), probably by IP_3 and possibly via Ca^{2+}-induced Ca^{2+}release. Once generated, IP_3 is promptly phosphorylated to inositol 1,3,4,5-tetrakisphosphate (*IP_4*) by an IP_3 kinase. It has been suggested that IP_4 may be responsible for opening plasmalemmal ion channels, thus permitting a low level of influx of extracellular Ca^{2+}. These Ca^{2+} ions may be responsible for maintaining $[Ca^{2+}]_i$ at, or slightly above, basal levels during the period of maintained tension. The *asterisks* indicate the points on the inner surface of the plasmalemma at which IP_4 might act. Additionally, IP_4 may contribute to the process involved in replenishing the store of Ca^{2+} in the IP_3-releasable pool. It has been postulated that C-kinase may be responsible for enhancing the sensitivity of the contractile apparatus to Ca^{2+}. The C-kinase pathway may, therefore, be involved in the maintenance of developed tension (tonic phase to contraction) at a time when $[Ca^{2+}]_i$ is low. Diacylglycerol is metabolised to phosphatidic acid (*PA*) which is then reincorporated into the membrane lipid pool. (Figure adapted from RODGER [1988])

tumour promoter thapsigargin (PUTNEY et al. 1989). Just how emptying of the Ca^{2+} pool in the sarcoplasmic reticulum influences the permeability of the plasmalemma, which may be somewhat remote, is not yet known. It is not inconceivable, however, that yet another intracellular chemical signalling process is responsible for this capacitive process. Another postulate is that one of the metabolites of IP_3, namely *myo*-inositol-1,3,4,5-tetrakisphosphate (IP_4) (IRVINE et al. 1986; NAHORSKI and BATTY 1986) may act as an intracellular regulator of a plasmalemmal calcium channel (Fig. 2). IP_4 has been shown to be formed under the action of a specific *myo*-inositol-1,4,5-trisphosphate-3-kinase which has a $K\hat{m}$(Ca) around 1 μM. Since the $[Ca^{2+}]_i$ easily achieves such levels during the initial, IP_3-triggered phase of a contraction this is an attractive hypothesis. At the present time, however, such a process has not been demonstrated in muscle cells.

There remains one further compelling reason for believing that extracellular Ca^{2+} must enter the cell during the plateau phase of a contraction. Numerous experiments have shown that repeated addition of contractile agonists to airway smooth muscle preparations bathed in Ca^{2+}-free medium causes a steady reduction in the size of the contractile response that can be induced. The simple explanation for this effect is that the intracellular Ca^{2+} stores are not able to be refilled in the interval between successive additions of the contractile agonist. In all likelihood, therefore, entry of extracellular Ca^{2+} is essential for the replenishment of the IP_3-releasable Ca^{2+} pool.

C. E/C Uncoupling Mechanisms

As outlined above, the relationship between the development and maintenance of tension and the $[Ca^{2+}]_i$ in airway smooth muscle is not a simple one. Furthermore, the long-held belief that relaxant drug action is accompanied by a fall in $[Ca^{2+}]_i$ has recently been challenged by experiments using Ca^{2+} fluorophores (TAKUWA et al. 1988; FELBEL et al. 1988).

Despite these difficulties, this section of the review will examine whether the actions of drugs which relax airway smooth muscle can be attributed to their interference with the cellular handling of Ca^{2+}. Discussion will focus on the two principal groups of drugs currently used as bronchodilators in the treatment of bronchial asthma, namely β-adrenoceptor agonists and alkylxanthines. In promoting relaxation of airway smooth muscle these drugs may exert a direct or indirect reduction in the Ca^{2+} sensitivity or responsiveness of the contractile machinery and/or a reduction in the cytosolic concentration of free Ca^{2+}. If they act to produce a reduction in the cytosolic free Ca^{2+} concentration, this could be achieved by: (a) inhibition of the cellular influx of Ca^{2+}, (b) inhibition of the release of Ca^{2+} from intracellular stores, (c) promotion of the sequestration of Ca^{2+} into intracellular stores or (d) promotion of Ca^{2+} extrusion from the cell. The relative importance of these different ways of interfering with the intracellular action of Ca^{2+} or its disposition with respect to the relaxant actions of β-adrenoceptor agonists and alkylxanthines is discussed below.

I. β-Adrenoceptor Agonists

1. β-Agonist-Induced Relaxation In Vitro: General Properties

β-Adrenoceptor agonists can suppress the tone of airway smooth muscle in vitro both when the tone is of spontaneous origin and when induced by an exogenous spasmogen. Such spasmogens include KCl, TEA and muscarinic cholinoceptor and histamine H_1 receptor agonists. As described earlier, these spasmogens may utilise different sources of activator Ca^{2+} to initiate tension development. The relaxant action of β-adrenoceptor agonists is, therefore, independent of the source of Ca^{2+} utilised to initiate contraction.

There is growing evidence (reviewed by FEDAN et al. 1988) that the airway epithelium may release a factor(s) which can modulate the responsiveness of airway smooth muscle, principally to spasmogens but also to relaxant agents. However, the tracheal relaxant potencies of β-adrenoceptor agonists are either unchanged or increased by epithelial removal (BARNES et al. 1986; FARMER et al. 1986; GOLDIE et al. 1986; HOLROYDE 1986). These findings suggest that the relaxant actions of β-adrenoceptor agonists in airway smooth muscle do not depend on the release of an epithelium-derived relaxant factor.

2. Role of Cyclic Adenosine Monophosphate (cAMP) Accumulation in the Relaxant Actions of β-Adrenoceptor Agonists

There is abundant evidence (reviewed by TORPHY and GERTHOFFER 1986) to suggest that the relaxant effects of β-adrenoceptor agonists in airway smooth muscle are mediated by the intracellular accumulation of cAMP. Activation of the β_2-adrenoceptor alters the working of a regulatory G-protein (Gs) so that the activity of adenylate cyclase is increased. This enzyme catalyses the intracellular conversion of ATP to form cAMP. The subsequent binding of cAMP to the regulatory subunit of cAMP-dependent protein kinase releases and activates the catalytic subunit. Activated cAMP-dependent protein kinase (A-kinase) then phosphorylates specific proteins and thereby sets in motion a variety of biochemical changes that result in relaxation.

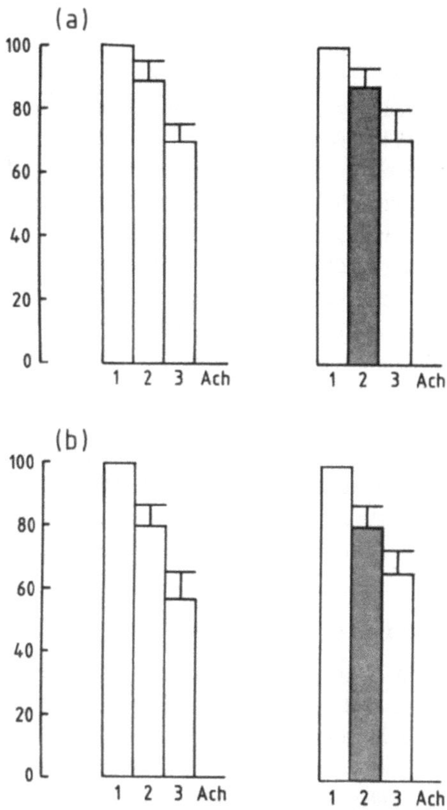

Fig. 3a,b. Guinea-pig isolated trachealis muscle skinned of its plasmalemmal membranes using Triton X-100. The effects of aminophylline and isoprenaline on responses of skinned trachealis to Ca^{2+}. The *abscissa scales* indicate the first, second and third (*1,2,3*) challenges with 20 μM Ca^{2+} or with 100 μM acetylcholine (*ACh*). *Left-hand panels*, time-matched control tissues; *right-hand* panels, test tissues. *Shaded columns* indicate the presence of **a** aminophylline 1 m*M* or **b** isoprenaline 1 μM. Each column represents the means of at least six experimental values and *vertical lines* indicate the SEM. There was no statistically significant difference ($p > 0.05$) between column 2 means in either **a** or **b** (two-tailed, unpaired *t* test). (From ALLEN et al. [1986])

3. β-Agonists and the Ca^{2+} Sensitivity (or Responsiveness) of the Contractile Machinery

The possibility that β-adrenoceptor agonists might directly reduce the Ca^{2+} responsiveness or sensitivity of the intracellular contractile machinery has been tested in trachealis chemically-skinned of its plasmalemmal membranes using Triton X-100 or saponin. In feline and guinea-pig skinned trachealis isoprenaline failed to reduce the sensitivity or responsiveness of the tissue to Ca^{2+} (ITO and ITOH 1984b; ALLEN et al. 1986) (Fig. 3). This suggests that the β-adrenoceptor agonists do not directly reduce the Ca^{2+} responsiveness of the intracellular contractile machinery. However, it is possible that, in the intact cell, they may do so by causing the intracellular accumulation of cAMP.

In trachealis skinned using Triton X-100 or saponin, cAMP and the catalytic subunit of cAMP-dependent protein kinase depress Ca^{2+}-induced spasm (ITO and ITOH 1984b; SPARROW et al. 1984; BRYSON and RODGER 1987). Since Triton X-100 destroys both the plasmalemma and the endoplasmic reticulum (MEISHERI et al. 1985) these effects of cAMP and A-kinase cannot represent interference with the cellular influx of Ca^{2+} or with Ca^{2+} sequestration into the sarcoplasmic reticulum. Instead they represent interference with Ca^{2+} activation of the contractile machinery. This may involve phosphorylation of myosin light chain kinase (BRYSON and RODGER 1987) and hence (RODGER 1986) reduced sensitivity of this enzyme to activation by Ca^{2+}-calmodulin.

In summary, in the intact cell, it is likely that β-adrenoceptor agonists can reduce the Ca^{2+} sensitivity of the contractile machinery by mechanisms involving the intracellular accumulation of cAMP and hence the activation of A-kinase. The importance of this to the overall relaxant activity of β-adrenoceptor agonists remains to be established.

4. β-Agonists and Inhibition of Ca^{2+} Influx

In bovine and canine trachealis certain β-adrenoceptor agonists (e.g. adrenaline, isoprenaline and procaterol) suppress spamogen-induced mechanical tone. This effect is associated with hyperpolarisation of the muscle cells (SUZUKI et al. 1976; KIRKPATRICK 1981; ITO and TAJIMA 1982; CAMERON et al. 1983; FUJIWARA et al. 1988). Isoprenaline, terbutaline and adrenaline, tested in guinea-pig trachealis (ALLEN et al. 1985a; HONDA et al. 1986), suppressed spontaneous mechanical tone and spontaneous electrical slow waves and caused hyperpolarisation (Fig. 4). Isoprenaline caused similar electrical changes in human trachealis (HONDA and TOMITA 1987).

The ability of β-adrenoceptor agonists to hyperpolarise airway smooth muscle is shared by the adenylate cyclase activator forskolin and by the membrane-permeable cyclic nucleotide dibutyryl cyclic AMP (HONDA et al. 1986). It therefore seems likely that the hyperpolarisation induced by β-adrenoceptor agonists may be mediated by the intracellular accumulation of cAMP.

The hyperpolaristion induced by β-adrenoceptor agonists seems to represent an increase in membrane conductance because electrotonic potentials induced by the passage of transmembrane current are reduced in amplitude during the hyperpolarisation (ITO and TAJIMA 1982; CAMERON et al. 1983). Early evidence that an increase in membrane K^+ conductance might underlie the hyperpolarisation induced by isoprenaline included the observation that the K^+ channel inhibitors procaine and TEA could suppress isoprenaline-induced hyperpolarisation (ALLEN et al. 1985a). Recent patch clamp studies (KUME et al. 1989) have revealed that isoprenaline increases the

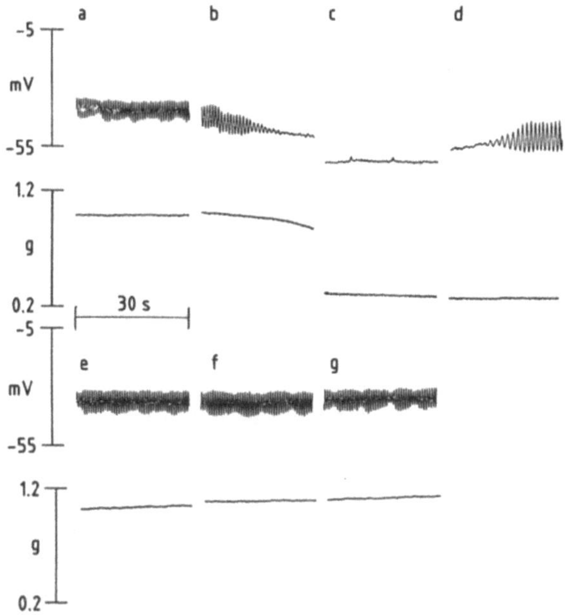

Fig. 4. The effects of propranolol (1 μ*M*) on the electrical and mechanical responses of guinea-pig isolated trachealis to isoprenaline (0.1 μ*M*).

In all records the *upper trace* represents membrane potential and the *lower trace* the mechanical activity of a contiguous segment of trachea. All electrical records are taken from the same cell. Activity was recorded (*a*) before, (*b*) 1 min and (*c*) 4 min after the initial challenge with isoprenaline (0.1 μ*M*); (*d*) 2 min and (*e*) 10 min after washout of isoprenaline using Krebs solution containing propranolol (1 μ*M*). Activity was also recorded (*f*) 1 min and (*g*) 4 min after a second challenge with isoprenaline (0.1 μ*M*) in the presence of propranolol. Note the ability of propranolol to abolish both the electrical and mechanical responses to isoprenaline. (From ALLEN et al. [1985a])

open state probability of a Ca^{2+}-dependent K^+ channel in rabbit trachealis cells. A-kinase mimics this action of isoprenaline and the actions of both agents are augmented by the protein phosphatase inhibitor okadaic acid. These data suggest, therefore, that A-kinase phosphorylates a Ca^{2+}-dependent K^+ channel (or a channel-related protein) resulting in an increase in its open-state probability and hence cellular hyperpolarisation (KUME et al. 1989).

Whatever the underlying biochemical mechanisms may be, the hyperpolarisation of airway smooth muscle induced by β-adrenoceptor agonists results in a reduced probability of the opening of the Ca^{2+} VOCs known (MARTHAN et al. 1989) to exist in this tissue. Ca^{2+} influx is consequently reduced. Whilst hyperpolarisation-induced inhibition of Ca^{2+} influx may contribute to the relaxant actions of β-adrenoceptor agonists, several findings suggest that it plays only a minor role.

In guinea-pig trachealis, K^+ channel blockade by procaine or TEA inhibits isoprenaline-induced hyperpolarisation without affecting the relaxant action (Allen et al. 1985a). Furthermore, isoprenaline retains relaxant activity in a K^+-rich (120 m*M*) medium (Kumar 1987; Allen et al. 1985a; Small et al. 1989). In this situation the K^+ equilibrium potential and the membrane potential of the cell are virtually coincident. Accordingly, K^+ channel opening cannot evoke sufficient hyperpolarisation to raise the membrane potential to a level where closure of Ca^{2+} VOCs occurs.

In summary, electrophysiological and mechanical studies have revealed that β-adrenoceptor agonists hyperpolarise airway smooth muscle cells and should therefore inhibit Ca^{2+} influx by reducing the opening of Ca^{2+} VOCs. However, this mechanism seems to play a supportive rather than a crucial role in the process by which relaxation is achieved.

While electrophysiological studies indicate that β-adrenoceptor agonists should reduce Ca^{2+} influx through VOCs, some studies involving the measurement of cytosolic Ca^{2+} with fluorophores suggest quite the opposite. In bovine trachealis, for example, isoprenaline and salbutamol have been observed to increase the cytosolic concentration of free Ca^{2+} (Felbel et al. 1988; Takuwa et al. 1988). In cat trachealis, however, procaterol failed to produce a similar effect (Fujiwara et al. 1988). The rise in cytosolic Ca^{2+} induced by β-adrenoceptor agonists in bovine trachealis was reduced both by a Ca^{2+}-free, EGTA-containing medium and by inhibitors of Ca^{2+} influx such as nitrendipine and D 600 (Felbel et al. 1988; Takuwa et al. 1988). This has let to the suggestion that β-adrenoceptor agonists promote the influx of Ca^{2+} through VOC (Felbel et al. 1988).

The immediate question is how can this apparent conflict between the results of electrophysiological studies and studies with Ca^{2+} fluorophores be resolved? When β-adrenoceptor agonists are administered following challenge with a muscarinic agonist, they reduce the sustained phase of the Ca^{2+} signal elicited by the muscarinic agonist and promote relaxation (Felbel et al. 1988; Fujiwara et al. 1988; Takuwa et al. 1988). This suggests that, in this situation, the sympathomimetics act to reduce Ca^{2+} influx. Felbel et al. (1988) observed that the rise in cytosolic Ca^{2+} induced by isoprenaline was inhibited by pertussis toxin, whereas this toxin was ineffective against the action of isoprenaline in suppressing the sustained phase of the Ca^{2+} signal induced by carbachol. These authors therefore proposed that β-adrenoceptor agonists could open Ca^{2+} VOCs by a mechanism involving a pertussis toxin-sensitive G protein (Gi) but could also close the Ca^{2+} channels by a cAMP-dependent process that was insensitive to pertussis toxin. This suggestion needs verification by further experimentation. However, in situations where the channel opening and closing actions occur simultaneously, the net effect of the β-adrenoceptor agonists on Ca^{2+} influx could be quite small. Such an explanation would be consistent with the results of the electrophysiological and mechanical studies described above which suggest that inhibition of Ca^{2+} influx through VOCs plays a supportive but not a crucial role in the process by which β-adrenoceptor agonists evoke relaxation.

5. β-Agonists and Inhibition of Ca^{2+} Release from Intracellular Stores

While there is some evidence to the contrary (LANGLANDS et al. 1989), structurally-specific spasmogens are widely believed to initiate tension development in airway smooth muscle via an IP_3-generated release of Ca^{2+} from intracellular stores (see earlier). The question of whether β-adrenoceptor agonists prevent Ca^{2+} release from intracellular stores by interfering with the production or action of IP_3 is addressed in this section.

Early studies using bovine trachealis treated with Li^+ showed that isoprenaline failed to inhibit inositol phosphate accumulation induced by carbachol, 5-hydroxytryptamine (5HT) or histamine (GRANDORDY and BARNES 1986; GRANDORDY et al. 1987). However, subsequent studies have shown that, in canine and bovine trachealis, β-adrenoceptor agonists (noradrenaline, isoprenaline and salbutamol) can inhibit inositol phosphate accumulation induced by histamine but not that induced by muscarinic cholinoceptor agonists (MADISON and BROWN 1988; HALL and HILL 1988).

The effects of the β-adrenoceptor agonists in inhibiting histamine-induced inositol phosphate accumulation can be mimicked by dibutyryl cAMP and by agents (e.g. forskolin) which act to promote the intracellular accumulation of cAMP. Accordingly, it has been suggested that the ability of β-adrenoceptor agonists to suppress histamine-induced inositol phosphate accumulation is related to the ability of these agents to stimulate cAMP production (MADISON and BROWN 1988; HALL et al. 1989).

The site of action of cAMP in preventing histamine-induced inositol phospholipid hydrolysis is at present unknown. However, it has been suggested (MADISON and BROWN 1988; HALL et al. 1989) that cAMP may inhibit phosphatidylinositol kinase, the enzyme which catalyses the conversion of phosphatidylinositol into phosphatidylinositol monophosphate, the precursor of PIP_2. Hence a rise in intracellular cAMP concentration may reduce the supply of substrate for hydrolysis by phospholipase C and hence impair the generation of IP_3.

Muscarinic receptor agonists have been shown to inhibit adenylate cyclase in airway smooth muscle (JONES et al. 1987). It might therefore be argued that the failure of β-adrenoceptor agonists to suppress inositol phosphate accumulation induced by muscarinic agonists reflects functional antagonism at the level of adenylate cyclase. By opposing β-adrenoceptor-mediated stimulation of adenylate cyclase, muscarinic agonists attenuate the rise in cAMP. Hence less cAMP (and A-kinase) is available to depress the activity of phospholipase C. However, dibutyryl cAMP attenuates histamine-induced inositol phosphate accumulation but not that induced by muscarinic agonists (MADISON and BROWN 1988). This suggests that inhibition of cAMP synthesis by muscarinic agonists does not explain the lack of effect of β-adrenoceptor agonists on phosphatidylinositol hydrolysis induced by muscarinic receptor agonists.

It is well-established that β-adrenoceptor agonists suppress mechanical responses of airway smooth muscle to a wide variety (e.g. KCl, TEA, histamine and muscarinic agonists) of spasmogens. Accordingly, the selective action of the β-agonists in inhibiting histamine-induced Ca^{2+} release (by suppressing PIP_2 hydrolysis) cannot represent the sole or even the most important mechanism by which these agents evoke relaxation.

This conclusion is supported by recent experiments in guinea-pig trachealis where the cGMP phosphodiesterase inhibitor M&B 22498 suppressed IP_3 production evoked by methacholine and by histamine. However, M&B 22498 failed to modify the spasmogenic activity of these agents (Langlands et al. 1989). This sugests that, in guinea-pig airway smooth muscle, the production of IP_3 may not be crucial for the process by which muscarinic and H_1 receptor agonists evoke tension development.

There have been several reports (Spilker and Minatoya 1975; Karlsson and Persson 1981; Allen et al. 1986; Small et al. 1989) that the potency of β-agonists as trachealis relaxants is reduced more greatly when tone is induced by muscarinic agonists than when equivalent tone is induced by histamine. It is tempting to suggest that these potency differences might reflect the reported (Hall and Hill 1988; Madison and Brown 1988) ability of the β-agonists to suppress inositol phosphate accumulation induced by histamine but not that induced by muscarinic agonists. However, this suggestion is not tenable if the claim of Langlands et al. (1989), that IP_3 production is not essential to the spasmogenic actions of histamine and muscarinic agonists, is substantiated.

6. β-Agonists and Promotion of Ca^{2+} Sequestration by Intracellular Stores

Working with feline trachealis, Ito and Itoh (1984b) obtained evidence to suggest that isoprenaline could promote the intracellular sequestration of Ca^{2+}. Acetylcholine was applied to the tissue in a Ca^{2+}-free, EGTA-containing medium and the resulting contraction was attributed to the release of Ca^{2+} from the intracellular store. A subsequent challenge with caffeine was used as an index of the degree of store depletion caused by acetylcholine. When isoprenaline was applied to the tissue prior to and during an acetylcholine challenge, the acetycholine-induced spasm was not reduced. However, a subsequent challenge with caffeine yielded spasm greater than that observed in control tissue not exposed to isoprenaline. This result was interpreted in terms of isoprenaline acting to promote the sequestration of Ca^{2+} into intracellular stores. However, further work is clearly merited in this area before we can be confident that part of the relaxant action of β-adrenoceptor agonists involves the intracellular sequestration of Ca^{2+}.

Ito and Itoh (1984b) suggested that the intracellular stores of airway smooth muscle are normally saturated with Ca^{2+}. If this is true, promotion of Ca^{2+} sequestration within the cell cannot be envisaged to play a powerful role in β-agonist-induced suppression of spontaneous tone. Furthermore β-agonists can suppress the spasm evoked when airway smooth muscle is exposed to high concentrations (120 mM) of KCl (Kumar 1978; Allen et al.

1985a; SMALL et al. 1989). In this situation Ca^{2+} floods into cells through VOCs (RAEBURN and RODGER 1984; SMALL and FOSTER 1986, 1988). It is difficult to envisage that the capacity of the intracellular store to sequester Ca^{2+} is so great that intracellular Ca^{2+} sequestration can explain the relaxant action of the β-agonists in this situation.

In summary, therefore, there is currently little evidence to suggest that β-adrenoceptor agonists can promote the intracellular sequestration of Ca^{2+} in airway smooth muscle. Any such effect cannot be envisaged to play an important role in the relaxant actions of the β-agonists observed in the presence of K^+-rich media.

7. β-Agonists and Ca^{2+} Extrusion

To our knowledge there have, as yet, been no reports of an action of β-adrenoceptor agonists in promoting the extrusion of Ca^{2+} from airway smooth muscle. This is an important potential mechanism of action, however, since an effect of this type could cause the suppression of pre-existing spasm in a manner which was independent of the source of activator Ca^{2+}.

II. Alkylxanthines

1. Alkylxanthine-Induced Relaxation In Vitro: General Properties

Like the β-adrenoceptor agonists alkylxanthines can suppress the tone of airway smooth muscle in vitro either when the tone is of spontaneous origin or when induced by any one of a variety of exogenous spasmogens. Accordingly it has been suggested (SMALL et al. 1988b) that the relaxant action of the alkylxanthines, too, is independent of the source of activator Ca^{2+} utilised to initiate tension development.

The tracheal relaxant potencies of aminophylline and theophylline are not altered by epithelial removal (GOLDIE et al. 1986; BEWLEY and CHAPMAN 1988; SMALL et al. 1989). Thus, the relaxant actions of alkylxanthines in airway smooth muscle, in all likelihood, do not depend upon the synthesis of an epithelium-derived relaxant factor.

2. Role of cAMP Accumulation in the Relaxant Actions of Alkylxanthines

PARSONS et al. (1988) have suggested that alkylxanthines may activate adenylate cyclase by interfering with the action of the inhibitory regulatory protein G_i. However, inhibition of cAMP-dependent phosphodiesterase (PDE) is commonly assumed to underlie the relaxant effects of alkylxanthines in isolated airway smooth muscle. The resultant intracellular accumulation of cAMP may then stimulate processes by which relaxation is achieved.

Theophylline and other alkylxanthines have been shown to inhibit cAMP-dependent PDE derived either from trachealis smooth muscle of the ox, guinea-pig and dog (LOHMANN et al. 1977; NEWMAN et al. 1978; POLSON et al. 1979, 1982; SMALL et al. 1989) or from lung tissue of man and guinea-pigs

(BERGSTRAND 1980; PERSSON 1985). Enzyme inhibition is observed at concentrations of alkylxanthines below those producing maximal relaxant effects (POLSON et al. 1979, 1982; SMALL et al. 1989) (Fig. 5). Moreover, a good correlation exists between the rank order of potency of alkylxanthines in relaxing trachealis muscle and their rank order of potency in inhibiting cAMP-dependent PDE (POLSON et al. 1982; PERSSON 1985).

Whilst most authors agree that alkylxanthines can inhibit cAMP-dependent PDE derived from respiratory tract tissue, findings with regard to the ability of alkylxanthines of modify tissue cAMP content are less consistent. In guinea-pig or bovine trachealis not exposed to exogenous spasmogens, theophylline (1-10 m*M*) was observed to increase the tissue content of cAMP (KATSUKI and MURAD 1977; KOLBECK et al. 1979). In guinea-pig trachealis isobutylmethylxanthine caused an increase in cAMP content but only at concentrations maximally effective as regards relaxation (BRYSON and RODGER 1987). KOLBECK et al. (1979) found no theophylline-induced change in the cAMP content of canine trachealis and LOHMANN et al. (1977) were unable to confirm that theophylline increased the cAMP content of bovine trachealis. The latter workers also failed to observe an increase in the cAMP content of bovine trachealis during theophylline-induced suppression of spasm evoked by carbachol.

The failure of alkylxanthines to consistently raise the cAMP content of airway smooth muscle has led to critical reappraisal (KOLBECK et al. 1979; PERSSON 1985) of the role of cAMP in mediating alkylxanthine-induced

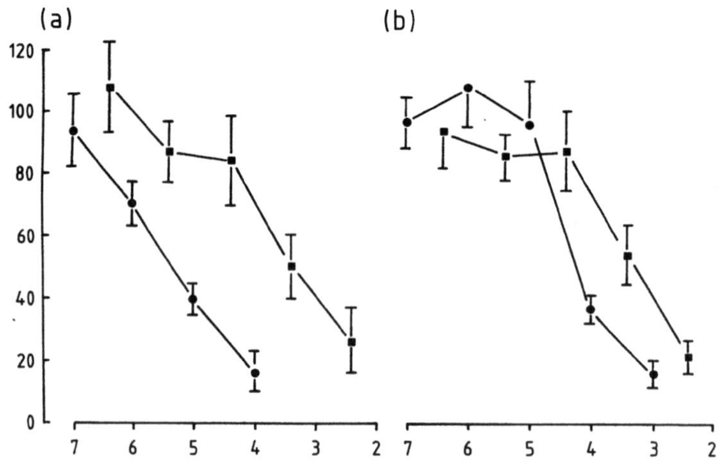

Fig. 5. Inhibition by AH 21-132 (●) and by theophylline (■) of **a** cAMP-dependent and **b** cGMP-dependent phosphodiesterase (PDE) of muscle-rich strips of guinea-pig trachea. *Abscissae*, Log molar concentration of inhibitor; *Ordinate scale*, enzyme activity (cAMP-PDE in **a**; cGMP-PDE in **b**) expressed as a percentage of control activity. Each data point indicates the mean of values from six experiments; *vertical bars* indicate SEM. (From SMALL et al. [1989])

relaxation. However, it is possible that very small changes in the cellular content of cAMP mediate the relaxant action or that the relevant increase in cAMP concentration occurs in a limited subcellular compartment (LOHMANN et al. 1977). Such changes would not be detected in experiments involving measurements of the total tissue content of cAMP.

The above-mentioned reappraisal of the role of cAMP in mediating alkylxanthine-induced relaxation has also been stimulated by the observation that other agents with the ability to inhibit cAMP-dependent PDE are ineffective in relaxing airway smooth muscle. Disodium cromoglycate, for example, is without relaxant effect on human isolated airway smooth muscle (COX 1967) yet inhibits cAMP-dependent PDE obtained from human lung (BERGSTRAND 1980).

It is possible, however, than anomalies of this kind may be explained by the existence in airway smooth muscle of several different isoenzymes of PDE. Amongst these the type Ic isoenzyme (which is Ca^{2+}- and calmodulin-dependent) accounts for the majority (85%) of the total cAMP hydrolytic activity of tissue homogenates. However, this isoenzyme may not be important for regulating the cAMP content and mechanical activity of intact tissue. In contrast the type III and type IV isoenzymes, while accounting for only 5% and 10% respectively, of the total cAMP hydrolytic activity of tissue homogenates, may nevertheless be important regulators of cAMP content and mechanical activity in intact tissue (TORPHY 1988).

The case for rejecting cAMP-dependent PDE inhibition as the mechanism underlying alkylxanthine-induced relaxation of airway smooth muscle will therefore only become strong if it can subsequently be shown that there is a poor correlation between the potency of alkylxanthines as relaxants and their potency as inhibitors of the mechanically-relevant isoenzymes of PDE.

3. Alkylxanthines and Reduction in the Ca^{2+} Sensitivity (or Responsiveness) of the Contractile Machinery

The possibility that alkylxanthines might directly reduce the responsiveness of the intracellular contractile machinery to cytosolic free Ca^{2+} has been tested in trachealis chemically-skinned of its plasmalemmal membranes. Neither aminophylline (ALLEN et al. 1986) (see Fig. 3) nor isobutylmethylxanthine (BRYSON and RODGER 1987) were able to depress the maximum spasmogenic response to Ca^{2+} in Triton X-100 skinned trachealis. This suggests that alkylxanthines do not directly reduce the Ca^{2+} responsiveness of the intracellular contractile machinery. However, it is possible that, in the intact cell, alkylxanthines (like β-adrenoceptor agonists) may do so by promoting the intracellular accumulation of cAMP.

4. Alkylxanthines and Inhibition of Ca^{2+} Influx

In guinea-pig trachealis aminophylline and theophylline have relaxant effects that are associated with the suppression of electrical slow waves and plasmalemmal hyperpolarisation (ALLEN et al. 1986; HONDA et al. 1986). As discussed above, these electrical changes can be mimicked by forskolin and by dibutyryl cAMP (HONDA et al. 1986) and hence may result from the intracellular accumulation of cAMP.

By causing cellular hyperpolarisation, the alkylxanthines (like the β-adrenoceptor agonists) would be expected to inhibit Ca^{2+} influx through VOCs. However, several observations suggest that inhibition of Ca^{2+} influx, consequent upon cellular hyperpolarisation, is relatively unimportant for, expression of the relaxant effects of the alkylxanthines. Firstly, low concentrations of aminophylline produce relaxation in the absence of membrane potential changes (ALLEN et al. 1986). Secondly, aminophylline and theophylline retain relaxant activity in a K^+-rich (120 m*M*) medium where hyperpolarisation cannot be achieved by K^+ channel opening because the resting membrane potential and the K^+ equilibrium potential are virtually coincident (ALLEN et al. 1986; SMALL et al. 1989). Thirdly, the K^+ channel inhibitors TEA and procaine significantly reduce aminophylline-induced hyperpolarisation without reducing its relaxant effects (ALLEN et al. 1986).

Evidence is now emerging that alkylxanthines may inhibit Ca^{2+} influx into smooth muscle more directly. In uterine smooth muscle from the pregnant rat, caffeine (> 1 m*M*) has been observed to reduce the inward Ca^{2+} current evoked by a depolarising voltage step. This action of caffeine was independent of the activation state of the Ca^{2+} VOCs and was independent of the Ca^{2+} content of the intracellular store (MARTIN et al. 1989). It has yet to be shown that alkylxanthines can similarly inhibit Ca^{2+} VOCs in airway smooth muscle. However, an action of this kind could certainly explain the ability of the alkylxanthines to suppress the spasm of airway smooth muscle evoked by K^+-rich (120 m*M*) media.

In summary, alkylxanthines, by causing membrane hyperpolarisation, can inhibit Ca^{2+} influx through VOCs. However, hyperpolarisation-induced inhibition of Ca^{2+} influx plays a minor role in the process by which alkylxanthine-induced relaxation of airway smooth muscle is achieved. It is possible that high concentrations (> 1 m*M*) of alkylxanthines may directly inhibit Ca^{2+} influx through VOCs.

5. Alkylxanthines and Inhibition of Ca^{2+} Release from Intracellular Stores

Early experiments using bovine trachealis treated with Li^+ (GRANDORDY and BARNES 1986; GRANDORDY et al. 1987) showed that theophylline was without effect on the accumulation of inositol monophosphate induced by carbachol, histamine or 5HT. More recently, however, it has been shown that, in bovine trachealis and in cultured canine trachealis cells, alkylxanthines can suppress histamine-induced inositol phosphate accumulation (MURRAY et al. 1988; HALL et al. 1989). Like the actions of the β-adrenoceptor agonists, the effects of the alkylxanthines against the histamine-induced inositol phosphate signal appear to be related to their ability to stimulate the intracellular accumulation of cAMP (HALL et al. 1989).

The ability of the alkylxanthines to suppress histamine-induced inositol phosphate production may therefore represent a cAMP-dependent action of these drugs in preventing Ca^{2+} release from intracellular stores. This cAMP-dependent action does not necessitate depletion of the store. In contrast, however, there is also evidence that alkylxanthines can prevent spasm of airway smooth muscle by a mechanism which is cAMP-independent and which does involve store depletion.

Xanthine and several of its alkylated derivatives cause spasm of guinea-pig trachealis treated with indomethacin and maintained at 12°C (SMALL et al. 1988a, 1989) (Fig. 6). Caffeine causes spasm of feline and guinea-pig trachealis maintained in a Ca^{2+}-free medium containing EGTA (ITO and ITOH 1984a, b; SMALL et al. 1988a). Furthermore, the use of the Ca^{2+} fluorophore aequorin has revealed that caffeine increases the cytosolic concentration of Ca^{2+} in bovine trachealis exposed to a Ca^{2+}-free medium containing EGTA (TAKUWA et al. 1987). Collectively these findings suggest that the alkylxanthines can release Ca^{2+} from intracellular stores in airway smooth muscle.

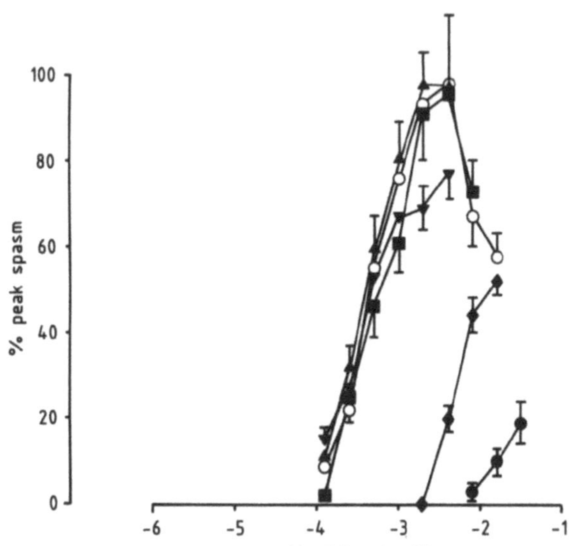

Fig. 6. The spasmogenic effects of some xanthine derivatives as observed in guinea-pig isolated trachealis maintained at 12°C in Krebs solution containing indomethacin (2.8 μ*M*). The *abscissa* represents the log molar concentration of the xanthine derivative. The *ordinate axis* represents the peak spasm expressed as a percentage of the peak spasm evoked by caffeine (10 μ*M*). The illustrated log-concentration effect curves are for enprofylline (■), theophylline (▲), caffeine (○), theobromine (▼), xanthine (♦) and 1,3,7,9-tetramethyl-xanthinium (•). Data indicate means of values from at least six tissues; SEM shown by *vertical bars*. (From SMALL et al. [1988a])

Caffeine-induced spasm can be observed in smooth muscle cells whose cell membranes have been rendered permeable to large molecules by treatment with saponin (HAEUSLER et al. 1981; SAIDA 1982; MEISHERI et al. 1985). The spasmogenic effects of alkylxanthines in guinea-pig trachealis at 12°C are not mimicked by the adenylate cyclase activator forskolin (SMALL et al. 1988a) or by the PDE inhibitor AH 21–132 (SMALL et al. 1989). Furthermore the rank order of potency of xanthine derivatives in causing spasm of guinea-pig trachealis at 12°C differs from that observed for suppression of spontaneous tone at 37°C (SMALL et al. 1988a). Taken together these observations suggest that alkylxanthine-induced release of Ca^{2+} from the intracellular store does not rely on the ability of the alkylxanthines to promote the intracellular accumulation of cAMP.

Tested in a Ca^{2+}-free medium containing EGTA, caffeine has been shown to cause a large, transient rise in cytosolic free Ca^{2+} of bovine trachealis and subsequently to prevent carbachol from producing a similar effect (TAKUWA et al. 1987). In feline trachealis exposed to similar medium, treatment with caffeine causes spasm and subsequently prevents acetylcholine-induced spasm (ITO and ITOH 1984a). Accordingly it seems that alkylxanthines can cause sufficient release of Ca^{2+} from the intracellular store to cause store depletion. Hence the actions of spasmogens, which depend on the integrity of the intracellular Ca^{2+} store, may be prevented by prior challenge with an alkylxanthine.

It seems unlikely, however, that depletion of intracellular Ca^{2+} stores can wholly explain the relaxant actions of alkylxanthines as observed in a medium containing physiological concentrations of Ca^{2+}. In the presence of Ca^{2+} (2.6 mM) enprofylline and theophylline evoke relaxation at concentrations well below those required to evoke spasm attributable to Ca^{2+} release from intracellular stores (SMALL et al. 1988a). Furthermore, depletion of Ca^{2+} stores does not explain the relaxant effects of alkylxanthines when tested against spasmogens (e.g. KCl) whose actions result from Ca^{2+} influx rather than release from intracellular stores.

6. Alkylxanthines and Promotion of Ca^{2+} Sequestration by Intracellular Stores

As mentioned above, ITO and ITOH (1984b) used caffeine-induced spasm as an index of the content of the intracellular Ca^{2+} store in feline trachealis. These workers obtained evidence to suggest that isoprenaline could promote the intracellular sequestration of Ca^{2+}. If this action of isoprenaline depends on the intracellular accumulation of cAMP, a similar effect might be expected of the alkylxanthines. However, as yet, there is no experimental evidence that alkylxanthines have such an action in airway smooth muscle. Furthermore, the store-depleting actions of the alkylxanthines described above would certainly make alkylxanthine-induced Ca^{2+} sequestration difficult to demonstrate experimentally.

Like the β-adrenoceptor agonists, the alkylxanthines can suppress the spasm of trachealis muscle evoked by high (120 m*M*) concentrations of KCl (ALLEN et al. 1986; SMALL et al. 1989). Again it is difficult to envisage that the capacity of the intracellular store is adequate to completely sequester Ca^{2+} entering the cells under such conditions. We therefore conclude that if alkylxanthines promote the intracellular sequestration of Ca^{2+}, such an effect might be offset by their store-depleting action. However, this action cannot be envisaged to play a major role when alkylxanthines suppress spasm evoked by exposure of airway smooth muscle to a K^+-rich medium.

7. Alkylxanthines and Ca^{2+} Extrusion

To our knowledge there have been no reports that alkylxanthines can promote Ca^{2+} extrusion from airway smooth muscle. However, such an effect could make an important contribution to the relaxant action of these drugs.

D. Future Trends in the Development of Bronchodilator Drugs

There is little doubt that β_2-adrenoceptor agonists form the mainstay of bronchodilator therapy in asthma. These drugs, however, possess several unwanted side effects including skeletal muscle tremor, nervous tension, headache, peripheral vasodilatation and tachycardia. Despite the fact that such side effects can be minimised when the bronchodilators are administered by inhalation they are known to be a source of patient noncompliance. Furthermore, many of the currently used β_2-adrenoceptor agonists have a short duration of action. Accordingly, their control of asthmatic episodes has been far from ideal. This is particularly so in the case of nocturnal asthma. The recent development of sustained release formulations of salbutamol may provide improved therapy in such circumstances. The future introduction into clinical practice of long-acting β_2-adrenoceptor agonists, such as salmeterol, offers potential further benefit. Whether the incidence of the unwanted side effects is improved correspondingly is much less certain.

The unwanted effects of theophylline include anorexia, nausea, vomiting, headache, tachycardia, anxiety, insomnia and (in high doses) cardiac arrhythmias and convulsions. While the introduction of sustained release formulations of theophylline has reduced both the incidence and severity of these side effects, the drug still possesses a narrow therapeutic index. Whether the new alkylxanthines currently under development will provide an improved therapy of asthma and a more favourable therapeutic index remains to be established.

Theophylline is a relatively non-selective inhibitor of the isoenzymes of cyclic nucleotide PDE. Thus, it has been argued that a more fruitful approach might be to develop compounds that are selective inhibitors of those PDE isoenzymes that are most physiologically relevant to the control of airway smooth muscle tone. At the present time no such isoenzyme-selective PDE inhibitor is in clinical use although several are in advanced stages of development.

Certain K^+ channel-opening drugs, typified by cromakalim, are currently undergoing clinical evaluation as novel bronchodilators. Since compounds of this kind frequently possess pronounced vasodilator activity it is essential that some form of selectivity (either pharmacological or distributional) for airway smooth muscle is achieved if they are to offer an advance over existing bronchodilator therapy. Notwithstanding, cromakalim has been shown to provide some benefit in the treatment of nocturnal asthma at doses that apparently do not cause significant vasodilatation. This K^+ channel-opener has a long biological half-life in man. Thus, taken orally late at night, it can attenuate the "morning dip" in lung function that characterises nocturnal asthma. Whether this effect is peculiar to cromakalim or simply a consequence of a prologed bronchodilator effect that may well be reproduced by other long-acting bronchodilator drugs such as salmeterol remains to be established.

It is evident from this concluding summary that several different lines of approach are being pursued in an attempt to develop improved bronchodilators for the therapy of asthma. One must be cognisant, however, of the fact that asthma is now recognised as not simply a disorder of airway smooth muscle that leads to exaggerated bronchoconstriction. The realisation that the disease has a pronounced inflammatory component is a major advance in our understanding. Thus, it could be convincingly argued that the ideal drug for the therapy of asthma would be one which incorporated both bronchodilator and anti-inflammatory activity. The bronchodilator component would provide relief from the bronchoconstriction whilst the anti-inflammatory component would address the underlying pathological changes. Such projections are clearly idealistic but in today's climate of drug discovery and development by no means unattainable.

Acknowledgements. IWR gratefully acknowledges the financial support of his research provided by the National Asthma Campaign, Medical Research Council, Wellcome Trust, Scottish Hospitals Endowment Research Trust (SHERT), Bayer UK and Organon Laboratories. RCS is grateful to the National Asthma Campaign, British Lung Foundation, Wellcome Trust, Glaxo Group Research, Napp Laboratories and Sandoz AG for providing support of his work. Both authors are indebted to the many colleagues who have contributed to the experimental work cited in this article.

References

Advenier C, Naline E, Renier A (1986) Effects of BAY k 8644 on contraction of the human isolated bronchus and guinea-pig trachea. Br J Pharmacol 88:33–39

Ahmed F, Foster RW, Small RC, Weston AH (1984) Some features of the spasmogenic actions of acetylcholine and histamine in guinea-pig isolated trachealis. Br J Pharmacol 83:227–233

Ahmed F, Foster RW, Small RC(1985) Some effects of nifedipine in guinea-pig isolated trachealis. Br J Pharmacol 84:861–869

Aksoy MO, Murphy RA, Kamm KE (1982) Role of Ca^{2+} and myosin light chain phosphorylation in regulation of smooth muscle. Am J Physiol 242:C109–C116

Aksoy MO, Mras S, Kamm KE, Murphy RA (1983) Ca^{2+}, cAMP and changes in myosin phosphorylation during contraction of smooth muscle. Am J Physiol 245:C255–C270

Allen SL, Beech DJ, Foster RW, Morgan GP, Small RC (1985a) Electro-physiological and other aspects of the relaxant action of isoprenaline in guinea-pig isolated trachealis. Br J Pharmacol 86:843–854

Allen SL, Foster RW, Small RC, Towart R (1985b) The effects of the dihydropyridine BAY k 8644 in guinea-pig isolated trachea. Br J Pharmacol 86:171–180

Allen SL, Cortijo J, Foster RW, Morgan GP, Small RC, Weston AH (1986) Mechanical and electrical aspects of the relaxant action of aminophylline in guinea-pig isolated trachealis. Br J Pharmacol 88:473–483

Baba K, Kawanishi M, Satake T, Tomita T (1985) Effects of verapamil on the contractions of guinea-pig tracheal smooth muscle induced by Ca, Sr and Ba. Br J Pharmacol 84:203–211

Baron CB, Cunningham M, Strauss JF, Coburn RF (1984) Pharmacomechanical coupling in smooth muscle may involve phosphatidylinositol metabolism. Proc Natl Acad Sci USA 81:6899–6903

Baron CB, Pring M, Coburn RF (1989) Inositol lipid turnover and compartmentation in canine trachealis smooth muscle. Am J Physiol 256:C375–C383

Barnes PJ, Cuss FM, Palmer JB (1986) The effect of airway epithelium on smooth muscle contractility in bovine trachea. Br J Pharmacol 86:685–691

Bergstrand H (1980) Phosphodiesterase inhibition and theophylline. Eur J Respir Dis [Suppl 109] 61:37–44

Berridge MJ, Irvine RF (1984) Inositol trisphosphate, a novel second messenger in cellular signal transduction. Nature 312:315–321

Berridge MJ, Irvine RF (1989) Inositol phosphates and cell signalling. Nature 341:197–203

Bewley JS, Chapman IE (1988) AH 21-132 a novel relaxant of airways smooth muscle. Br J Pharmacol 93:52P

Bolton TB (1979) Mechanisms of action of transmitters and other substances on smooth muscle. Physiol Rev 59:606–718

Bryson SE, Rodger IW (1987) Effects of phosphodiesterase inhibitors on normal and chemically-skinned isolated airways smooth muscle. Br J Pharmacol 92:673–681

Cameron AR, Kirkpatrick CT (1977) A study of excitatory neuromuscular transmission in bovine trachea. J Physiol (Lond) 270:733–745

Cameron AR, Johnson CF, Kirkpatrick CT, Kirkpatrick MCA (1983) The quest for the inhibitory neurotransmitter in bovine tracheal smooth muscle. Q J Exp Physiol 68:413–426

Chatterjee M, Tejada M (1986) Phorbol ester-induced contraction in chemically skinned vascular smooth muscle. Am J Physiol 251:C356–C361

Chilvers ER, Challis J, Barnes PJ, Nahorski SR (1989) Mass changes of inositol(1,4,5)trisphosphate in trachealis muscle following agonist stimulation. Eur J Pharmacol 614:587–590

Coburn RF (1979) Electromechanical coupling in canine trachealis muscle: acetylcholine contractions. Am J Physiol 236:C177–C184

Coburn RF, Yamaguchi T (1977) Membrane potential-dependent and -independent tension in canine trachealis. J Pharmacol Exp Ther 201:276–284

Cox JSG (1967) Disodium cromoglycate (FPL 670) ('Intal'): a specific inhibitor of reaginic antibody-antigen mechanisms. Nature 216:1328–1329

Creese BR (1983) Calcium ions, drug action and airways obstruction. Pharmacol Ther 20:357–375

Dale MM, Obianime AW (1985) Phorbol myristate acetate causes in guinea pig lung parenchymal strip a maintained spasm resistant to isoprenaline. FEBS Lett 190:6–10

Dale MM, Obianime AW (1987) 4β-PDBu contracts parenchymal strip and synergises with raised cytosolic calcium. Eur J Pharmacol 141:23–32
DeLanerolle P, Condit JR, Tanenbaum M, Adelstein RS (1982) Myosin phosphorylation, agonist concentration and contraction of tracheal smooth muscle. Nature 298:871–872
Duncan RA, Krzanowski JJ, Polson JB, Coffey RG, Szentivanyi A (1987) Polyphosphoinositide metabolism in canine tracheal smooth muscle (CTSM) in response to a cholinergic stimulus. Biochem Pharmacol 36:307–310
Farley JM, Miles PR (1977) Role of depolarisation in acetylcholine-induced contractions of dog trachealis muscle. J Pharmacol Exp Ther 201:199–205
Farmer SG, Fedan JS, Hay DWP, Raeburn D (1986) The effects of epithelium removal on the sensitivity of guinea-pig isolated trachealis to bronchodilator drugs. Br J Pharmacol 89:407–414
Fedan JS, Hay DWP, Farmer SG, Raeburn D (1988) Epithelial cells. Modulation of airway smooth muscle reactivity. In: Barnes PJ, Rodger IW, Thompson NC (eds) Asthma: basic mechanisms and clinical management. Academic, London, pp. 143–162
Felbel J, Trockur B, Ecker T, Landgraf W, Hofmann F (1988) Regulation of cytosolic calcium by cAMP and cGMP in freshly isolated smooth muscle cells from bovine trachea. J Biol Chem 263:16764–16771
Foster RW, Small RC, Weston AH (1983a) The spasmogenic action of potassium chloride in guinea-pig trachealis. Br J Pharmacol 80:553–559
Foster RW, Small RC, Weston AH (1983b) Evidence that the spasmogenic action of tetraethylammonium in guinea-pig trachealis is both direct and dependent upon the cellular influx of calcium ions. Br J Pharmacol 79:255–263
Foster RW, Okpalugo BI, Small RC (1984) Antagonism of Ca^{2+} and other actions of verapamil in guinea-pig isolated trachealis. Br J Pharmacol 81:499–507
Fujiwara T, Suminoto K, Itoh T, Kuriyama H (1988) Relaxing actions of procaterol, a beta2-adrenoceptor stimulant, on smooth muscle cells of the dog trachea. Br J Pharmacol 93:199–209
Gerthoffer WT (1986) Calcium dependence of myosin phosphorylation and airway smooth muscle contraction and relaxation. Am J Physiol 250:C597–C604
Gerthoffer WT, Murphy RA (1983) Myosin phosphorylation and regulation of cross-bridge cycle in tracheal smooth muscle. Am J Physiol 244:C182–C187
Ghosh TK, Eis PS, Mullaney JM, Ebert CL, Gill DL (1988) Competitive, reversible, and potent antagonism of inositol 1,4,5-trisphosphate-activated calcium release by heparin. J Biol Chem 263:11075–11079
Ghosh TK, Mullaney JM, Tarazi FM, Gill DL (1989) GTP-activated communication between distinct inositol 1,4,5-trisphosphate-sensitive and -insensitive calcium pools. Nature 340:236–239
Giembycz MA, Rodger IW (1987) Electrophysiological and other aspects of excitation-contraction coupling and uncoupling in mammalian airways smooth muscle. Life Sci 41:111–132
Goldie RG, Papadimitriou JM, Paterson JW, Self H, Spina D (1986) Influence of the epithelium on responsiveness of guinea-pig isolated trachea to contractile and relaxant agonists, Br J Pharmacol 87:5–14
Grandordy BM, Barnes PJ (1986) Effects of antiasthma drugs on membrane phospholipid breakdwon in airway smooth muscle. Am Rev Respir Dis 133(4):A179
Grandordy BM, Barnes PJ (1987) Phosphoinositide turnover in airway smooth muscle. Am Rev Respir Dis 136:S17–S20.
Grandordy BM, Cuss FM, Sampson AS, Palmer JB, Barnes PJ (1986) Phosphatidylinositol response to cholinergic agonists in airway smooth muscle: relationship to contraction and muscarinic receptor occupancy. J Pharmacol Exp Ther 238:273–279
Grandordy BM, Cuss FM, Barnes PJ (1987) Breakdown of phosphoinositides in

airway smooth muscle: lack of influence of antiasthma drugs. Life Sci 41:1621–1627

Haeusler G, Richards JG, Thorens A (1981) Noradrenaline contractions in rabbit mesenteric arteries skinned with saponin. J Physiol (Lond) 321:537–556

Hai C-M, Murphy RA (1988) Cross-bridge phosphorylation and regulation of the latch state in smooth muscle. Am J Physiol 254:C99–C106

Hai C-M, Murphy RA (1989) Ca^{2+}, crossbridge phosphorylation and contraction. Annu Rev Physiol 51:285–298

Hall IP, Hill SJ (1988) β_2-Adrenoceptor stimulation inhibits histamine-stimulated inositol phospholipid hydrolysis in bovine tracheal smooth muscle. Br J Pharmacol 95:1204–1212

Hall IP, Hill SJ (1989) Inhibition of histamine-stimulated inositol phospholipid hydrolysis by agents which increase cyclic AMP levels in bovine tracheal smooth muscle. Br J Pharmacol 97:603–613

Hall IP, Donaldson J, Hill SJ (1989) Inhibition of histamine-stimulated inositol phospholipid hydrolysis by agents which increase cyclic AMP levels in bovine tracheal smooth muscle. Br J Pharmacol 97:603–613

Hashimoto T, Hirata M, Ito Y (1985) A role for inositol 1,4,5-trisphosphate in the initiation of agonist-induced contractions of dog tracheal smooth muscle. Br J Pharmacol 86:191–199

Holroyde MC (1986) The influence of epithelium on responsiveness of guinea-pig isolated trachea. Br J Pharmacol 87:501–507

Honda K, Tomita T (1987) Electrical activity in isolated human tracheal muscle. Jpn J Physiol 37:333–336

Honda K, Satake T, Takagi K, Tomita T (1986) Effects of relaxants on electrical and mechanical activities in the guinea-pig tracheal muscle. Br J Pharmacol 87:P 665–671

Irvine RF, Letcher AJ, Heslop JP, Berridge MJ (1986) The inositol tris/tetrakisphosphate pathway – demonstration of Ins (1,4,5)P_3-kinase activity in animal tissues. Nature 320:631–633

Ito Y, Itoh T (1984a) The roles of stored calcium in contraction of cat tracheal smooth muscle produced by electrical stimulation, acetylcholine and high K^+. Br J Pharmacol 83:677–676

Ito Y, Itoh T (1984b) Effects of isoprenaline on the contraction-relaxation cycle in the cat trachea. Br J Pharmacol 83:677–686

Ito Y, Tajima K (1982) Dual effects of catecholamines on pre- and post-junctional membranes in the dog trachea. Br J Pharmacol 75:433–440

Jones CA, Madison JM, Tom-Moy M, Brown JK (1987) Muscarinic cholinergic inhibition of adenylate cyclase in airway smooth muscle. Am J Physiol 253:C97–C104

Kamm KE, Stull JT (1985a) The function of myosin and myosin light chain kinase phosphorylation in smooth muscle. Annu Rev Pharmacol Toxicol 25:593–620

Kamm KE, Stull JT (1985b) Myosin phosphorylation, force and maximal shortening velocity in neurally-stimulated tracheal smooth muscle. Am J Physiol 249:C238–C247

Kamm KE, Stull JT (1986) Activation of smooth muscle contraction: correlation between myosin phosphorylation and stiffness. Science 232:80–82

Kannan MS, Jager LP, Daniel EE, Garfield RE (1983) Effects of 4-aminopyridine and tetraethylammonium chloride on the electrical activity and cable properties of canine tracheal smooth muscle. J Pharmol Exp Ther 227:706–715

Karlsson JA, Persson CGA (1981) Influence of tracheal contraction on relaxant effects in vitro of theophylline and isoprenaline. Br J Pharmacol 74:73–79

Katsuki S, Murad F (1977) Regulation of adenosine cyclic 3,5-monophosphate and guanosine cyclic 3,5-monophosphate levels and contractility in bovine tracheal smooth muscle. Mol Pharmacol 13; 330–341

Kirkpatrick CT (1981) Tracheobronchial smooth muscle. In: Bulbring E, Brading AF,

Jones AW, Tomita T (eds) Smooth muscle: an assessment of current knowledge. Arnold, London, pp 385–395
Kirkpatrick CT, Jenkinson HA, Cameron AR (1975) Interaction between drugs and potassium-rich solutions in producing contraction in bovine tracheal smooth muscle: Studies in normal and calcium-depleted tissues. Clin Exp Pharmacol Physiol 2:559–570
Kitazawa T, Kobayashi S, Horiuti K, Somlyo AV, Somlyo AP (1989) Receptor-coupled, permeabilised smooth muscle: role of the phosphatidylinositol cascade, G-proteins, and modulation of the contractile response to Ca^{2+}. J Biol Chem 264:5339–5342
Kobayashi S, Somlyo AV, Somlyo AP (1988) Heparin inhibits the inositol 1,4,5-trisphosphate-dependent, but not the independent, calcium release induced by guanine nucleotide in vascular smooth muscle. Biochem Biophys Res Commun 153:625–631
Kolbeck RC, Speir WA, Carrier GO, Bransome ED (1979) Apparent irrelevance of cyclic nucleotides to the relaxation of tracheal smooth muscle induced by theophylline. Lung 156:173–183
Kotlikoff MI (1988) Calcium currents in isolated canine airway smooth muscle cells. Am J Physiol 254:C793–C801
Kotlikoff MI, Murray RK, Reynolds EE (1987) Histamine-induced calcium release and phorbol antagonism in cultured airway smooth muscle cells. Am J Physiol 245:C561–C566
Kroeger EA, Stephens NL (1975) Effect of tetraethylammonium on tonic airway smooth muscle: initiation of phasic electrical activity. Am J Physiol 228:633–636
Kumar MA (1978) The basis of beta-adrenergic bronchodilation. J Pharmacol Exp Ther 206:528–534
Kume H, Takai A, Tokuno H, Tomita T (1989) Regulation of Ca^{2+}-dependent K^+-channel activity in tracheal myocytes by phosphorylation. Nature 341:152–154
Langland JM, Rodger IW, Diamond J (1989) The effect of M and B 22948 on methacholine- and histamine-induced contraction and inositol 1,4,5-trisphosphate levels in guinea-pig tracheal tissue. Br J Pharmacol 98:336–338
Lohmann SM, Miech RP, Butcher FR (1977) Effects of isoproterenol, theophylline and carbachol on cyclic nucleotide levels and relaxation in bovine tracheal smooth muscle. Biochim Biophys Acta 499:238–250
Madison JM, Brown JK (1988) Differential inhibitory effects of forskolin, isoproterenol and dibutyryl cyclic adenosine monophosphate on phosphoinositide hydrolysis in canine tracheal smooth muscle. J Clin Invest 82:1462–1465
Marthan RJ, Savineau JP, Mironneau J (1985) Acetylcholine-induced contraction in human isolated bronchial smooth muscle: role of an intracellular calcium store. Respir Physiol 67:127–135
Marthan RJ, Armour CL, Johnson PRA, Black JL (1988) Extracellular calcium and human isolated airway muscle: ionophore A23187-induced contraction. Respir Physiol 71:157–168
Marthan RJ, Martin C, Amedee T, Mironneau J (1989) Calcium channel currents in isolated smooth muscle cells from human bronchus. J Appl Physiol 66:1706–1714
Martin C, Dacquet C, Mirroneau C, Mirroneau J (1989) Caffeine-induced inhibition of calcium channel current in cultured smooth muscle cells from pregnant rat myometrium. Br J Pharmacol 98:493–498
McCaig DJ, Rodger IW (1988) Electrophysiological effects of leukotriene D_4 in guinea-pig airway smooth muscle. Br J Pharmacol 94:729–736
McCaig DJ, Souhrada JF (1980) Alteration of electrophysiological properties of airway smooth muscle from sensitized guinea-pigs. Respir Physiol 41:49–60
Meisheri KD, Ruegg JC, Paul RJ (1985) Studies on skinned fiber preparations. In: Grover AK, Daniel EE (eds) Calcium and contractility: smooth muscle. Humana, Clifton, pp 191–224

Menkes H, Baraban JM, Snyder SH (1986) Protein kinase C regulates smooth muscle tension in guinea-pig trachea and ileum, Eur J Pharmacol 122; 19–28

Merritt J, Armstrong WP, Hallam TJ, Jaxa-Chamiec A, Leigh BK, Moores KE, Rink TJ (1989) SK and F 96365, a novel inhibitor of receptor-mediated calcium entry and aggregation in Quin 2-loaded human platelets. Br J Pharmacol 98:674P

Meurs H, Roffel AF, Postema JB, Timmermans A, Elzinga CRA, Kauffman HF, Zaagsma J (1988) Evidence for a direct relationship between phosphoinositide metabolism and airway smooth muscle contraction induced by muscarinic agonists. Eur J Pharmacol 156:271–274

Meurs H, Timmermans S, Van Amsterdam RGM, Brouwer F, Kauffman HF, Zaagsma J (1989) Muscarinic receptors in human airway smooth muscle are coupled to phosphoinositide metabolism. Eur J Pharmacol 164:369–371

Miller JR, Hawkins DJ, Wells JN (1986) Phorbol diesters alter contractile responses of porcine conronary artery. J Pharmacol Exp Ther 239:38–42

Miller-Hance WC, Miller JR, Wells, JN, Stull JT, Kamm KE (1988) Biochemical events associated with activation of smooth muscle contraction. J Biol Chem 263:13979–13982

Mong S, Hoffman K, Wu H-L, Crooke ST (1987) Leukotriene-induced hydrolysis of inositol lipids in guinea-pig lung. Mechanism of signal transduction for leukotriene D_4 receptors. Mol Pharmacol 31:35–41

Mong S, Miller J, Wu H-L, Crooke ST (1988) Leukotriene D_4 receptor-mediated hydrolysis of phosphoinositide and mobilization of calcium in sheep tracheal smooth muscle cells. J Pharmacol Exp Ther 244:508–515

Mullaney JM, Yu M, Ghosh TK, Gill DL (1988) Calcium entry into the inositol 1,4,5-trisphosphate-releasable calcium pool is mediated by a GTP-regulatory mechanism. Proc Natl Acad Sci USA 85:2499–2503

Murlas CG, Doupnik CA (1989) Electromechanical coupling of ferret airway smooth muscle in response to leukotriene D_4. J App Physiol 66:1533–1538

Murphy RA (1989) Contraction in smooth muscle cells. Annu Rev Physiol 51:275–283

Murray RK, Fluharty SJ, Kotlikoff MJ (1988) Phorbol ester blocks histamine-induced insoitol trisphosphate production in cultured airway smooth muscle. Am Rev Respir Dis 137:A309

Nahorski SR, Batty I (1986) Inositol tetrakisphosphate: recent developments in PI metabolism and receptor function. Trends Pharmacol Sci 7:83–85

Newman DJ, Colella DF, Spainhour CB, Brann EG, Zabko-Potapovich B, Wardell JR (1978) cAMP-Phosphodiesterase inhibitors and tracheal smooth muscle relaxation. Biochem Pharmacol 27:729–732

Nishizuka Y (1983) Calcium, phospholipid turnover and transmembrane signalling. Philos Trans R Soc Lond [Biol] 302:101–112

Nishizuka Y (1988) The molecular heterogeneity of protein kinase C and its implications for cellular regulation. Nature 334:661–665

Nouailhetas VLA, Lodge NJ, Twort CHC, Van Breemen C (1988) The intracellular calcium stores in rabbit trachealis. Eur J Pharmacol 157:165–172

Obianime AW, Hirst SJ, Dale MM (1988) Interactions between phorbol esters and agents which increase cytosolic calcium in guinea pig parenchymal strip: direct and indirect effects on the contractile response. J Pharmacol Exp Ther 247:262–270

Panettieri RA, Murray RK, DePalo LR, Yadvish PA, Kotlikoff MI (1989) A human airway smooth muscle cell line that retains physiological responsiveness. Am J Physiol 256:C329–C335

Park S, Rasmussen H (1986a) Activation of tracheal smooth muscle contraction: synergism between Ca^{2+} and activators of protein kinase C. Proc Natl Acad Sci USA 82:8835–8839

Park S, Rasmussen H (1986b) Carbachol-induced protein phosphorylation changes in bovine tracheal smooth muscle. J Biol Chem 261:15373–15379

Parsons WJ, Ramkumar V, Stiles GL (1988) Isobutylmethylxanthine stimulates

adenylate cyclase by blocking the inhibitory regulatory protein, G_i. Mol Pharmacol 34:37–41
Persechini A, Kamm KE, Stull JT (1986) Different phosphorylated forms of myosin in contracting tracheal smooth muscle. J Biol Chem 261:6293–6299
Persson CGA (1985) Experimental lung actions of xanthines In: Andersson KE, Persson CGA (eds) Anti-asthma xanthines and adenosine. Excerpta Medica, Amsterdam, pp 61–83
Polson JB, Krzanowski JJ, Anderson WH, Fitzpatrick DF, Hwang DPC, Szentivanyi A (1979) Analysis of the relationship between pharmacological inhibition of cyclic nucleotide phosphodiesterase and relaxation of canine tracheal smooth muscle. Biochem Pharmacol 28:1391–1395
Polson JB, Krzanowski JJ, Szentivanyi A (1982) Inhibition of a high affinity cyclic AMP phosphodiesterase and relaxation of canine tracheal smooth muscle. Biochem Pharmacol 31:3403–3406
Putney JW (1986) A model for receptor-regulated calcium entry. Cell Calcium 7:1–12
Putney JW, Takemura H, Hughes AR, Horstman DA, Thastrup O (1989) How do inositol phosphates regulate calcium signalling? FASEB J 3:1899–1905
Raeburn D, Rodger IW (1984) Lack of effect of leukoriene D_4 on Ca-uptake in airway smooth muscle. Br J Pharmacol 83:499–504
Raeburn D, Roberts JA, Rodger IW, Thomson NC (1986) Agonist-induced contractile responses of human bronchial muscle in vitro: effects of Ca^{2+} removal, La^{3+} and PY 108068. Eur J Pharmacol 121:251–255
Rasmussen H, Barrett PQ (1984) Calcium messenger system: an integrated view. Physiol Rev 64:938–984
Rasmussen H, Kojima I, Kojima K, Zawalich W, Appeldorf W (1984) Calcium as intracellular messenger: sensitivity modulation, C-kinase pathway and sustained cellular response, Adv Cyclic Nucleotide Res 18:159–193
Rasmussen H, Takuwa Y, Park S (1987) Protein kinase C in the regulation of smooth muscle contraction. FASEB J 1:177–185
Rodger IW (1986) Calcium ions and contraction of airways smooth muscle. In: Kay AB (ed) Asthma: clinical pharmacology and therapeutic progress. Blackwell, Oxford, pp 114–127
Rodger IW (1987) Calcium channels in airway smooth muscle. Am Rev Respir Dis 136:S15–S17
Rodger IW (1988) Biochemistry of airway smooth muscle contraction. In: Barnes PJ, Rodger IW, Thomson NC (eds) Asthma: basic mechanisms and clinical management. Academic, London, pp 57–79
Ross CA, Meldolesi J, Milner TA, Satoh T, Supattapone S, Snyder SH (1989) Inositol 1,4,5-trisphosphate receptor localised to endoplasmic reticulum in cerebellar Purkinje neurons. Nature 339:468–470
Saida K (1982) Intracellular Ca^{2+} release in skinned smooth muscle. J Gen Physiol 80:191–192
Saida K, Twort CHC, Van Breemen C (1988) The specific GTP requirement for inositol 1,4,5-trisphosphate-induced Ca^{2+} release from skinned vascular smooth muscle. J Cardiovasc Pharmacol 12:S47–S50
Silver PJ, Stull JT (1982) Regulation of myosin light chain and phosphorylase phosphorylation in tracheal smooth muscle. J Biol Chem 257:6145–6150
Silver PJ, Stull JT (1984) Phosphorylation of myosin light chain and phosphorylase in tracheal smooth muscle in response to KCl and carbachol. Mol Pharmacol 25:267–274
Small RC, Foster RW (1986) Airways smooth muscle: an overview of morphology, electrophysiology and aspects of the pharmacology of contraction and relaxation. In:Kay AB (ed) Asthma:clinical pharmacology and therapeutic progress. Blackwell, London, pp 101–113
Small RC, Foster RW (1987) Electrophysiologic behaviour of normal and sensitized airway smooth muscle. Am Rev Respir Dis 136:S7–S11
Small RC, Foster RW (1988) Electrophysiology of the airway smooth muscle cell.

In:Barnes PJ, Rodger IW, Thompson NC (eds) Asthma: basic mechanisms and clinical management. Academic, London, pp 35–56
Small RC, Boyle JP, Cortijo J, Curtis-Prior PB, Davies JM, Foster RW, Hofer P (1988a) The relaxant and spasmogenic effects of some xanthine derivatives acting on guinea-pig isolated trachealis muscle. Br J Pharmacol 94:1091–1100
Small RC, Foster RW, Boyle JP, Davies JM (1988b) The site and mechanism of the relaxant effects of aminophylline and other methylxanthines in isolated airways smooth muscle. In: Paton DM (ed) Adenosine and adenine nucleotides: physiology and pharmacology. Taylor and Francis, London. pp 271–280
Small RC, Boyle JP, Duty S, Elliott KRF, Foster RW, Watt AJ (1989) Analysis of the relaxant effects of AH 21-132 in guinea-pig isolated trachealis. Br J Pharmacol 97:1165–1173
Sparrow MP, Pfitzer G, Gagelmann M, Ruegg JC (1984) Effect of calmodulin, Ca^{2+} and cAMP protein kinase on skinned tracheal smooth muscle. Am J Physiol 246:C308–C314
Spat A, Bradford PG, McKinney JS, Rubin RP, Putney JW (1986) A saturable receptor for ^{32}P-inositol-1,4,5-trisphosphate in hepatocytes and neutrophils. Nature 319:514–516
Spilker B, Minatoya H (1975) The role of bronchoconstrictors in evaluating smooth muscle relaxant activity. Arch Int Pharmacodyn 27:201–217
Suzuki H, Morita K, Kuriyama H (1976) Innervation and properties of the smooth muscle of the dog trachea. Jpn J Physiol 26:303–320
Takuwa Y, Takuwa N, Rasmussen H (1986) Carbachol induces a rapid and sustained hydrolysis of polyphosphoinositide in bovine tracheal smooth muscle. Measurements of the mass of polyphosphoinositides, 1,2-diacylglycerol and phosphatidic acid. J Biol Chem 261:14670–14675
Takuwa Y, Takuwa N, Rasmussen H (1987) Measurement of cytoplasmic free Ca^{2+} concentration in bovine tracheal smooth muscle using aequorin. Am J Physiol 253:C817–C827
Takuwa Y, Takuwa N, Rasmussen H (1988) The effects of isoproterenol on intracellular calcium concentration. J Biol Chem 263:762–768
Taylor DA, Stull JT (1988) Calcium dependence of myosin light chain phosphorylation in smooth muscle cells. J Biol Chem 263:14456–14462
Taylor DA, Bowman BF, Stull JT (1989) Cytoplasmic Ca^{2+} is a primary determinant for myosin phosphorylation in smooth muscle cells. J Biol Chem 264:6207–6213
Thomas AP (1988) Enhancement of the inositol 1,4,5-trisphosphate-releasable Ca^{2+} pool by GTP in permeabilised hepatocytes. J Biol Chem 263:2704–2711
Torphy TJ (1988) Action of mediators on airway smooth muscle: functional antagonism as a mechanism for bronchodilator drugs. Agents Actions [Suppl] 23:37–53
Torphy TJ, Gerthoffer WT (1986) Biochemical mechansims of airway smooth muscle contraction and relaxation. In:Hollinger MA (ed) Current topics in pulmonary pharmacology and toxicology, vol 1. Elsevier, New York, pp 23–56
Twort CHC, Van Breemen C (1989) Human airway smooth muscle in cell culture: control of the intracellular calcium store. Pulm Pharmacol 2:45–53

CHAPTER 5

Neural Mechanisms in Asthma

P.J. BARNES

A. Introduction

In the seventeenth century it was commonly held that asthma was a nervous disease. Sir John Floyer believed that an attack of asthma occured when "nerves are filled with windy spirits" (FLOYER 1698). Hyde Salter, in his brilliant description of asthma more than 100 years ago, wrote that "in asthma, there is no peculiarity in the stimulus, the air breathed in is the same in the asthmatic and the non-asthmatic...nor, probably, is there any peculiarity in the irritability of bronchial muscle; the peculiarity is confined to the link that connects these two – the nervous system – and consists in its perverse sensibility, in its receiving and transmitting on to the muscle, as a stimulus to contraction, that of which it should take cognizance...it is clear that the vice in asthma consists not in the production of any special irritant but in the irritability of the part irritated" (SALTER 1868). Thus, asthma was seen as a disease of excessive irritability of the airways produced by neural mechanisms. Later, immunological theories of asthma gained favour and there was an emphasis on inflammatory mediators and mast cells. More recently there has been a re-awakening of interest in neurogenic mechanisms in asthma, now that it is clear that there is a complex interaction between inflammation and neural control of airways.

Autonomic nerves regulate many aspects of airway function and influence the tone of airway smooth muscle, airway secretions, blood flow and microvascular permeability (BARNES 1986a). In inflammatory diseases of the airways nerves may augment or modulate the inflammatory response.

Neural control of human airways is complex and the contribution of neurogenic mechanisms to the pathogenesis of airway disease is still uncertain (KALINER and BARNES 1988). Because changes in bronchomotor tone in asthma occur rapidly, it was suggested many years ago that there might be an abnormality in autonomic neural control in asthma with an imbalance between excitatory and inhibitory pathways, resulting in excessively "twitchy" airways (Fig. 1). The idea that autonomic control may be abnormal in asthma was first suggested by the observations of ALEXANDER and PADDOCK (1921), who observed that wheezing was precipated in asthmatics, but not in normal subjects, by injection of the cholinergic agonist pilocarpine and that this was reversed by injection of the β-adrenoceptor agonist adrenaline. This

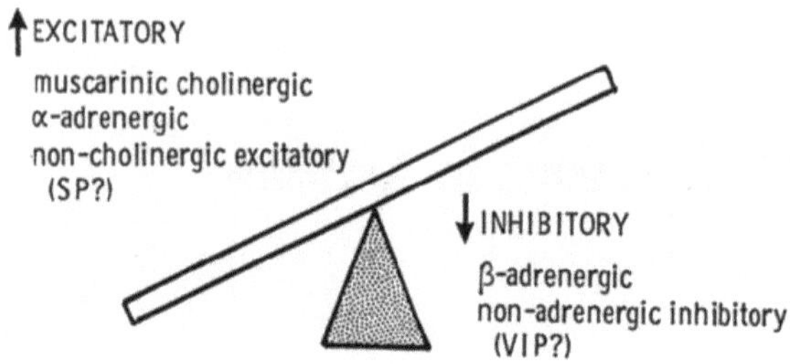

Fig. 1. Autonomic imbalance in asthma. There may be an excess of excitatory effects or a defect in inhibitory mechanisms (or both)

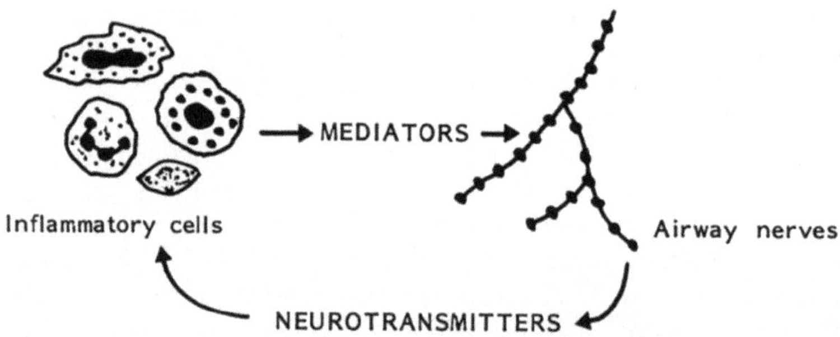

Fig. 2. Interaction between nerves and inflammatory cells. Inflammatory mediators may inhibit or facilitate the release of neurotransmitters from airway nerves, and in turn neurotransmitters may modulate or enhance the inflammatory response

suggested an imbalance between cholinergic and adrenergic nervous systems. Several types of autonomic defect have been proposed in asthma, including enhanced cholinergic, α-adrenergic and non-adrenergic non-cholinergic (NANC) excitatory mechanisms or reduced β-adrenergic and NANC inhibitory mechanisms. Various abnormalities in airway control have been described in asthma, although it now seems unlikely that these are primary defects but are secondary to the disease or its treatment. Nevertheless, it is likely that neural mechanisms contribute to the symptoms of asthma and other airway diseases. There is even evidence for autonomic abnormalities outside the airways (KALINER et al. 1982), although the magnitude of these defects is small and their significance is dubious.

The recognition that inflammation plays a key role in asthma raises the possibility that there may be some interaction between neural and inflammatory mechanisms (Fig. 2) (BARNES 1986d). Inflammatory mediators may modulate or facilitate the release of neurotransmitters from airway nerves, or may act on autonomic receptors. Similarly, neural mechanisms may contribute to the inflammatory reaction in the airway wall, and the concept of neurogenic inflammation, which is well established in skin and gut (SZOLCZANYI 1988), may also apply to the airways (BARNES 1986e; LUNDBERG and SARIA 1987).

The purpose of this chapter is to review some of the possible neural mechanisms which may be relevant to asthma. The recognition that, in addition to classical cholinergic and adrenergic pathways, there are NANC nerves and many neuropeptides which have potent effects on airway function has revived interest in neural control of airways (Fig. 3).

B. Cholinergic Mechanisms

Cholinergic nerves are the dominant neural bronchoconstrictor pathway in animal and human airways (RICHARDSON 1979; BARNES 1986a), and there has been considerable interest in whether cholinergic mechanisms are exaggerated in asthma or chronic obstructive airways disease (COAD). This view is supported by the observation that many stimuli which produce bronchospasm (such as sulphur dioxide, prostaglandins, histamine and cold air) also stimulate afferent receptors and may therefore lead to reflex cholinergic bronchoconstriction (NADEL 1980; NADEL and BARNES 1984; NADEL et al. 1986). Anticholinergic drugs are effective bronchodilators in COAD and acute asthma, although they are usually less useful in chronic asthma. There are several mechanisms which might contribute to cholinergic bronchoconstriction in asthma (BARNES 1987d; BOUSHEY 1984).

I. Increased Vagal Tone

There may be an increase in central vagal drive, although there is no direct evidence for this in asthma. Indirect evidence which may suggest such an increase in vagal tone is the enhanced vagal cardiac tone (as determined by the Valsalva manoeuvre and sinus arrhythmia) which has been demonstrated in asthma (KALLENBACH et al. 1985). There is also evidence that the increase in the sinus arrhythmia gap increases at night as bronchoconstriction increases, suggesting that an increase in vagal cholinergic tone may contribute to nocturnal asthma (POSTMA et al. 1985).

II. Reflex Bronchoconstriction

There may be increased reflex bronchoconstriction due to stimulation of sensory receptors in the airway (irritant receptors and C-fibre endings) by inflammatory mediators. Several mediators, such as histamine, prostaglandins and bradykinin, have been shown to stimulate sensory receptors (NADEL 1980), and it is possible that these receptors may be more easily triggered in

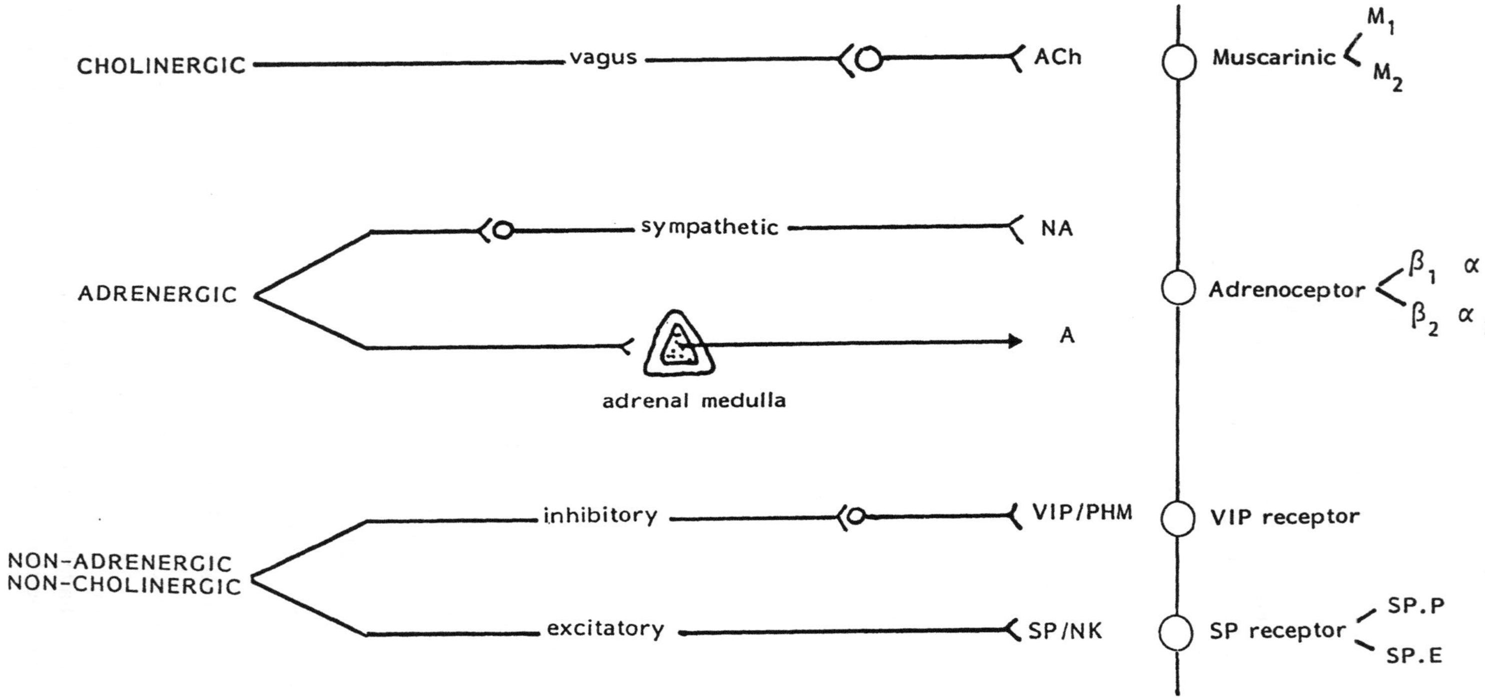

Fig. 3. Neural control of airways. At least three types of neural mechanisms are recognised. Cholinergic nerves release acetylchoine (*ACh*) which acts on muscarinic receptors (*M*); adrenergic nerves release noradrenaline (*NA*) and the adrenal me dulla releases adrenaline (*A*) which act on adrenoceptors. NANC nerves are also recognised, which may include several types of nerves, some bronchodilator and some bronchoconstrictor. The neurotransmitters of NANC nerves are likely to be peptides, such as vasoactive-intestinal peptide (*VIP*), peptide histidine methionine (PHM), substance P (*SP*), neurokinin A (*NKA*)

asthma, since airway epithelium may be damaged or sensory endings "sensitized" by inflammatory mediators such as bradykinin or prostaglandins (KAUFMAN et al. 1980; COLERIDGE and COLERIDGE 1984).

III. Increased Acetylcholine Release

There may be enhanced neurotransmission in cholinergic ganglia, perhaps becasue of release of other neurotransmitters or mediators or facilitation of acetylcholine release from postganglionic nerve terminals. Thus, a thromboxane mimetic, U 46619, facilitates acetylcholine release in canine airways (CHUNG et al. 1985), whereas tachykinins have a similar action in rabbit, ferret and guinea-pig airways (TANAKA and GRUNSTEIN 1986; HALL et al. 1989). Since adrenergic nerves may inhibit acetylcholine release, either via β-adrenoceptors (DANSER et al. 1987; RHODEN et al. 1988) or via α_2-receptors (GRUNDSTROM et al. 1981), it is possible that a defect in adrenergic nerves or receptors may be reflected by an increase in cholinergic tone. Neurotransmission through parasympathetic ganglia may be enhanced by inflammatory mediators, and there is evidence for facilitated transmission in sensitized animals exposed to allergen (MCCAIG 1987). Histamine, surprisingly, has an inhibitory effect on ganglionic neurotransmission and on postganglionic nerves in guinea-pig and human airways, which is mediated through specific H3-receptors (ICHINOSE et al. 1989; ICHINOSE and BARNES 1989). These receptors are activated by low concentrations of histamine and may represent a safety device to inhibit reflex bronchoconstriction, whereas larger concentrations of histamine may overide this mechanism and cause bronchoconstriction directly by acting on H1-receptors on airway smooth muscle.

IV. Muscarinic Receptors

There may be an increased effect of cholinergic stimulation of airway smooth muscle due to an increase in muscarinic receptor density or affinity or due to increased efficacy of signal transduction (BARNES 1987b). Asthmatic patients show an exaggerated bronchoconstrictor response to cholinergic agonsits but this is not specific, since a similar increased responsiveness is observed with other spasmogens, such as histamine, leukotrienes or prostaglandins (BOUSHEY et al. 1980), suggesting that an isolated enhancement of muscarinic receptors is unlikely. There are conflicting reports about the response to asthmatic airways in vitro to cholinergic agonists and other spasmogens, and an exaggerated response to cholinergic agonists has not consistently been demonstrated (reviewed in THOMSON 1987). In guinea-pigs, which become hyperresponsive to acetylcholine after intravenous or inhaled platelet-activating factor (PAF), there is no increae in the response of airways in vitro to acetylcholine, no change in muscarinic receptors as determined by direct receptor binding nor any change in coupling or biochemical consequences of receptor activation as measured by cholinergic stimulation of phosphoinositide turnover (ROBERTSON et al. 1988). This indicates that the increased cholinergic responsiveness cannot be explained by enhancement of muscarinic receptors in airway smooth muscle and perhaps is more likely to be explained by mechanical factors such as airway oedema induced by PAF.

V. Cholinergic Mechanisms in Asthma

Although there are serveral reasons why cholinergic mechanisms might be exaggerated in asthma, the evidence for this is not convincing, since cholinergic antagonists are generally less effective as bronchodilators in asthma than are beta agonsits, which reverse bronchoconstriction irrespective of the contractile stimulus (GROSS and SKORODIN 1984a; MANN and GEORGE 1985; GROSS 1988). This implies that airway smooth msucle is contracted mainly by mechanisms other than vagal tone, and presumably several mediators may act directly on airway smooth muscle. Thus, cholinergic antagonists would not be effective. Similarly, histamine bronchoconstriction is only weakly affected by anticholinergic treatment in human subjects (whereas in dog histamine bronchoconstriction is largely abolished by vagal nerve section or by atropine). Cholinergic mechanisms are likely to be more important in acute asthma exacerbations, since nebulised anticholinergics may be almost as effective as beta agonists as bronchodilators (WARD et al. 1981; REBUCK et al. 1987).

VI. Cholinergic Mechanisms in COAD

Cholinergic mechanisms appear to play a more important role in COAD, and anticholinergics may be as effective as beta agonists (GROSS and SKORODIN 1984b). This indicates that vagal tone is the only reversible element in the airways obstruction of COAD. There is no direct evidence for increased vagal cholinergic tone in COAD and the bronchodilator response may be explained entirely by geometric factors since vagal tone will have a relatively greater effect on airway resistance if the airway is narrowed (BARNES 1987d). Cholinergic stimulation may contribute to the increased mucus secretion in chronic bronchitis.

VII. Muscarinic Receptor Subtypes

At least three subtypes of muscarinic receptor have been recognised pharmacologically (MITCHELSON 1988) and five distinct receptor subtypes have now been cloned (BONNER et al. 1987, 1988). Muscarinic receptor subtypes have now also been recognised in airways (BARNES et al. 1988; BARNES 1989; MINETTE and BARNES 1989). Inhibitory muscarinic receptors (autoreceptors) have been demonstrated on cholinergic nerves of airways in animals in vivo

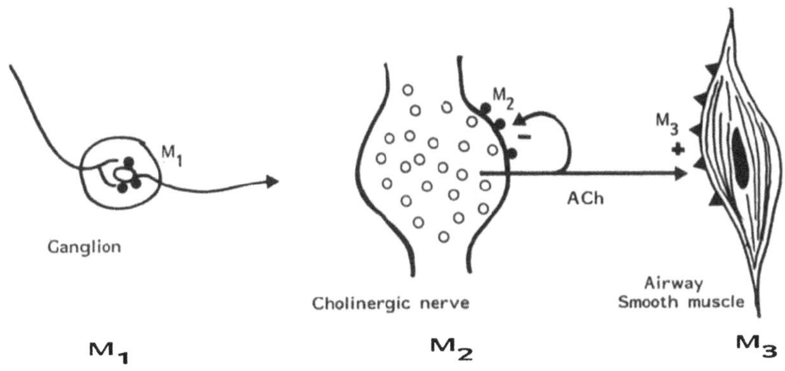

	M1	M2	M3
AGONIST	McN-A-343	Pilocarpine	—
ANTAGONIST	Pirenzepine	Gallamine AF-DX 116 Methoctramine	4-DAMP Hexahydrosila-difenidol

Fig. 4. Muscarinic receptor subtypes in airways

(Fryer and Maclagan 1984; Blaber et al. 1985). Autoreceptors have also been demonstrated on human bronchi in vitro (Minette and Barnes 1988). These receptors inhibit acetylcholine release and therefore may serve to limit vagal bronchoconstriction. These receptors are of the M2-subtype and clearly differ pharmacologically from M3-receptors on airway smooth muscle (Fig. 4). Drugs such as atropine and ipratropium bromide, which block both prejunctional receptors and those on smooth muscle with equal affinity, therefore increase acetylcholine release which may then overcome the post-junctional blockade. This means that such drugs will not be as effective against vagal bronchoconstriction as against cholinergic agonists, so it may be necessary to re-evaluate the contribution of cholinergic nerves when drugs which are selective for the muscarinic receptors on airway smooth muscle (M3-antagonists) are developed. Indeed, in animals low doses of ipratropium bromide increase vagally mediated bronchoconstriction (Fryer and Maclagan 1987).

Recently, the presence of muscarinic autoreceptors has been demonstrated in human subjects in vivo. A cholinergic agonist, pilocarpine, which selectively activates these M2-receptors, inhibits cholinergic reflex bronchoconstriction induced by sulphur dioxide in normal subjects, but such an inhibitory mechanism does not appear to operate in asthmatic subjects, suggesting that there may be dysfunction of these autoreceptors (Minette et al. 1988). Such a defect in muscarinic autoreceptors may then result in exaggerated cholinergic reflexes in asthma, since the normal feedback inhibition of acetylcholine release may be lost. This might also explain the sometimes catastrophic bronchoconstriction which occurs with beta blockers in asthma which, at least in mild asthmatics, appears to be mediated by

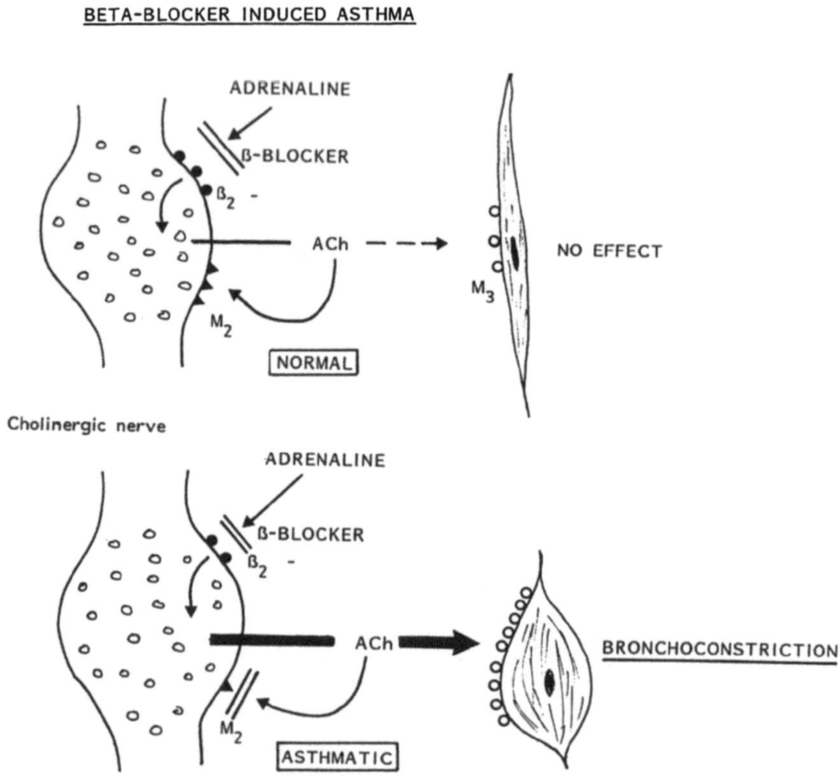

Fig. 5. Possible mechanism of β-blocker-induced bronchoconstriction. β-Blockers may inhibit the modulatory effect of circulating adrenaline on β_2 receptors of cholinergic nerves, thus increasing acetylcholine (*ACh*) release. In normal people this may act at M_2 autoreceptors to inhibit further ACh release, so no effect on airway smooth muscle is seen. In asthmatic patients, if there is a defect in M_2 autoreceptors the increased *ACh* cannot switch itself off. Released *ACh* also has a greater effect on airway smooth muscle because of bronchial hyperreactivity

cholinergic pathways (IND et al. 1989). Antagonism of inhibitory β-receptors on cholinergic nerves would result in an increased release of acetylcholine which could not switch itself off in the asthmatic subejct (BARNES 1989) (Fig. 5).

M1-receptors, which are excitatory, are present in airway ganglia of animals and may be inhibited by pirenzepine (BLOOM et al. 1987, 1988). The function of these M1-receptors role in regulation of airway tone is not yet certain, but they may play an important role in the "setting" of vagal tone, since these receptors appear to be important in the chronic regulation of ganglionic transmission, whereas the classical nicotinic receptors may be more important in rapid neurotransmission, as in reflexes. Similar receptors are also likely to be present in human parasympathetic ganglia, since pirenzepine inhibits reflex bronchoconstriction at doses which do not inhibit the direct effect of cholinergic agonists on airway tone (LAMMERS et al. 1989a). It is hoped that the development of more selective muscarinic drugs and the application of molecular biology techniques with cDNA probes for muscarinic subtypes will help to elucidate the role of muscarinic receptor subtypes in asthma.

C. Adrenergic Mechanisms

Since adrenergic agonists have a dramatic effect in relieving asthmatic bronchoconstriction, it was logical to suggest that there might be a defect in adrenergic control in asthma. SZENTIVANYI (1968) proposed that there was a fundamental defect in β-adrenoceptor function in asthma. Adrenergic mechanisms involve sympathetic nerves, circulating catecholamines and α and β-adrenoceptors. Several defects might be present in asthma (BARNES 1988).

I. Adrenergic Innervation

Adrenergic nerves do not control human airway smooth muscle directly (RICHARDSON 1979; BARNES 1986a) but could influence cholinergic neurotransmission via prejunctional adrenoceptors (DANSER et al. 1987; RHODEN et al. 1988; GRUNDSTROM et al. 1981). There is no evidence to suggest that sympathetic neurotransmission may be abnormal in asthma, but it is possible that inflammatory mediators, such as histamine, might impair the release of noradrenaline from adrenergic nerves. Adrenergic neural control of the bronchial vasculature may also be important in asthma, particularly during exercise and cold air challenge, when conditioning of inspired air may be important.

II. Circulating Catecholamines

Since adrenergic nerves do not directly control airway smooth muscle, it seems probable that circulating catecholamines may play a more important role in regulation of bronchomotor tone (BARNES 1986b). Although the catecholamines noradrenaline, adrenaline and dopamine are present in the circulation, only adrenaline has physiological effects and is secreted by the adrenal medulla (CRYER 1980). Since beta blockers cause bronchoconstriction in asthmatic patients but not in normal subjects, this suggests that there is an increase in adrenergic drive to the airways, and in the absence of adrenergic innervation this might be provided by circulating adrenaline. However, plasma adrenaline concentrations are not elevated in asthmatic patients

(BARNES et al. 1982b), even in those who bronchoconstrict with intravenous propranolol (IND et al. 1985b). Even during acute exacerbations of asthma there is no elevation of plasma adrenaline (IND et al. 1985b), suggesting that severe bronchoconstriction is not a stimulus for adrenaline release. Furthermore, plasma adrenaline is not elevated during bronchoconstriction induced by isocapnic hyperventilation (BARNES et al. 1981a) or inhaled histamine and allergen (WARREN et al. 1984; SANDS et al. 1985; LARSSEN et al. 1985). A degree of exercise sufficient to precipitate exercise-induced asthma fails to elevate plasma adrenaline, although such a rise is seen in normal subjects who perform the same degree of exercise (BARNES et al. 1981a; WARREN et al. 1982). This suggests that there may be a problem in mobilisation of adrenaline in asthma, but this is unlikely to be due to a defect in adrenal medullary secretion, since hypoglycaemia stimulates a large rise in plasma adrenaline, as in normal subjects (IND et al. 1983). The nature of this impaired secretory response in asthma remains uncertain.

III. β-Adrenoceptors

The possibility that β-receptors are abnormal in asthma has been extensively investigated. The original suggestion that there was a primary defect in β-receptor function in asthma (SZENTIVANYI 1968) has not been substantiated (TATTERSFIELD et al. 1983) and any defect in β-receptors is likely to be secondary to the disease, perhaps as a result of inflammation or as a result of adrenergic therapy. Recent studies have demonstated that airways from asthmatic patients fail to relax normally in response to isoprenaline, suggesting a possible defect in β-receptor function in airway smooth muscle (CERRINA et al. 1986; GOLDIE et al. 1986). Whether this is due to a reduction in β-receptors, a defect in receptor coupling or some abnormality in the biochemical pathways leading to relaxation is not yet known. However, the fact that asthmatic airways relax normally in response to theophylline suggests that the defect may be specific for β-receptors (GOLDIE et al. 1986).

It is possible that inflammation might lead to such abnormalities in β-receptor function. β-Receptors are widely distributed in lung and are localised to many different cell types (BARNES et al. 1982a, 1983a; CARSTAIRS et al. 1985). Incubation of airway smooth muscle with activated macrophages causes reduced β-adrenergic responses, which appear to be due to the release of oxygen radicals (ENGELS et al. 1985). Airways from guinea-pig which are hyperresponsive after PAF exposure show a reduced bronchodilator response to isoprenaline in vivo but relax normally in vitro and there is no reduction in β-receptor density or affinity (BARNES et al. 1987). Recent studies have suggested that stimulation of phosphoinositide hydrolysis (PI) may lead to down-regulation and uncoupling of β-receptors in airway smooth muscle, probably via the activation of protein kinase C which occurs in response to diacylglycerol generated by PI hydrolysis (GRANDORDY et al. 1989) (Fig. 6).

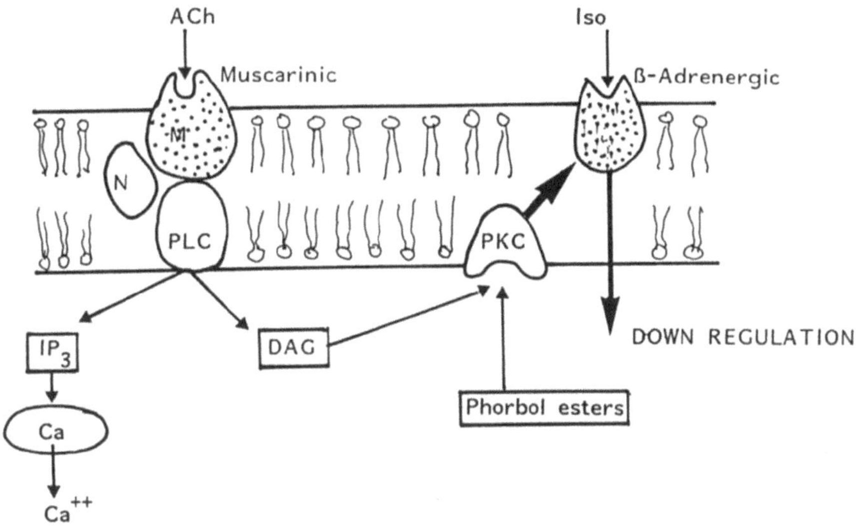

Fig. 6. Receptor transmodulation in airway smooth muscle. Stimulation of receptors linked to phosphoinositide hydrolysis activates phospholipase C (*PLC*) leading to the formation of inositol trisphosphate (*IP_3*), which releases intracellular calcium ions, and diacylglcerol (*DAG*) which activates protein kinase C (*PKC*). PKC activation may then phosphorylate β-receptors or their coupling proteins, thus leading to down-regulation of receptors or uncoupling of responses, resulting in an impaired β-adrenergic response

Since many inflammatory mediators and spasmogens may stimulate PI hydrolysis in airway smooth muscle cells (GRANDORDY and BARNES 1987; HALL and CHILVERS 1989), this may provide an explanation for impaired β-receptor function in asthma. Thus, stimulation of one type of receptor may affect another receptor (receptor transmodulation). Since such a defect would be proportional to the severity of inflammation it might only be relevant in more severe asthmatic patients.

IV. α-Adrenoceptors

α-Receptors, which mediate bronchoconstriction, have been demonstrated in airways of several species, including humans, but α-adrenoceptor mediated contraction may only be demonstrated under certain conditions (KNEUSSL and RICHARDSON 1978). In canine trachea histamine and serotonin facilitate α-adrenergic bronchoconstriction in vitro (KNEUSSL and RICHARDSON 1978; BARNES et al. 1983c) and in vivo (BROWN et al. 1983), an effect which is not due to any change in α-receptor density or affinity (BARNES et al. 1983b). In sensitized guinea pigs exposed to allergen aerosol there is an increase in α-receptor density (BARNES et al. 1980). If α-adrenergic facilitation was important in asthma then α-blocking drugs should be beneficial, yet specific α-blockers such as prazosin have little or no effect in asthma (BARNES et al. 1981b,c; BAUDOUIN et al. 1988). It is difficult to understand how α-receptors on airway smooth muscle would be activated in the absence of direct adrenergic innervation, and in any case noradrenaline causes bronchodilation rather than bronchoconstriction in asthmatic patients (LARSSEN et al. 1986).

The situation is complicated by the demonstration of inhibitory α_2-receptors on cholinergic nerves which would have the opposite effect of α-receptors on airway smooth muscle (ANDERSON et al. 1986), and the demonstration that alpha agonists reduce microvascular leakiness in airways (BOSCHETTO et al. 1989) and may therefore have a beneficial anti-inflammatory effect in asthma by reducing airway mucosal oedema. Indeed, it is this property of adrenaline to blanch the skin which first suggested, at the beginning of this century, that it might be useful in asthma, which was then viewed as a disease of vascular instability.

D. NANC Nerves and Neuropeptides

The role of NANC nerves in regulation of airway calibre remains uncertain, since the neurotransmitters involved have not been conclusively identified and no specific blockers are available (BARNES 1986c). There is increasing evidence that neuropeptides may act as neurotransmitters of NANC nerves and many neuropeptides, which have potent effects on many aspects of airway function, have now been identified in human airways (Table 1) (BARNES 1987a,c,e; LUNDBERG and SARIA 1987; BARNES and LUNDBERG 1990).

I. NANC Inhibitory Nerves

A defect in non-adrenergic inhibitory nerves has been proposed in asthma, since this is the only neural bronchodilator pathway in human airways (RICHARDSON and BELAND 1976). This could arise from an intrinsic defect in these nerves (which seems unlikely), or a secondary abnormality such as increased breakdown of neurotransmitter (which may be vasoactive intestinal peptide, VIP) or a defect in NANC inhibitory receptors. An absence of VIP-immunoreactive nerves has been described in asthmatic airways (OLLERENSHAW et al. 1989b). VIP is susceptible to enzymatic degradation by a variety of enzymes (CAUGHEY et al. 1988). Inflammatory cells release a variety of peptidases (including tryptase), which may lead to more rapid breakdown of VIP in asthmatic airways (BARNES 1987a). Mast cell tryptase is able to block NANC inhibitory responses in canine airways in vitro and to increase airway responsiveness (SEKIZAWA et al. 1989). NANC bronchodilation has been demonstrated in human airways in vivo (MICHOUD et al. 1987; LAMMERS et al. 1988; ICHINOSE et al. 1988), but does not appear to be defective in mild asthma (LAMMERS et al. 1989b). However, this does not preclude a secondary defect in more severe asthma, when inflammation is more pronounced. VIP is a potent dilator of human bronchi in vitro (PALMER

Table 1. Neuropeptides in airways

Vasoactive intestinal peptide
Peptide histidine isoleucine/methionine
Peptide histindine valine
Substance P
Neurokinin A
Neuropeptide K
Calcitonin gene-related peptide
Neuropeptide Y
Galanin
Gastrin releasing peptide (mammalian bombesin)
Cholecystokinin octapeptide
Somatostatin
Enkephalins

et al. 1986). VIP is a cotransmitter with acetylcholine in airway cholinergic nerves (LAITINEN et al. 1985) and may act as "braking" mechanism to cholinergic bronchoconstriction, since VIP acts as a functional antagonist to acetylcholine-induced contraction of airway smooth muscle. If VIP is more rapidly degraded in asthma, this may lead to exaggerated cholinergic bronchoconstriction (Fig. 7).

II. Increased NANC Excitatory Nerves

Perhaps a more likely abnormality is an increase in activity of NANC excitatory nerves. Evidence in favour of this is the observation that bronchoconstriction, induced by sulphur dioxide and metabisulphite, is believed to act on sensory nerve endings and is only minimally reduced by anticholinergic treatment in asthmatics (TAN et al. 1982; NICHOL et al. 1989). Evidence from animal studies suggests that NANC bronchoconstriction is due to release of neuropeptides, such as substance P and neurokinin A from C-fibre sensory nerve endings (ANDERSSON and GRUNDSTROM 1983; LUNDBERG et al. 1988). If these sensory endings are activated in asthma by exposure of epithelial nerve endings and release of mediators, such as bradykinin and prostaglandins, then an axon reflex may be triggered (BARNES 1986e). Release of tackykinins would then lead to bronchoconstriction, microvascular leakage and mucus hypersecretion.

III. Tachykinins

Substance P (SP) has long been recognised as a bronchoconstrictor (NILSSON et al. 1977) and also contracts human and animal airways in vitro (LUNDBERG

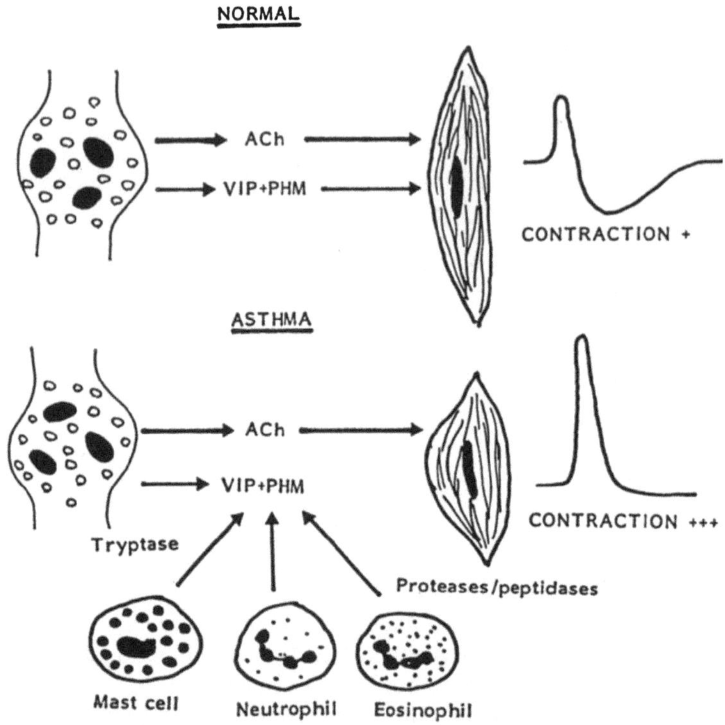

Fig. 7. VIP-cotransmission in asthma. *VIP* and peptide histidine methionine (*PHM*) acts as a cotransmitter with acetylcholine (*ACh*) in airway cholinergic nerves and may act as a functional "brake" for cholinergic nerve bronchoconstriction. In asthma peptidases from inflammatory cells (e.g. tryptase from mast cells) may rapidly degrade *VIP*, leading to unopposed cholinergic bronchoconstriction

et al. 1983a). It also has a number of other effects on airway function which mimic asthma (BARNES 1987; LUNDBERG et al. 1988). Other tackykinins have now been identified in airway nerves, including neuropeptide K and neurokinin A (NKA) (MARTLING et al. 1987). SP constricts human airways in vitro (LUNDBERG et al. 1983a) but has no effect when given by infusion or inhalation in vivo (FULLER et al. 1987b). By contrast, neurokinin A is more potent as a constrictor of human bronchi in vitro (MARTLING et al. 1987; ADVENIER et al. 1987; PALMER and BARNES 1987) and also causes bronchoconstriction in vivo either by infustion (EVANS et al. 1988) or by inhalation (JOOS et al. 1987).

Tachykinins are degraded predominantly by neutral metalloendopeptidase (enkephalinase) (SKIDGEL et al. 1984; SEKIZAWA et al. 1987) which is localised mainly to airway epithelium (JOHNSON et al. 1985). Epithelium removal greatly exaggerates the constrictor effect of tachykinins (GRANDORDY et al. 1988; FROSSARD et al. 1989). Specific inhibitors of neutral endopeptidase, such as phosphoramidon and thiorphan, enhance the contractile response to NKA to the same extent as epithelial removal, and no further effect of epithelium removal can be demonstrated, suggesting that metabolism of NKA by epithelial enzymes accounts for this effect (FROSSARD et al. 1989). For SP it is slightly more complex, since it may in addition release prostaglandins and a relaxant factor from airway epithelium via SP-preferring (NK-1)receptors (Fig. 8). These observations may be of relevance to asthma since if epithelium is shed then tachykinins, if released from sensory nerves, would not be degraded as effectively and this would exaggerate their effects, such as bronchoconstriction, microvascular leakage and mucus secretion.

SP is a potent inducer of microvascular leakage in airways (LUNDBERG et al. 1983b) and is more potent than NKA in this respect (ROGERS et al. 1988). SP also stimulates mucus secretion both in animal (COLES et al. 1984) and human bronchi in vitro (ROGERS et al. 1989). SP is again more potent than NKA. These last two responses are therefore mediated by NK-1 receptors.

Recent studies suggest that SP-immunoreactive nerves may be increased in the airways of asthmatics (OLLERENSHAW et al. 1989a). This could be the result of chronic inflammation.

IV. Calcitonin Gene-Related Peptide

Calcitonin gene-related peptide (CGRP) is also localised to sensory nerves in airways (UDDMAN et al. 1985; CADIEUX et al. 1986), but perhaps its most striking effect is potent and long-lasting vasodilation (BRAIN et al. 1985). CGRP is a potent dilator of bronchial vessels in vitro (MCCORMACK et al. 1988) and causes a long-lasting increase in blood flow in canine trachea (SALONEN et al. 1988). Receptors for CGRP are localised predominantly to

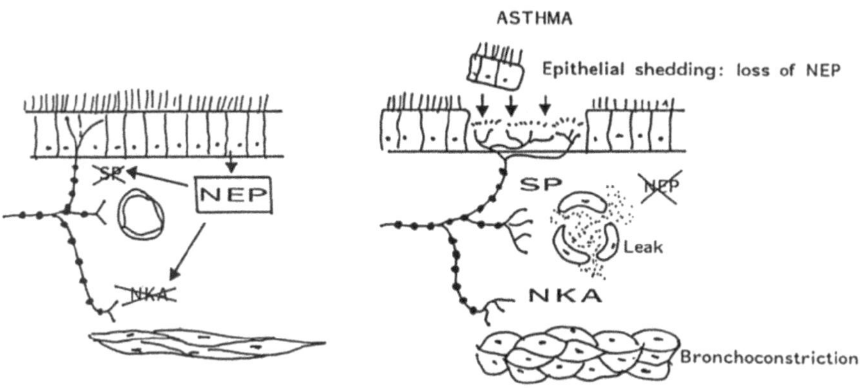

Fig. 8. Tachykinin, such as substance P (*SP*) and neurokinin A (*NKA*) released from sensory nerves are broken down by neutral endopeptidase (*NEP*) which is predominantly localised to airway epithelium. If epithelium is shed in asthma and *NEP* is lost this may lead to enhanced effects of these tachykinins

airway blood vessels in animal and human airways (Mak and Barnes 1988). This suggests that CGRP may account for some of the hyperaemia which is characteristic of asthmatic airways.

V. Axon Reflex Mechanisms

Shedding of airway epithelium may expose sensory nerve endings and these may be triggered or "sensitized" by certain mediators. Bradykinin may be particularly effective in this respect, and it iduces a marked dyspnoea which is reminiscent of an asthma attack (Fuller 1987a). Bradykinin could be formed from plasma which exudes into the asthmatic airway lumen as a result of microvascular leakage. Stimulation of the exposed sensory nerves may then cause their antidromic stimulation, with release of their neuropeptides (Fig. 9). Such neurogenic inflammation is well-described in skin and gut (Szolczanyi 1988) and has also been demonstrated in rodent lung (Lundberg et al. 1988). Mediators such as bradykinin release tachykinins from sensory nerves in lung (Saria et al. 1988). In some animal models of hyperreactivity depletion of sensory neuropeptides, by prior treatment with capsaicin, results in a reduction in responsiveness (Thompson et al. 1987). Whether such neurogenic inflammation is involved in human asthma has not yet been

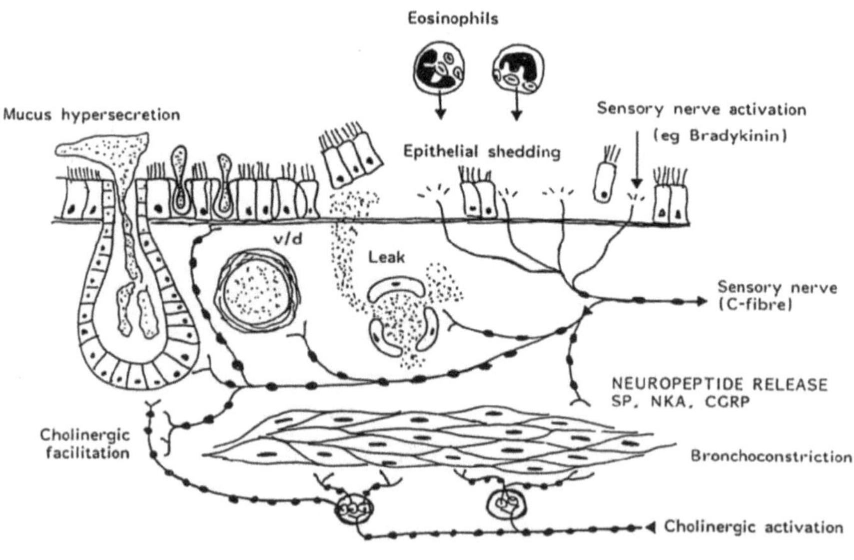

Fig. 9. Axon reflex mechanisms in asthma. Neuropeptides in sensory nerves, such as substance P (*SP*), neurokinin A (*NKA*), and calcition gene-related peptide (*CGRP*), may be released from collateral branches of sensory nerves to augment the inflammatory response, with bronchroconstriction, cholinergic activation vasodilations (*v/d*), mucus hypersecretion, and microvascular leak

demonstrated, since there are no specific blockers of tachykinins available for clinical use. However, the increase in SP-immunoreactive nerves described in asthmatic airways may provide evidence that such local reflexes become important in disease. Axon reflex mechanisms, and possibly also local reflexes mediated by sensory nerve input to parasympathetic ganglia, provide a mechanism for amplifying the inflammatory response in asthmatic airways. These nerves would spread the inflammatory response from sites of epithelial damage and therefore amplify the inflammatory process. It follows that drugs which reduce neurogenic inflammation might be useful in asthma therapy.

VI. Modulation of NANC Bronchoconstriction

In the central nervous system opiates inhibit the release of SP, and enkephalins act as the endogenous modulators of SP-ergic neurones. Opiates are effective in inhibiting NANC constriction of guinea-pig airways in vitro (Frossard and Barnes 1987) and in vivo (Belvisi et al. 1988). This is mediated by μ-opioid receptors on sensory nerve endings which inhibit the release of tachykinins. Similarly, opioids inhibit neurogenic microvascular leakage in guinea-pig airways (Belvisi et al. 1989b), and mucus secretion induced by capsaicin in human bronchi (Rogers and Barnes 1989). The central inhibitory neurotransmitter, α-aminobutyric acid, is similarly effective on NANC bronchoconstriction (Belvisi et al. 1989a).

E. Conclusions

Neural control of airway calibre is far from simple and it is likely to contribute to airway narrowing and bronchial hyperresponsiveness in asthma. Complex interactions may occur between the different components of autonomic control, and inflammatory mediators released from cells in the inflamed airway may profoundly alter the balance of neural control. Further elucidation of these interactions should shed light on the mechanisms of airway disease and may lead to therapeutic advances. The discovery of many potent neuropeptides in human airway nerves complicates matters further, but may lead to novel approaches to treatment in the future.

Acknowledgement. I am very grateful to Madeleine Wray for her careful preparation of this manuscript.

References

Advenier C, Naline E, Drapeau G, Regoli D (1987) Relative potencies of neurokinins in guinea-pig and human bronchus. Eur J Pharmacol 139:133–137

Alexander HL, Paddock R (1921) Bronchial asthma: response to pilocarpine and epinephrine. Arch Intern Med 27:184–191

Andersson RGG, Grundstrom N (1983) The excitatory non-cholinergic, non-adrenergic nervous system of the guinea-pig airways. Eur J Respir Dis 64:141–157

Andersson RGG, Fugner A, Lundgren BR, Muacevic G (1986) Inhibitory effects of clonidine on bronchospasm induced in guinea-pig by vagal stimulation or antigen challenge. Eur J Pharmacol 123:181–185

Barnes PJ (1986a) State of art. Neural control of human airways in health and disease. Am Rev Respir Dis 134:1289–1314

Barnes PJ (1986b) Endogenous catecholamines and asthma. J Allergy Clin Immunol 77:791–795

Barnes PJ (1986c) Non-adrenergic non-cholinergic neural control of human airways. Arch Int Pharmacodyn 280 [Suppl]:208–228

Branes PJ (1986d) Airway inflammation and autonomic control. Eur J Respir Dis 69 [Suppl 147]:80–87

Barnes PJ (1986e) Asthma as an axon reflex. Lancet i:242–245

Barnes PJ (1987a) Airway neuropeptides and asthma. Trends Pharm Sci 8:24–27

Barnes PJ (1987b) Muscarinic receptors in lung. Postgrad Med J 63: [Suppl]:13–19

Barnes PJ (1987c) Neuropeptides in the lung: localization, function and pathophysiologic implications. J Allergy Clin Immunol 79:285–295

Barnes PJ (1987d) Cholinergic control of airway smooth muscle. Am Rev Respir Dis 136:S42–45

Barnes PJ (1987e) Neuropeptides in human airways: function and clinical implications. Am Rev Respir Dis 136:S77–83

Barnes PJ (1988) Adrenergic regulation of airway function. In:Kaliner MA, Barnes PJ (eds) The airways: neural control in health and disease. Dekker, New York, pp 57–85

Barnes PJ (1989) Muscarinic receptor subtypes: implications for lung disease. Thorax 44:161–167

Barnes PJ, Lundberg JM (1990) Airway neuropeptides and asthma. In: Kaliner M, Barnes PJ, Persson CGA (eds) Asthma and bronchial hyperresponsiveness: pathogenesis and treatment. Dekker, New York (in press)

Barnes PJ, Dollery CT, MacDermot J (1980) Increased pulmonary α-adrenergic and reduced β-adrenergic receptors in experimental asthma. Nature 285:569–571

Barnes PJ, Brown MJ, Silverman M, Dollery CT (1981a) Circulating catecholamines in exercise and hyperventilation-induced asthma. Thorax 36:435–440

Barnes PJ, Ind PW, Dollery CT (1981b) Inhaled prazosin in asthma. Thorax 36:378–381

Barnes PJ, Wilson NM, Vickers H (1981c) Prazosin, an alpha1-adrenoceptor antagonist partially inhibits exercise-induced asthma. J Allergy Clin Immunol 68:411–419

Barnes PJ, Basbaum CB, Nadel JA, Roberts JM (1982a) Localization of β-adrenoceptors in mammalian lung by light microscopic autoradiography. Nature 299:444–447.

Barnes PJ, Ind PW, Brown MJ (1982b) Plasma histamine and catecholamines in stable asthmatic subjects. Clin Sci 62:661–665

Barnes PJ, Basbaum CB, Nadel JA (1983a) Autoradiographic localization of autonomic receptors in airway smooth muscle: marked differences between large and small airways. Am Rev Respir Dis 127:758–762

Barnes PJ, Skoogh B-E, Nadel JA, Roberts JM (1983b) Postsynaptic α_2-adrenoceptors predominate over alpha1-adrencoptors in canine tracheal smooth muscle and mediate neuronal and humoral alpha-adrenergic contraction. Mol Pharmacol 23:570–575

Barnes PJ, Skoogh B-E, Brown JK, Nadel JA (1983c) Activation of alpha-adrenergic responses in tracheal smooth muscle: a post-receptor mechanism. J Appl Physiol 54:1469–1476

Barnes PJ, Grandordy BM, Page CP, Rhoden KJ, Robertson DN (1987) The effect of platelet activating factor on pulmonary beta-adrenoceptors. Br J Pharmacol 90:709–715

Barnes PJ, Minette PA, Maclagan J (1988) Muscarinic receptor subtypes in lung. Trends Pharmacol Sci 9:412–416

Baudouin SV, Altman TJ, Johnson AJ (1988) Prazosin in the treatment of chronic asthma. Thorax 43:385–387
Belvisi MG, Chung KF, Jackson DM, Barnes PJ (1988) Opioid modulation of non-colinergic neural bronchoconstriction in guinea-pig in vivo. Br J Pharmacol 95:413–418
Belvisi MG, Ichinose M, Barnes PJ (1989a) Evidence for GABA B receptors on non-adrenergic non-cholinergic nerves in guinea-pig airways. Br J Pharmacol 97:1225–1231
Belvisi MG, Rogers DF, Barnes PJ (1989b) Neurogenic plasma extravasation:inhibition by morphine in guinea pig airways in vivo. J Appl Physiol 66:268–272
Blaber LC, Fryer AD, Maclagan J (1985) Neuronal muscarinic receptors attenuate vagally-induced contraction of feline bronchial smooth muscle. Br J Pharmacol 86:723–728
Bloom JW, Yamamura HI, Baumgartner C, Halonen M (1987) A muscarinic receptor with high affinity for pirenzepine mediates vagally induced bronchoconstriction. Eur J Pharmacol 133:21–27
Bloom JW, Baumgartener-Folkerts C, Palmer JD, Yamamura HI, Halonen M (1988) A muscarinic receptor subtype modulates vagally stimulated bronchial contraction. J Appl Physiol 85:2144–2150
Bonner TI, Buckley NJ, Young AC, Bann MR (1987) Identification of a family of muscarinic acetylcholine receptor genes. Science 237:527–532
Bonner TI, Young AC, Brann MR, Buckley NJ (1988) Cloning and expression of the human and rat M5 muscarinic acetylcholine receptor genes. Neuron 1:403–410
Boschetto P, Roberts NM, Rogers DF, Barnes PJ (1989) Effect of anti-asthma drugs on microvascular leakage in guinea-pig airways. Am Rev Respir Dis 139:416–421
Boushey HA (1984) The role of the parasympathetic system in the regulation of bronchial smooth msucle. Eur J Respir Dis 65 [Suppl]:80–90
Boushey HA, Hotzman MJ, Sheller JR, Nadel JA (1980) Bronchial hyperreactivity. Am Rev Respir Dis 121:389–413
Brain SD, Williams TJ, Tippins JR, Morris HR, MacIntyre I (1985) Calcitonin gene-related peptide is a potent vasodilator. Nature 313:54–56
Brown JK, Shields R, Jones C, Gold WM (1983) Augmentation of alpha-adrenergic responsiveness in trachealis muscle of living dogs. J Appl Physiol 54:1558–1566
Cadieux A, Springall DR, Mulderry PK, Rodrigo J, Ghatei MA, Terenghi G, Bloom SR, Polak JM (1986) Occurrence, distribution and ontogeny of CGRP-immunoreactivity in the rat lower respiratory tract: effect of capsaicin treatment and surgical denervations. Neuroscience 19:605–627
Carstairs JR, Nimm AJ, Barnes PJ (1985) Autoradiographic visualization of beta-adrenoceptor subtypes in human lung. Am Rev Respir Dis 132:541–547
Caughey GH, Leidig F, Viro NF, Nadel JA (1988) Substance P and vasoactive intestinal peptide degradation by mast cell tryptase and chymase. J Pharmacol Exp Ther 244:133–137
Cerrina J, Ladurie ML, Labat C, Raffestin B, Bayol A, Brink C (1986) Comparison of human bronchial muscle responses to histamine in vivo with histamine and isoprotenerol agonists in vitro. Am Rev Respir Dis 134:57–61
Chung KF, Evans TW, Graf PD, Nadel JA (1985) Modulation of cholinergic neurotransmission in canine airways by thromboxane-mimetic U 46619. Eur J Pharmacol 117:373–375
Coleridge JCG, Coleridge HM (1984) Afferent vagal C fibre innervation of the lung and airways and its functional significance. Rev Physiol Biochem Pharmacol 99:1–110
Coles SJ, Neill KH, Reid LM (1984) Potent stimulation of glycoprotein secretion in canine trachea by substance P. J Appl Physiol 57:1323–1327
Cryer PE (1980) Physiology and pathophysiology of the human sympathoadrenal neuroendocrine system. N Engl J Med 303:436–444

Danşer AHJ, van den Ende R, Lorenz RR, Flavahan NA, Vanhoutte PM (1987) Prejunctional betal-adrenoceptors inhibit cholinergic neurotransmission in canine bronchi. J Appl Physiol 62:785–790
Engels F, Oosting RS, Nijkamp FP (1985) Pulmonary macrophages induce deterioration of guinea-pig tracheal β-adrenergic function through release of oxygen radicals. Eur J Pharmacol 111:143–144
Evans TW, Dixon CMS, Clarke B, Conradson T-B, Barnes PJ (1988) Comparison of neurokinin A and substance P on cardiovascular and airway function in man. Br J Clin Pharmacol 25:273–275
Floyer J (1968) A treatise of the asthma. Wilkins, London
Frossard N, Barnes PJ (1987) μ-Opioid receptors modulate non-cholinergic constrictor nerves in guinea-pig airways. Eur J Pharmacol 141:519–522
Frossard N, Rhoden KJ, Barnes PJ (1989) Influence of epithelium on guinea pig airway responses to tachykinins: role of endopeptidase and cyclooxygenase. J Pharmacol Exp Ther 248:292–299
Fryer AD, Maclagan J (1984) Muscarinic inhibitory receptors in pulmonary parasymathetic nerves in the guinea-pig. Br J Pharmacol 83:973–978
Fryer AD, Maclagan J (1987) Ipratropium bromide potentiates bronchoconstriction induced by vagal nerve stimulation in the guinea-pig. Eur J Pharmacol 139:187–191
Fuller RW, Dixon CMS, Cuss FMC, Barnes PJ (1987a) Bradykinin-induced bronchoconstriction in man: mode of action. Am Rev Respir Dis 135:176–180
Fuller RW, Maxwell DL, Dixon CMS, McGregor GP, Barnes VF, Bloom SR, Barnes PJ (1987b) The effects of substance P on cardiovascular and respiratory function in human subjects. J Appl Physiol 62:1473–1479
Goldie RG, Spina D, Henry PJ, Lulich KM, Paterson JW (1986) In vitro responsiveness of human asthmatic blockers to carbachol, histamine, β-receptor agonists and theophylline. Br J Clin Pharmacol 22:669–676
Grandordy BM, Barnes PJ (1987) Phosphoinositide turnover in airway smooth muscle. Am Rev Respir Dis 136 [Suppl]:S17–20
Grandordy BM, Frossard N, Rhoden KJ, Barnes PJ (1988) Tachykinin-induced phosphoinositide breakdown in airway smooth muscle and epithelium: relationship to contraction. Mol Pharmacol 33:515–519
Grandordy B, Rhoden K, Barnes PH (1990) Effects of protein kinase C activation on adrenoceptors in airway smooth muscle. Biochem Biophys Res Commun (in press)
Gross NJ (1988) Ipratropium bromide. N Engl J Med 319:486–494
Gross NJ, Skorodin MS (1984a) Anticholinergic antimuscarinic bronchodilators. Am Rev Respir Dis 129:856–870
Gross NJ, Skorodin MS (1984b) Role of the parasympathetic system in airway obstruction due to emphysema. N Engl J Med 311:321–325
Grundstrom N, Andersson RGG, Wikberg JES (1981) Prejunctional alpha2-adrenoceptors inhibit contraction of tracheal smooth muscle by inhibitory cholinergic neurotransmission. Life Sci 28:2981–2986
Hall AK, Barnes PJ, Meldrum LA, Maclagan J (1989) Facilitation by tachykinins of neurotransmission in guinea-pig pulmonary parasympathetic nerves. Br J Pharmacol 97:274–280
Hall I, Chilvers ER (1989) Inositol phosphates and airway smooth muscle. Pulm Pharmacol 2:113–120
Ichinose M, Barnes PJ (1989) Inhibitory histamine H3-receptors on cholinergic nerves in human airways. Eur J Pharmacol 163:383–386
Ichinose M, Inoue H, Miura M, Takishima T (1988) Nonadrenergic bronchodilation in normal subjects. Am Rev Respir Dis 138:31–34
Ichinose M, Stretton CD, Schwartz J-C, Barnes PJ (1989) Histamine H3-receptors inhibit cholinergic neurotransmission in guinea pig airways. Br J Pharmacol 97:13–15

Ind PW, Brown MJ, Barnes PJ (1983) Sympathoadrenal responses in asthma. Thorax 38:702
Ind PW, Barnes PJ, Brown MJ, Dollery CT (1985a) Plasma histamine concentration during propranolol induced bronchoconstriction. Thorax 40:903–909
Ind PW, Causon RC, Brown MJ, Barnes PJ (1985b) Circulating catecholamines in acute asthma. Br Med J 290:267–279
Ind PW, Dixon CMS, Fuller RW, Barnes PJ (1989) Anticholinergic blockade of beta-blocker induced bronchoconstriction. Am Rev Respir Dis 139:1390–1394
Johnson AR, Asthon J, Schulz WW Erdos EG (1985) Neutral metalloendopeptidases in human lung tissue and cultured cells. Am Rev Respir Dis 132:564–568
Joos G, Pauwels R, van der Straeten M (1987) Effect of inhaled substance P and neurokinin A on the airways of normal and asthmatic subjects. Thorax 42:779–783
Kaliner MA, Barnes PJ (eds) (1988) The airways: neural control in health and disease. Dekker, New York, pp 657 (Lung biology in health and disease series, vol 33)
Kaliner M, Shelhamer J, Davis PB, Smith LJ, Venter JC (1982) Autonomic nervous system abnormalities and allergy. Ann Intern Med 96:349–357
Kallenbach JM, Webster T, Dowdeswell R, Reinach SG, Scott Millar RN, Zwi S (1985) Reflex heart rate control in asthma. Chest 87:644–648
Kaufman MP, Coleridge HM, Coleridge JCG, Baker DG (1980) Bradykinin stimulates afferent vagal C-fibers in intrapulmonary airways of dogs. J Appl Physiol 48:551–517
Kneussl MP, Richardson JB (1978) Alpha-adrenergic receptors in human and canine tracheal and bronchial smooth muscle. J Appl Physiol 45:307–311
Laitinen A, Partanen M, Hervonen A, Peto-Huikko M, Laitinen LA (1985) VIP-like immunoreactive nerves in human respiratory tract. Light and election micrsoscopic study. Histochemistry 82:313–319
Lammers J-W, Minette P, McCusker MT, Chung KF, Barnes PJ (1988) Nonadrenergic bronchodilator mechanisms in normal human subjects in vivo. J Appl Physiol 64:1817–1822
Lammers J-W, Minette P, McCusker M, Barnes PJ (1989a) The role of pirenzepine-sensitive (M1) muscarinic receptors in vagally mediated bronchoconstriction in humans. Am Rev Respir Dis 139:446–449
Lammers J-W, Minette P, McCusker MT, Chung KF, Barnes PJ (1989b) Capsaicin-induced bronchodilation in asthmatic patients: role of the nonadrenergic inhibitory system. J Appl Physiol 67:856–861
Larssen K, Grunneberg R, Hjemdahl P (1985) Bronchodilatation and inhibition of allergen-induced bronchoconstriction by circulatory epinephrine in asthmatic subjects. J Allergy Clin Immunol 75:586–593
Larssen K, Martinsson A, Hjemdahl P (1986) Influence of circulating 4a-adrenoceptor agonists on pulmonary function and cardiovascular variables in patients with exercise induced asthma and healthy subjects. Thorax 41:522–528
Lundberg JM, Saria A (1987) Polylpeptide-containing neurons in airway smooth muscle. Ann Rev Physiol 49:557–572
Lundberg JM, Martling C-R, Saria A (1983a) Substance P and capsaicin-induced contraction of human bronchi. Acta Physiol Scand 119:49–53
Lundberg JM, Saria A, Brodin E, Rusells S, Folkers R (1983b) A substance P antagonist inhibits vagally induced increase in vascular permeability and bronchial smooth muscle contraction in the guinea-pig. Proc Natl Acad Sci USA 80:1120–1124
Lundberg JM, Saria A, Lundblad L, Angaard A, Martling C-R, Theodorsson-Norheim E, Stjarne P, Hokfelt T (1988) Bioactive peptides in capsaicin-sensitive C-fiber afferents of the airways: functional and pathophysiological implications. In: Kaliner MA, Barnes PJ (eds) The airways: neural control in health and disease. Dekker, New York, pp 417–445

Mak JCM, Barnes PJ (1988) Autoradiographic localization of calcitonin gene-related peptide binding sites in human and guinea pig lung. Peptides 9:957–964
Mann JS, George CF (1985) Anticholinergic drugs in the treatment of airways disease. Br J Dis Chest 79:209–228
Martling C-R, Theordorsson-Norheim E, Lundberg JM (1987) Occurrence and effects of multiple tachykinins: substance P, neurokinin A, neuropeptide K in human lower airways. Life Sci 40:1633–1643
McCaig DJ (1987) Comparison of autonomic responses in the trachea isolated from normal and albumin sensitive guinea-pigs. Br J Pharmacol 92:809–816
McCormack DG, Salonen RO, Widdicombe JG, Barnes PJ (1988) Sensory neuropeptides are potent vasodilators of canine bronchial arteries in vitro. Am Rev Respir Dis 137 [Suppl]:369
Michoud M-C, Amyot R, Jeanneret-Grosjean A, Couture J (1987) Reflex decrease of histamine-induced bronchoconstriction after laryngeal stimulation in humans. Am Rev Respir Dis 136:618–622
Minette PA, Barnes PJ (1988) Prejunctional inhibitory muscarinic receptors on cholinergic nerves in human and guinea-pig airways. J Appl Physiol 64:2532–2537
Minette PA, Barnes PJ (1989) Muscarinic receptor subtypes in airways: function and clinical significance. Am Rev Respir Dis 141:8162–8165
Minette PA, Lammers J-W, Barnes PJ (1988) Is there a defect in inhibitory muscarinic receptors in asthma? Am Rev Respir Dis 137:239
Mitchelson F (1988) Muscarinic receptor differentiation. Pharmacol Ther 37:357–423
Nadel JA (1980) Autonomic regulation of airway smooth muscle. In: Nadel JA (ed) Physiology and pharmacology of the airways. Dekker, New York, pp 217–257
Nadel JA, Barnes PJ (1984) Autonomic regulation of the airways. Annu Rev Med 35:451–467
Nadel JA, Barnes PJ, Holtzman MJ (1986) Autonomic factors in the hyperreactivity of airway smooth muscle. In: Handbook of Physiology: the respiratory system, vol III. American Physiological Society, Besthesda, pp 693–702
Nichol GM, Nix A, Chung KF, Barnes PJ (1989) Characterisation of bronchoconstrictor responses to sodium metabisulphite aerosol in atopic asthmatic and non-asthmatic subjects. Thorax 44:1009–1014
Nilsson G, Dahlberg K. Brodin E, Sundler F, Strandberg K (1977) Distribution and constrictor effect of substance P in guinea pig tracheobronchial tissue. In: von Euler US, Pernow B (eds) Substance P. Raven Press, New York, pp 57–61
Ollerenshaw SL, Jarvis DL, Woolcock AJ, Scheibner T, Sullivan CE (1989a) Substance P immunoreactive nerve fibers in airways from patients with and without asthma. Am Rev Respir Dis 139:A237
Ollerenshaw S, Jarvis D, Woolcock A, Sullivan C, Scheibner T (1989b) Absence of immunoreactive vasoactive intestinal polypeptide in tissue from the lungs of patients with asthma. N Engl J Med 320:1244–1248
Palmer JBD, Barnes PJ (1987) Neuropeptides and airway smooth muscle function. Am Rev Respir Dis 136:S50–54
Palmer JB, Cuss FMC, Barnes PJ (1986) VIP and PHM and their role in non-adrenergic inhibitory responses in isolated human airways. J Appl Physiol 61:1322–1328
Postma DS, Keyzer JJ, Koeter GH, Sluiter HJ, De Vries K (1985) Influence of the parasympathetic and sympathetic nervous systems on nocturnal bronchial obstruction. Clin Sci 69:251–258
Rebuck AS, Chapman KR, Abboud R, Pare PD, Kreisman H, Wolkove N, Vickerson P (1987) Nebulized anticholinergic and sympathomimetic treatment of asthma and chronic obstructive airways disease in the emergency room. Am J Med 82:59–64
Rhoden KJ, Meldrum LA, Barnes PJ (1988) Inhibition of cholinergic neurotransmission in human airways by beta2-adrenoceptors. J Appl Physiol 65:700–705

Richardson JB (1979) State of art. Nerve supply to the lungs. Am Rev Respir Dis 119:785–802
Richardson J, Beland J (1976) Nonadrenergic inhibitory nervous system in human airways. J Appl Physiol 41:764–771
Robertson DN, Rhoden KJ, Grandordy B, Page CP, Barnes PJ (1988) The effect of platelet activating factor on histamine and muscarinic receptor function in guinea-pig airways. Am Rev Respir Dis 137:1317–1322
Rogers DF, Barnes PJ (1989) Opioid inhibition of neurally mediated mucus secretion in human bronchi. Lancet i:930–932
Rogers DF, Belvisi MG, Aursudkij B, Evans TW, Barnes PJ (1988a) Effects and interactions of sensory neuropeptides on airway microvascular leakage in guinea pig. Br J Pharmacol 95:1109–1116
Rogers DF, Aursudkij B, Barnes PJ (1989) Effects of tachykinins on mucus secretion in human bronchi in vitro. Eur J Pharmacol 174:283–286
Salone RO, Webber SE, Widdicombe JG (1988) Effects of neuropeptides and capsaicin on the canine tracheal vasculature in vivo. Br J Pharmacol 95:1262–1270
Salter HH (1868) On asthma: its pathology and treatment, 2nd edn. Churchill, London
Sands MF, Douglas FL, Green J, Banner AS, Robertson GL, Leff AR (1985) Homeostatic regulation of bronchomotor tone by sympathetic activation during bronchoconstriction in normal and asthmatic humans. Am Rev Respir Dis 132:993–998
Saria A, Martling C-R, Yan Z, Theordorson-Norheim E, Gamse R, Lundberg JM (1988) Release of multiple tachykinins from capsaicin-sensitive sensory nerves in the lung by bradykinin, histamine, dimethyulphenyl piperazinium and vagal nerve stimulation. Am Rev Respir Dis 137:1330–1335
Sekizawa K, Tamaoki J, Nadel JA, Borson DB (1987) Enkephalinase inhibitor potentiates substance P and electrically induced contraction in ferret trachea. J Appl Physiol 63:1401–1405
Sekizawa K, Caughey GH, Lazarus SK, Gold WM, Nadel JA (1989) Mast cell tryptase causes airway smooth muscle hyperresponsiveness in dogs. J Clin Invest 83:175–179
Skidgel RA, Engelbrecht A, Johnson AR, Erdos EG (1984) Hydrolysis of substance P and neurokinins by converting enzyme and neutral endoproteinase. Peptides 5:769–776
Szentivanyi A (1968) The beta adrenergic theory of the atopic abnormality in bronchial asthma. J Allergy 42:203–232
Szolcsanyi J (1988) Antidromic vasodilatation and neurogenic inflammation. Agents Actions 23:5–11
Tan WC, Cripps E, Douglas N, Sudlow MF (1982) Protective effects of drugs on bronchoconstriction induced by sulphur dioxide. Thorax 37:671–676
Tanaka DT, Grustein MM (1986) Effect of substance P on neurally mediated contraction of rabbit airway smooth muscle. J Appl Physiol 60:458–463
Tattersfield AE, Holgate ST, Harvey JE, Gribbin HR (1983) Is asthma due to partial beta-blockade of airways. Agents Actions 13:265–271
Thompson JE, Scypinski LA, Gordon T, Sheppard D (1987) Tachykinins mediate the increase in airway responsiveness caused by toluene diisocynante. Am Rev Respir Dis 136:43–49
Thomson NC (1987) In vivo versus in vitro human airway responsiveness to different pharmacologic stimuli. Am Rev Respir Dis 136:S58–62
Uddman R, Luts A, Sundler F (1985) Occurrence and distribution of calcitonin gene-related peptide in the mammalian respiratory tract and middle ear. Cell Tissue Res 241:551–555
Ward MJ, Fentem PH, Roderick Smith WH, Davies D (1981) Ipratropium bromide in acute asthma. Br Med J 282:590–600

Warren JB, Keynes RJ, Brown MJ, Jenner DA, McNicol MW (1982) Blunted sympathoadrenal response to exercise in asthmatic subjects. Br J Dis Chest 76:147–150

Warren JB, Dalton N, Turner C, Clark TJH (1984) Protective effect of circulating epinephrine within the physiological range on the airway response to inhaled histamine in non-asthmatic subjects. J Allergy Clin Immunol 74:683–686

CHAPTER 6

Pharmacology and Therapeutics of β-Adrenoceptor Agonists

R.G. GOLDIE, J.W. PATERSON, and K.M. LULICH

A. Introduction

Asthma is a disease involving reversible airway obstruction (PATERSON et al. 1979; NORN and SKOV 1980) with an associated non-specific bronchial hyperreactivity to bronchoconstrictor stimuli (BOUSHEY et al. 1980; PATERSON et al. 1984; WOOLCOCK 1986). Airway obstruction in asthma is produced by a combination of bronchospasm and inflammation (WOOLCOCK 1986; BARNES 1987). The inflammatory component is characterized by mucosal and bronchial wall oedema, lymphocyte and eosinophil infiltration, desquamation of the airway epithelium and the presence of mucus plugs within the airway lumen (REED 1986). In recent years, there has been a greater appreciation of the inflammatory nature of the disease and, as a result, the therapeutic emphasis in asthma has moved away from bronchodilator drugs, i.e. those drugs inducing relaxation of constricted airway smooth muscle (REED 1986; WOOLCOCK 1986).

The β-adrenoceptor agonist bronchodilators have been the mainstay of treatment for asthma for many years (PATERSON et al. 1979). Agents selective for β_2-adrenoceptors (which predominate in human lung) can be delivered by inhalation and provide rapid and effective reversal of acute airway obstruction caused by bronchoconstriction without causing significant side effects (LULICH et al. 1986; BARNES 1987). It is now recognized that β-adrenoceptors are widely distributed throughout the lung, raising the possibility of therapeutically beneficial actions of β-agonists at sites in addition to airway smooth muscle (BARNES 1984; 1986).

This review describes the pharmacology and therapeutics of β-adrenoceptor agonists in asthma in the light of this more recent knowledge.

B. The Molecular Nature of the β-Adrenoceptor and Its Relationship to Adenylate Cyclase

The β-adrenoceptor is a glycoprotein moiety embedded in the cell plasma membrane (STILES et al. 1984). The agonist/β-adrenoceptor complex interacts with membrane bound adenylate cyclase via regulatory proteins, inducing

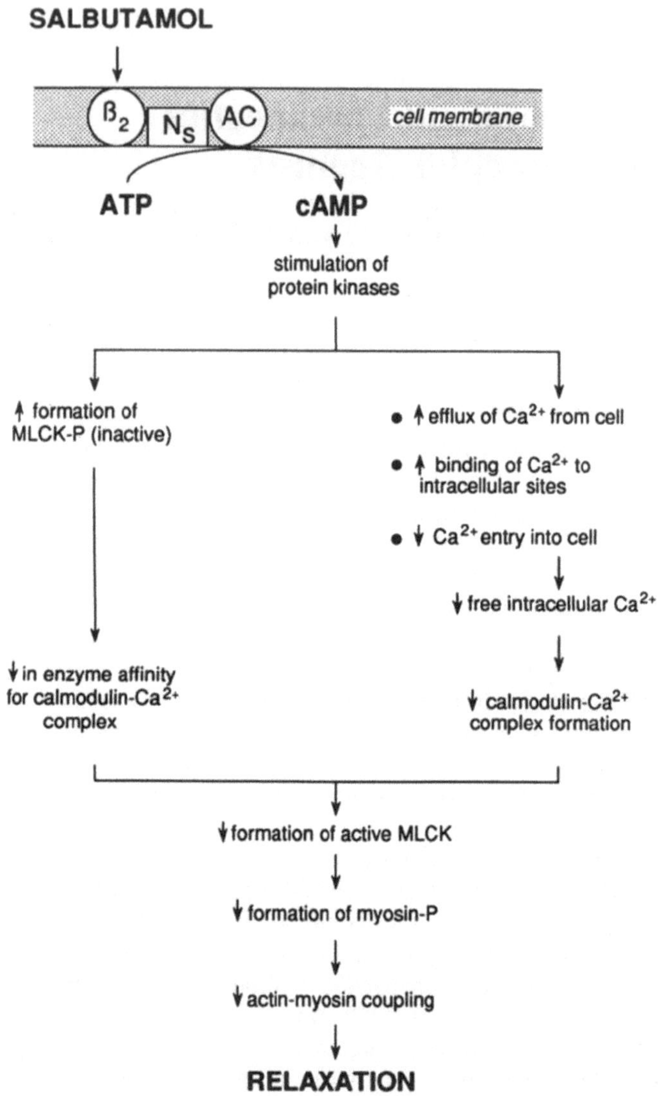

Fig. 1. Cellular processes involved in β-adrenoceptor-induced relaxation of airway smooth muscle

the production of the second messenger substance 3′,5′,cyclic adenosine monophosphate (cAMP) (Fig. 1).

I. The β-Adrenoceptor

Compelling pharmacological and biochemical evidence indicates the existence of two functionally distinct β-adrenoceptor subtypes, i.e. two different polypeptides arising from the expression of two distinct genes (STRADER et al. 1987). The β_2-adrenoceptor isolated from hamster lung has a molecular weight of approximately 64000 (CARON et al. 1985; DIXON et al. 1986) and is a single polypeptide chain consisting of 418 amino acids. There is a high degree of homology between β_2-adrenoceptors found in hamster and in humans (CARON et al. 1988).

II. Regulatory Proteins

Agonist-bound β-adrenoceptors activate adenylate cyclase through guanyl nucleotide binding stimulatory proteins (Ns or Gs) (BENOVIC et al. 1985; CARON et al. 1985). The Ns proteins are heterotrimers. In order of decreasing mass, the subunits are designated α, β and γ (CARON et al. 1985). The α-subunit of Ns contains the binding site for guanosine 5′-triphosphate (GTP), while the β,γ-subunit complex inhibits the activation of Ns (GILMAN 1987). The binding of agonist causes the β-adrenoceptor to form a complex with Ns (STILES et al. 1984). This promotes the displacement of GDP by GTP from the α-subunit of the Ns trimer and, as a result, the α-subunit sheds the inhibitory β, γ-subunit complex (Gilman 1987). The activated form of Ns (Ns(α)-GTP) then stimulates adenylate cyclase. The activation cycle is terminated by the enzymatic hydrolysis of α-subunit-bound GTP to GDP (CARON et al. 1985; GILMAN 1987).

III. Adenylate Cyclase

Adenylate cyclase appears to be a single polypeptide glycoprotein (GILMAN 1987). Its catalytic moiety converts ATP to cAMP which then acts as an intracellular messenger by activating specific kinases that are composed of tetramers containing two regulatory and two catalytic subunits (R_2C_2) (LOHMANN and WALTER 1984). The cAMP-dependent protein kinases are activated by cAMP binding to the regulatory subunits, causing them to dissociate from the catalytic subunits as illustrated below:

$$R_2C_2 \text{ (inactive)} + 4\text{cAMP} \rightarrow 2(R(\text{cAMP})_2) + 2C \text{ (active)}$$

Protein kinases activated by cAMP then mediate effects, such as relaxation of airway smooth muscle, by phosphorylating myosin light chain kinase (MLCK) and a variety of proteins involved in other enzymatic activities or ion transport (CAUVIN et al. 1984).

IV. β-Adrenoceptor Desensitization

A characteristic feature of the β-adrenoceptor/adenylate cyclase system is the desensitization produced by prolonged exposure to a β-adrenoceptor agonist (STILES et al. 1984). The decrease in the number of functional β-adrenoceptors appears to be the result of phosphorylation of agonist-bound β-adrenoceptors which is mediated by the cytosolic enzyme β-adrenoceptor kinase (BARK) (Benovic et al. 1986, 1988). Evidence suggests that the BARK is able to phosphorylate adenylate cyclase coupled receptors other than the β-adrenoceptor (Strasser et al. 1986). Specific desensitization of the β-adrenoceptor occurs because only agonist-occupied receptors are substrates for this kinase (Lefkowitz et al. 1986). Binding of agonist to the β-adrenoceptor produces a conformational change in the receptor such that normally encrypted phosphorylation sites at the C-terminus on the cytoplasmic face of the cell membrane are exposed to BARK. Phosphorylation of the β-adrenoceptor is associated with uncoupling from adenylate cyclase. Continued exposure to β-adrenoceptor agonist results in the sequestration of β-adrenoceptors within the cell and thus a decrease in the number of β-adrenoceptors in the cell membrane. Sequestrated β-adrenoceptors may be dephosphorylated, reactivated and recycled to the plasma membrane surface to become recoupled to adenylate cyclase and fully functional once again (Lefkowitz et al. 1986; Strasser et al. 1986). Alternatively, internalized β-adrenoceptors may be degraded and new receptors synthesized and seeded into the plasma membrane where they may couple to adenylate cyclase.

C. Density and Distribution of β-Adrenoceptors in the Lung

I. Radioligand Binding Studies

The study of pulmonary β-adrenoceptors has been facilitated by the advent of highly specific and selective β-antagonist radioligands including, [^{3}H] dihydroalprenolol (DHA) (Caron and Lefkowitz 1976) and ^{125}-iodocyanopindolol (I-CYP) (Engel et al. 1981). These radioligands associate with binding sites in membrane homogenates of whole lung, allowing quantification of receptor number and affinity.

Rugg et al. (1978), using rabbit lung membranes, and Szentivanyi (1979), using human lung membrane preparations, were among the first to demonstrate high levels of binding of DHA to pulmonary β-adrenoceptors. Similarly high numbers of β-adrenoceptors were revealed in guinea-pig (Barnes et al. 1980; Engel et al. 1981), rat (Rugg et al. 1978) and hamster lung (Benovic et al. 1983). Importantly, such studies demonstrated that the peripheral lung contained heterogeneous populations of β-adrenoceptors. For example, guinea-pig parenchymal membranes possessed both β_1- and β_2-adrenoceptors in the approximate proportions 20% and 80% respectively (Engel et al. 1981; Carswell and Nahorski 1983). Similar values were obtained in hamster (Benovic et al. 1983), rat (Dickinson et al. 1981; Minneman et al. 1979) and human lung membranes (Brown et al. 1985). In sharp contrast, β_1-adrenoceptors predominated in rabbit lung with estimates varying between 60% (Rugg et al. 1978; Barnett et al. 1978) and 80%

(BRODDE et al. 1983). However, no information concerning the cellular locations of these receptors was obtained from such membrane homogenate studies.

II. Autoradiography

The lung and trachea contain over 40 different cell types (SOROKIN 1970). These include airway smooth muscle, epithelial and submucosal gland cells, alveolar type I and type II pneumocytes, endothelial cells and interstitial cells. The functions of some of these cells are modulated via β-adrenoceptors. Light microscopic autoradiography (YOUNG and KUHAR 1979) has enabled assessments of β-adrenoceptor numbers and subtypes on different cell types in both central and peripheral lung. β-Adrenoceptors have been localized in mammalian trachea and lung from various species including ferret (BARNES et al. 1982), guinea-pig (GOLDIE et al. 1986b), rabbit (BARNES et al. 1984), rat (FINKEL et al. 1984), pig (GOLDIE et al. 1986a) and humans (CARSTAIRS et al. 1985).

1. Trachea

β-Adrenoceptors have been detected in ferret trachea using DHA. Large numbers of β-adrenoceptors were associated with epithelial cells, submucosal glands and airway smooth muscle (BARNES and BASBAUM, 1983). The highest density of β-adrenoceptors was associated with the epithelium > mucous gland cells > serous gland cells > airway smooth muscle. In the guinea-pig trachea, the binding of I-CYP to β-adrenoceptors was greatest over epithelium and submucosal cells with significantly lower numbers of β-adrenoceptors associated with airway smooth muscle (GOLDIE et al. 1986b). In mouse trachea, the number of β-adrenoceptors was similar in epithelium and airway smooth muscle (HENRY et al. 1990).

2. Parenchyma

β-Adrenoceptors have been detected by light microscopic autoradiography throughout the lungs of seveal animal species as well as humans. The greatest density of β-adrenoceptors was found in alveolar septae in mouse (HENRY et al.1990), rat (FINKEL et al. 1984; CONNER and REID 1984), guinea-pig (CARSWELL and NAHORSKI 1983), pig (GOLDIE et al.1986a) and human lung (CARSTAIRS et al. 1984). High numbers of β-adrenoceptors were also detected in ferret lung (BARNES et al. 1983). In rabbit lung, bronchioles contained greater numbers of β-adrenoceptors than alveoli (BARNES et al. 1984). These studies have also demonstrated significant numbers of β-adrenoceptors in bronchial and bronchiolar epithelium and smooth muscle and in vascular endothelium and smooth muscle.

Volume density measurements in human lung revealed that approximately 78% of the total lung tissue volume consisted of alveolar tissue, 8%

was vascular smooth muscle and 3% was airway smooth muscle. The remaining 11% consisted of connective tissue and cartilage (Bertram et al. 1983). β-Adrenoceptor numbers were three times higher over alveolar cells than over bronchial smooth muscle and nearly one and a half times higher than over bronchiolar smooth muscle. Thus, approximately 96% of the β-adrenoceptor population was located in alveolar tissue. Similar calculations in pig lung indicated that approximately 95% of lung β-adrenoceptors existed in alveoli (Goldie 1986a). This is consistent with data in mouse lung (Henry et al. 1990) and in rat lung where 97% of the β-adrenoceptor population resided over alveolar tissue (Conner and Reid 1984). In pig lung (Goldie et al. 1986a) and in mouse lung (Henry et al. 1990), β_1- and β_2-adrenoceptors co-existed in the alveolar wall, with the β_2-subtype predominating.

D. Therapeutic Sites of Action

I. Airway Smooth Muscle

It is generally accepted that β-adrenoceptor agonists reverse airway obstruction in asthmatics, primarily by relaxing airway smooth muscle (Paterson et al. 1979; Barnes 1986). β-Agonists are potent relaxants of human isolated bronchial preparations (Svedmyr et al. 1976; Goldie et al. 1984; Zaagsma et al. 1984), bronchioles (Zaagsma et al. 1984) and peripheral lung strips (Goldie et al. 1982a,b). Thus in humans, functional β-adrenoceptors are found in airway smooth muscle from the upper airways to terminal bronchioles. Functional studies indicate that β_2-adrenoceptors are responsible for β-agonist induced bronchodilation in humans. This is consistent with the fact that intravenous prenalterol (β_1-selective) increased the pulse rate of asthmatics by 25 beats per min but failed to cause significant bronchodilation (Lofdahl and Svedmyr 1982), while terbutaline (β_2-selective) caused pronounced bronchodilation in the same subjects. Autoradiographic studies have confirmed that human bronchial β-adrenoceptors are entirely of the β_2-subtype (Carstairs et al. 1985).

Autoradiographic studies in animals and humans indicate that the density of β-adrenoceptors in airway smooth muscle increases progressively from trachea to terminal bronchioles (Barnes et al. 1983; Carstairs et al. 1985). The functional importance of this still remains to be clarified. The predominant site of action of β-agonists in asthma may be either the central airways or peripheral airways. However, both levels may be affected to a similar extent, depending upon the route of administration, dose of inhaled drug or the principal site of airways obstruction (De Troyer et al. 1978; Fairshter and Wilson 1980; Chu 1984).

β-Agonists can relax airway smooth muscle by increasing cAMP and activating the cAMP-dependent protein kinase that phosphorylates myosin light chain kinase (MLCK) (Silver and Stull 1982). Phosphorylated MLCK exhibits decreased affinity for the calmodulin/calcium complex causing a decrease in the phosphorylation of myosin and a reduction in actin/myosin coupling (CAUVIN et al. 1984; LULICH et al. 1988). However, isoprenaline may also inhibit airway contraction produced by various spasmogens mainly by enhancing the sequestration of free calcium ions into intracellular storage sites and/or extrusion of calcium ions into the extracellular space (ITO and ITOH 1984). β-Adrenoceptor activation may also relax smooth muscle by inhibition of calcium entry into the cells (CAUVIN et al. 1984). Thus, an increase in the level of intracellular cAMP produced by β-adrenoceptor stimulation can relax airway smooth muscle by increasing the phosphorylation of MLCK and/or decreasing the free calcium ion concentration as illustrated in Fig. 1.

II. Secretory Cells and Cilia

Abnormalities of airway mucociliary function are a key feature of asthma and widespread mucus plugging is commonly observed in status asthmaticus (WANNER et al. 1979). The major source of mucus or tracheobronchial secretions are serous and mucous cells located in submucosal glands in the bronchi and trachea, goblet cells, Clara cells and type II pneumocytes in the alveoli (KALINER et al. 1984; PAVIA et al. 1987b). Some reports have indicated that β-agonists had no effect on the output of secretory glycoproteins from submucosal glands in human isolated bronchi (BOAT and KLEINERMAN 1975; SHELHAMER et al. 1980). However, in these studies, sampling occured 4 h after treatment with β-agonists and thus transient increases in secretion would not have been detected. When sampling occurred 15 min after treatment, it was shown that β-agonists stimulated the output of radiolabeled secretory glycoproteins from human isolated bronchi (PHIPPS et al. 1982). β-Agonists may promote secretion in glands by triggering the phosphorylation of proteins involved in stimulus/secretion coupling (BASBAUM et al. 1988). This class of bronchodilator appears to act on mucous cells rather than on serous cells to produce viscous secretions (BASBAUM et al. 1981; NADEL 1983; LEIKAUF et al. 1984). This is consistent with autoradiographic data indicating that β-adrenoceptors are located predominantly over mucous rather than serous cells in ferret trachea (BARNES and BASBAUM 1983). The β-adrenoceptors present in submucosal glands in human lung are predominantly of the β_2-subtype (approximately 90%) (CARSTAIRS et al. 1985).

Pulmonary surfactant is a complex lipoprotein secreted by alveolar type II cells which stabilizes alveoli at low lung volumes by lowering surface tension following compression (HOLLINGSWORTH and GILFILLAN 1984). In addition, surfactant reduces the adhesiveness of airway secretions (ZIMENT 1987). The major constituent of alveolar surfactant is phosphatidylcholine. Both in vivo studies (EKELUND et al. 1983; YOUNG and SILBAJORIS 1985) and in vitro studies with alveolar type II cells (DOBBS and MASON 1979; METTLER et al. 1981) indicate that β_2-agonists increase surfactant secretion. Since surfactant acts as an anti-adhesive on cilia/mucus coupling, an increase in the production of surfactant may improve the transport of mucus in the airway lumen (SCHLIMMER et al. 1983; ALLEGRA et al. 1985). Secretion from Clara cells has also been reported to be stimulated by β-agonists (MASSARO et al. 1981). Lipid secretions from Clara cells may have a beneficial anti-adhesive action similar to that of surfactant from type II alveolar cells, but the relevance of this to asthma is uncertain.

Respiratory secretions separate into periciliary fluid (sol phase) adjacent to the luminal surface of the epithelial lining cells and a more superficial mucus (gel phase) (KALINER et al. 1984) The thickness of the periciliary layer is critical for the effectiveness of mucus propulsion. If this fluid layer becomes too thin, the efficiency of ciliary beating in clearing mucus is impaired (SLEIGH et al. 1983). If mucus propulsion is halted due to thinning of the periciliary fluid, luminal transport of water will help to restore this fluid layer. It has been shown both in canine and feline trachea that terbutaline stimulated the secretion of chloride ions and water towards the surface of epithelial cells without altering sodium secretion (DAVIS et al. 1979; PHIPPS et al. 1980). High concentrations of isoprenaline activate β-adrenoceptors to produce a concentration dependent increase in the ciliary beat frequency of cultured epithelial cells from rabbit trachea (VERDUGO et al. 1980) and rat tracheal explants (LOPEZ-VIDRIERO et al. 1985). Thus β-agonists also directly stimulate airway ciliary cells.

Therapeutic doses of β-agonists have been shown to stimulate mucociliary transport in patients with asthma in a manner that does not appear to be related to their bronchodilator action (SANTA CRUZ et al. 1974; MOSSBERG et al. 1976). This suggests that promotion of mucus transport may be a significant therapeutic action in asthma. The increase in surfactant secretion produced by β-agonists may contribute towards improving mucus transport as a result of the secretions becoming less adhesive. β_2-Agonists can also improve mucus clearance by promoting ciliary function by a direct action on cilia and indirectly by promoting ion and water transport across bronchial epithelium. However, in more severe asthma, where there may be extensive disruption or desquamation of the bronchial epithelium (LAITINEN et al. 1985; REED 1986), the action of β-agonists on ciliary cells and water flux across epithelium may become inoperative. Whether the action of β-agonists on mucous cells exacerbates plugging of the airways by promoting mucus secretion or changes the physical properties of airway mucus so it is cleared more effectively has not been resolved.

III. Tracheobronchial Microvessels

Exudation of plasma from tracheobronchial microvessels causes oedema of the tracheobronchial mucosa which contributes towards airway narrowing and possibly other changes observed in asthma such as epithelial shedding and even bronchial hyperresponsiveness (PERSSON 1986). Histamine and other mediators of asthma produce microvascular leakage in the tracheobronchial circulation by contracting postcapillary venular endothelial cells and thus causing the formation of gaps between these cells (PERSSON 1987). Pretreatment with a β_2-agonist inhibits mediator-induced microvascular extravasation in guinea-pig airways by prior relaxation of post-capillary endothelial cells so that gap formation is prevented (PERSSON et al. 1986).

Where tracheobronchial microvascular baseline leakage has already been increased by airway injury, pretreatment with a β_2-agonist does not reduce the leakage of molecules induced by a further inflammatory stimulus (ERJEFALT et al. 1985; PERSSON 1987). Furthermore, established oedema in the tracheobronchial model associated with airway inflammation may not be readily reversed by the addition of a β_2-agonist (BARNES 1986; BARNES et al. 1987). Hence, the therapeutic importance of the relaxant effect of β_2-agonists on postcapillary venular endothelium still remains to be established.

IV. Cholinergic Nerves

The parasympathetic nervous system is the predominant neural pathway producing bronchoconstriction in asthmatic airways. In canine bronchi and trachea, β_2-agonists act prejunctionally, either at cholinergic ganglia or postsynaptic nerve terminals, to reduce cholinergic neurotransmission (VERMEIRE and VANHOUTTE 1979; ITO and TAJIMA 1982). RHODEN et al. (1987) reported that isoprenaline was approximately 80 times more potent in inhibiting cholinergic neurotransmission than in reducing acetylcholine-induced tone in human isolated bronchi. Results from this study also suggested that this inhibitory action on cholinergic neurotransmission was mediated by the β_2-adrenoceptor subtype. However, the extent to which the inhibitory action of β_2-agonists on the cholinergic pathway contributes to their bronchodilator effect in asthma still remains to be quantified.

V. Mast Cells

It is well-established that β_2-agonists are potent inhibitors of antigen-induced release of mast cell mediators from human lung fragments (CHURCH and YOUNG 1983; HUGHES et al. 1983) and dispersed human lung mast cells in vitro (PETERS et al. 1982; CHURCH and HIROI 1987). β-Adrenoceptors mediating inhibition of the anaphylactic release of histamine and slow reacting substance-A (SRS-A) (a mixture of leukotrienes C_4, D_4 and E_4; MARONE 1985), have been shown to be of the β_2 subtype (BUTCHERS et al. 1980; HUGHES et al. 1983). Thus, there is strong evidence for the presence of inhibitory β_2-adrenoceptors on mast cells in human lung.

While salbutamol is a potent inhibitor of histamine release from human lung fragments, disodium cromoglycate is a relatively weak inhibitor of histamine release in this system (BUTCHERS et al. 1979, CHURCH and YOUNG

1983). CHURCH and HIROI (1987) found that salbutamol was between 2000 and 30000 times more potent than disodium cromoglycate in inhibiting histamine release from enzymatically dispersed human lung mast cells. In asthmatics, salbutamol had a greater inhibitory effect on the antigen-induced increase in plasma histamine than a dose of disodium cromoglycate which was 100 times greater (HOWARTH et al. 1985).

It would be expected that, when an inhaled β_2-agonist is used to alleviate an acute attack of asthma, its inhibitory effect on lung mast cells would be less important than its effect on airway smooth muscle, since mast cells mediators would already have been released. The inhibitory action of a β_2-agonist on lung mast cells may be more relevant if it is used prophylactically.

The more serious morbidity which can result from the inhalation of allergens and other stimuli appears to result from the inflammatory processes involved in the late asthmatic response and subsequent bronchial hyperreactivity (CHUNG 1986; O'BYRNE et al. 1987). Pretreatment with a β_2-agonist before allergen challenge abolishes the early asthmatic response but does not prevent the late response (BOOIJ-NOORD et al. 1972; COCKCROFT and MURDOCH 1987). The late asthmatic response in asthmatics is inhibited by pretreatment with disodium cromoglycate (COCKCROFT and MURDOCH 1987) even though disodium cromoglycate is very much less potent than salbutamol in inhibiting histamine release from human mast cells (BUTCHERS et al. 1979; CHURCH and YOUNG 1983; CHURCH and HIROI 1987), Disodium cromoglycate may be effective in preventing the activation of eosinophils and other inflammatory leukocytes recruited during the late response (MOQBEL et al. 1986; KAY et al. 1987). This suggests that the mast cell stabilizing properties of β_2-agonists are not clinically important.

VI. Inflammatory Cells

Since the eosinophil has a key role in asthma (REED 1986; WARDLAW and KAY 1987), the effect of β_2-agonists on the eosinophil count in peripheral blood could be important. In addition, the effect of β_2-agonists on isolated lymphocyte, polymorphonuclear leukocyte, macrophage and platelet cell preparations may be relevant, because these cells appear to play an important role in the inflammation associated with asthma (KAY 1986).

1. Eosinophils

The eosinophil generates mediators which can produce many of the pathological features of asthma, including epithelial damage, mucus production, oedema and bronchospasm, and there is considerable evidence implicating the eosinophil as the major effector cell in asthma (DE MONCHY et al. 1985; FRIGAS and GLEICH 1986; WARDLAW and KAY 1987). It has been suggested that asthma be renamed chronic desquamating eosinophilic bronchitis in order to describe its histopathology (REED 1986). In normal volunteers, it was found that an intramuscular injection of 0.5 mg of adrenaline caused a drop in the eosinophil count from 214 ± 19.8 cells/mm^3 (mean ± sem) to 76.7 ± 6.3 cells/mm^3 within 3.5 h (KOCH-WESER 1968). Since the mean drop in circulating eosinophils of 62% was completely blocked by oral pretreatment with 40 mg propranolol hydrochloride, this eosinopaenic action of adrenaline appeared to be mediated directly by β-adrenoceptors in a manner yet to be clarified. OHMAN et al. (1972) reported that isoprenaline reduced the blood eosinophil count by 34%, from 149 to 95 cells/mm^3, in healthy male volunteers. REED et al. (1970) observed a reduced eosinopaenic effect of adrenaline in asthmatic subjects compared to normals. They found that adrenaline (10 μg/kg) reduced the blood eosinophil count from 391 to 245 cells/mm^3 (i.e. a decrease of 37%) in asthmatics, in contrast to a reduction from 114 to 47 cells/mm^3 (a reduction of 59%) in normal subjects. It is important to note that these asthmatics were asymptomatic and had not needed any medication for at least a month before the study. In more severe asthmatics, blunting of the β_2-agonist-induced eosinopaenia may be much more pronounced due to a combination of impaired β-adrenoceptor function secondary to the disease (KOETER et al. 1982; GOLDIE et al. 1986c) and β_2-agonist-induced desensitization of β-adrenoceptors (CONOLLY and GREENACRE 1976; GALANT et al. 1978b; SANO et al. 1983). KRAAN et al. (1985) reported that after 4 weeks of prophylactic treatment with inhaled terbutaline (500 μg four times a day), the eosinophil count of 389 ± 72 cells/mm^3 was not reduced. Thus a decrease in the number of circulating eosinophils does not appear to be a consequence of inhaled β_2-agonists in asthmatics.

2. Lymphocytes

β_2-Adrenoceptors predominate on peripheral blood lymphocytes (WILLIAMS et al. 1976) and these receptors are linked to adenylate cyclase (KOETER et al. 1982; SANO et al. 1983). It has been proposed that the β-agonist-induced increase in intracellular cAMP in lymphocytes inhibits the proliferateive response of lymphocytes to allergen and inhibits lymphokine secretion (BOURNE et al. 1974; REED 1985). There is evidence of a reduction in β-adrenoceptor number or β-adrenoceptor function in asthmatic lymphocytes, both as a consequence of the disease (BROOKS et al. 1979; MEURS et al. 1982; MOTOJIMA et al. 1983) and following β-agonist-induced desensitization of β-adrenoceptors (CONOLLY and GREENACRE 1976; MOTOJIMA et al. 1983; SANO et al. 1983). MOTOJIMA et al. (1983) found that the number of β-adrenoceptors per lymphocyte in normal subjects, drug free asymptomatic asthmatics and asthmatics given β-agonist therapy was 1146 ± 98, 845 ± 114 and 582 ± 47 sites/cell (mean ± sem), respectively. CONOLLY and GREENACRE (1976) reported that the maximal increase in cAMP produced by isoprenaline in lymphocytes from normal subjects was 393% ± 44% but only 67.5% ± 24.2% in asthmatics on β-agonist therapy.

3. Polymorphonuclear Leukocytes

The neutrophil is the predominant polymorphonuclear leukocyte (PMNL) present in circulating blood, while eosinophils and basophils comprise a very small proportion. Stimulation of β_2-adrenoceptors (Galant et al. 1978a) in human PMNL stimulates cAMP formation (Galant et al. 1980; Davis et al. 1986; Nielson 1987). It has been demonstrated in vitro that pretreatment with a β-agonist inhibits lysosomal β-glucuronidase release, superoxide generation and leukotriene (LT) B_4 generation by non-asthmatic PMNLs activated by the calcium ionophore A23187 or opsonised zymosan particles (Mack et al. 1986; Busse and Sosman 1984; Nielsen 1987). In asymptomatic or mild asthmatics, there was no significant decrease in PMNL β-adrenoceptor number or adenylate cyclase function, but, in those asthmatics who were on β-agonist medication, there was a marked drop in β-adrenoceptor number and adenylate cyclase activity (Galant et al. 1980; Sano et al. 1983; Davis et al. 1986). Busse et al. (1979) reported that isoprenaline-induced inhibition of zymosan-induced release of β-glucuronidase from isolated PMNLs was decreased in asthma and this dysfunction was accentuated by the prophylactic use of β-agonist aerosol. The potential therapeutic benefit of the action of β_2-agonists on PMNLs in asthmatics needs to be further clarified.

4. Macrophages

Lung macrophages are derived from blood monocytes and they secrete numerous mediators including platelet activating factor (PAF) (Arnoux et al. 1982), which may be important in initiating and regulating inflammatory reactions (Kay 1986). Schenkelaars and Bonta (1985) have reported that activation of β_2-adrenoceptors in rat peritoneal macrophages by isoprenaline or salbutamol inhibited LTC_4-induced release of lysosomal β-glucuronidase. In contrast, β-agonists were shown to be without effect on activated macrophage tumoricidal function (Schultz et al. 1979) or the proliferation of committed macrophage progenitor cells (Kurland et al. 1977). Fuller et al. (1988) found that isoprenaline did not affect the release of thromboxane B_2, LTB_4, N-acetyl-β-D-glucosaminidase or superoxide from human alveolar macrophages that were activated by opsonised zymosan or IgE/anti-IgE complexes. These results suggest that human alveolar macrophages do not possess functional inhibitory β-adrenoceptors.

5. Platelets

Cook et al. (1987) have demonstrated β_2-adrenoceptors in intact human platelets, but these receptors do not appear to stimulate cAMP production. Thus, the β_2-adrenoceptor in platelets appears to be very weakly coupled to adenylate cyclase, if at all, and its function is not established.

6. Anti-Inflammatory Action in Asthmatic Airways

The overall effectiveness of a drug in suppressing airway inflammation may be assessed by studying its inhibitory effect on the late asthmatic response. The late asthmatic response is not inhibited by pretreatment with β_2-agonists (Booij-Noord et al. 1972; Cockcroft and Murdoch 1987) but it is inhibited by pretreatment with disodium cromoglycate or beclomethasone dipropionate (Cockcroft and Murdoch 1987) and by theophylline (Pauwels et al. 1985). This indicates that β_2-agonist pretreatment is ineffective in preventing the acute exacerbation of airway inflammation produced by allergen challenge in asthmatics. Infusion of PAF activates a number of inflammatory cells and induces a sustained inflammatory response as well as non-selective bronchial hyperresponsiveness in laboratory animals and humans (Cuss et al. 1986; Barnes et al. 1988). Mazzoni et al. (1985) found that isoprenaline failed to inhibit bronchial hyperreactivity produced by PAF infusion in the guinea-pig while disodium cromoglycate, hydrocortisone and aminophylline were effective in reducing PAF-induced hyperreactivity.

The severity of bronchial inflammation is an important determinant of the level of bronchial hyperreactivity to inhaled spasmogens (Woolcock 1986). Thus bronchial hyperreactivity tests with spasmogens may be used to ascertain indirectly the inflammatory status of the airways over an extended period of time. If prophylactic use of a drug significantly reduces the level of airway inflammation, an associated reduction in bronchial hyperreactivity would be expected. Kraan et al. (1985) examined the prophylactic action of inhaled terbutaline (500 μg four times a day) over 4 weeks and Kerribijn et al. (1987) examined inhaled terbutaline (500 μg three times a day) over 6 months. Both studies showed that inhaled terbutaline produced no reduction in hyperreactivity while chronic therapy with inhaled corticosteroid reduced bronchial hyperreactivity in asthmatics. β_2-Agonists do not appear to have a significant anti-inflammatory effect on the airways in asthma. Furthermore, the mast cell and leukocyte stabilizing properties of β-agonists may not be clinically useful in this disease.

E. The Status of β-Adrenoceptor Function in Asthma

Szentivanyi (1968) proposed that diminished responsiveness of the β-adrenoceptor/adenylate cyclase system was the major defect in asthma. It was suggested that both the origins and the symptoms of asthma could be accounted for in the light of such a defect. This hypothesis was supported by in vivo observations showing that several metabolic and cardiovascular responses to catecholamines were reduced in asthmatic patients (Grieco et al. 1968; Middleton and Finke 1968). However, it has been pointed out that interpretation of these studies is difficult in view of complicating factors, including metabolic status and cardiovascular reflex mechanisms (Paterson et al. 1984).

I. In Vivo Studies

Tattersfield et al. (1983) have shown that mild asthmatics and healthy subjects behave similarly in response to inhaled or infused β-adrenoceptor agonists. In contrast, Barnes and Pride (1983) demonstrated that the bronchodilator effect of inhaled salbutamol was reduced in asthmatics compared with non-diseased subjects. Furthermore, this reduced responsiveness to salbutamol was related to disease severity, as assessed by lung function measurements. Decreased lung function in asthmatic airways may have reduced access of the inhaled β-adrenoceptor agonist to airway smooth muscle (Barnes and Pride 1983), or elevated bronchial tone may have reduced the bronchodilator efficacy of salbutamol (Barnes 1986). Alternatively, β-adrenoceptor function may be directly impaired in severe asthma. It has been suggested that the clinical significance of β-adrenoceptor dysfunction is small considering that most asthmatics respond well to bronchodilator β-adrenoceptor agonists (Barnes 1986). However, it is well established that asthmatics exposed to β-adrenoceptor antagonists, such as propranolol, may suffer severe bronchoconstriction (Zaid and Beall 1966; Richardson and Sterling 1969; Schwartz et al. 1980), while non-asthmatics similarly exposed neither bronchoconstrict nor develop asthma symptoms (Richardson and Sterling 1969; Boushey et al. 1980; Paterson et al. 1984). Thus, maximal β-adrenoceptor function is not essential to the maintenance of adequate airway calibre in healthy subjects. In contrast, bronchial β-adrenoceptor function may be critical in asthma where obstruction and elevated levels of airway tone may exist (Paterson et al. 1984). To fully address the status of bronchial β-adrenergic function in asthma, in vitro approaches uncomplicated by the many problems encountered in vivo were required.

II. In Vitro Studies

1. Isolated Human Leukocytes

Taking mixed leukocytes and lymphocytes from peripheral blood, Parker and Smith (1973) demonstrated that the magnitude of the increase in cAMP in response to isoprenaline was blunted in cells taken from asthmatic subjects compared with that seen in cells from non-diseased subjects. However, the reduction in β-adrenoceptor responsiveness, may have been due to the desensitizing influence of β-adrenoceptor agonist therapy (for review, see Harden 1983; Stiles et al. 1984).

Several studies have demonstrated that isoprenaline-stimulated increases in cAMP levels in lymphocytes (Gillespie et al. 1974; Tashkin et al. 1982) and PMNLs (Galant et al. 1980) were no different in asthmatics not taking β-adrenoceptor agonist medication or in whom medication had ceased prior to the study than in non-asthmatic subjects. However, radioligand binding studies have also demonstrated a reduction in β-adrenoceptor density in lymphocyte membrane fragments from asthmatic patients not receiving β-adrenoceptor agonist medication (BROOKS et al. 1979; SZENTIVANYI 1979; KARIMAN 1980; Motojima et al. 1983; SANO et al. 1983). In contrast, other studies reported no significant difference in β-adrenoceptor density in lymphocytes (TASHKIN et al. 1982) or PMNLs (GALANT et al. 1978a,b; 1980) between non-diseased and asthmatic individuals.

The divergent results may be explained in light of the following studies. BROOKS et al. (1979) demonstrated that the more severe the disease the greater the loss in lymphocyte β-adrenoceptor density. These cells were obtained from asthmatic individuals who had ceased to take β-adrenoceptor agonist medication between 48 h and some weeks prior to the experiment. KOETER et al. (1982) demonstrated that before inhalation of house dust mite, lymphocyte cAMP responses in asthmatics not receiving β-adrenoceptor agonist medication were similar to those observed in non-diseased individuals. However, 24 h after allergen challenge, the maximal cAMP level in lymphocytes exposed to isoprenaline in vitro was reduced from 339% to 194% above baseline in asthmatics experiencing a late phase inflammatory reaction; in non-asthmatic patients and in subjects with no late phase reaction cAMP responses were unchanged. Furthermore, in asthmatic individuals who developed a late reaction, airway sensitivity to the bronchoconstrictor influence of propranolol was increased, again suggesting specific β-adrenoceptor dysfunction following exposure to allergen. MEURS et al. (1982) also showed that the density of β-adrenoceptors and the cAMP response to isoprenaline, NaF and guanyl-5-yl-imidodiphosphate (GppNHp) were only reduced in lymphocytes from asthmatic patients 24 h after a late reaction caused by inhalation of house dust mite. This reduced function may involve an uncoupling of the β-adrenoceptor from the guanine nucleotide regulatory protein and/or adenylate cyclase as well as some reduction in membrane receptor number.

2. Isolated Human Bronchi

Few studies have directly investigated the bronchodilator effect of β-adrenoceptor agonists in bronchi isolated from asthmatic lung. However, SVEDMYR et al. (1976) and WHICKER et al. (1988) found no significant difference in responsiveness to isoprenaline in bronchial preparations from non-asthmatic and asthmatic individuals. SVEDMYR et al. concluded that β-adrenoceptor dysfunction did not exist in asthma. Interestingly, the bronchial preparations tested in that study were from asthmatic subjects undergoing surgery for lung cancer. The fact that these asthmatic patients were well enough to undergo surgery suggests that they had stable asthma. Similarly, the bronchi tested by WHICKER et al. were from mild asthmatics. Thus, bronchial β-adrenoceptor hypofunction might not occur in mild airway disease. In a more recent study, CERRINA et al. (1986) showed that the effective concentration of isoprenaline producing 50% of the maximal relaxant response was tenfold greater in bronchi from asthmatics compared with that in preparations from non-diseased patients. Furthermore, a significant negative correlation was observed between asthma severity (as assessed by airway reactivity to inhaled histamine in vivo) and diminished bronchial responsiveness to isoprenaline in vitro.

GOLDIE et al. (1986c) also showed that when the relaxant potencies of various β-adrenoceptor agonists were compared in bronchial preparations from human non-diseased and asthmatic lung, responsiveness to isoprenaline, fenoterol and terbutaline was significantly attenuated in bronchial preparations from asthmatic lung of patients dying of severe asthma. Conversely, responsiveness to the non-β-adrenoceptor agonist, theophylline, was similar in all bronchi tested, suggesting selective dysfunction of the β-adrenergic system. These results clearly suggest bronchial β-adrenoceptor dysfunction in individuals with severe asthma. It could be argued that this defect may have occurred as a result of β-adrenoceptor agonist medication. However, sensitivity to isoprenaline in bronchi from two asthmatic subjects who did not receive regular β-agonist medication was also significantly attenuated, suggesting that hypofunction was related to disease rather than to inhalation of β-adrenoceptor agonists.

The lesion probably exists either at the level of the β-adrenoceptor or in the coupling with the guanine nucleotide regulatory protein and/or adenylate cyclase. An assessment of adenylate cyclase function can be obtained by using forskolin, a direct acting adenylate cyclase activator (SEAMON et al. 1981) or by measuring cyclic nucleotide levels in bronchial smooth muscle. Interestingly, in ovalbumin-sensitized guinea-pigs, β-adrenoceptor density was reduced as was forskolin-stimulated adenylate cyclase activity (TAKI et al. 1986). Similarly, cyclic nucleotide levels were reduced in lung parenchyma from patients who had obstructed airways compared with those in non-diseased subjects (KRZANOWSKI et al. 1979). These studies implicate alterations in β-adrenoceptor coupling in asthma, which may account for the observed hypofunction.

3. Radioligand Binding Studies and Autoradiography in Lung

SZENTIVANYI (1979) demonstrated that the amount of specific DHA (12 n*M*) bound was significantly reduced in lung from asthmatic patients compared with that in lung from healthy subjects. A 43% reduction in the level of cyclic nucleotides was observed in lung tissue from patients with obstructed airways compared with normals (KRZANOWSKI et al. 1979) and provides further evidence of β-adrenergic dysfunction. In contrast, a later study showed that the maximum specific binding level (B_{max}) for DHA in tissue from a healthy individual was well within the range of values obtained for lung tissue from six chronic bronchitis patients with mild obstructive disease (BARNES et al. 1980). Furthermore, an increase in this value has also been observed in patients with chronic obstructive lung disease compared with normal controls (RAAIJMAKERS et al. 1985), with no difference observed in the dissociation constant (K_d).

We have recently used lung samples obtained post-morten from healthy individuals and from subjects who suffered and/or died as a result of asthma to determine whether or not reduced responsiveness of bronchi from asthmatic lung to β-adrenoceptor agonists was reflected in a decrease in β-adrenoceptor number in the lung in general and in bronchial smooth muscle in particular (GOLDIE et al. 1987). In healthy human lung, specific I-CYP binding was saturable, involving a homogeneous class of non-interacting sites with high affinity. The K_d value for I-CYP binding was 47.8 p*M*. This value compared favourably with values obtained in binding studies using homogenates of human lung (10 p*M*; ENGEL 1980; ENGEL et al. 1981) and slide-mounted tissue sections of human lung (11 p*M*; CARSTAIRS et al. 1985). K_d values fro I-CYP binding, obtained using lung tissue from two severe asthmatics, were 48 and 115 p*M*. These estimates compare favourably with those determined in non-diseased lung and in pig lung (73 p*M*; GOLDIE et al. 1986a). B_{max} estimates in the two asthmatic lung samples were 81 and 149 fmol/mg protein, while B_{max} in non-diseased human lung was 56 ± 7 fmol/mg protein ($n = 8$). These data suggest that B_{max} was not reduced and may even be increased in severe asthma, although further lung samples need to be tested before a firm conclusion can be drawn.

The absence of reduced β-adrenoceptor density in the two asthmatic lungs studied is consistent with the notion that the β-adrenergic hypofunction in asthma is not necessarily associated with a loss in β-adrenoceptors. An increase in the density of autoradiographic grains representing specific I-CYP binding sites was observed over smooth muscle in bronchi from a severe asthmatic. However, in bronchial preparations from the same lung, isoprenaline and fenoterol were 10- and 13-fold less potent than in non-diseased lung. The relaxant potency of theophylline was unaltered. These data are also consistent with inflammation-induced uncoupling of the β-adrenoceptor/adenylate cyclase system in severe asthma (LULICH et al. 1988)

F. The Importance of β_2-Adrenoceptor Selectivity and Efficacy

I. β_2 Selectivity

In the mid 1960s, there was an increase in the death rate of asthmatics which was correlated with increased use of isoprenaline aerosols (LULICH et al. 1986). It is now considered that these deaths were probably caused by over-reliance on isoprenaline as a bronchodilator resulting in delays in the introduction of corticosteroid therapy in patients whose asthma was deteriorating because of worsening inflammation (SLY et al. 1985; LULICH et al. 1986). Initially, however, the increase in the number of deaths was attributed to isoprenaline-induced cardiac arrhythmias because this agonist does not discriminate between β_1-adrenoceptors in the heart and β_2-adrenoceptors in the airways (PATERSON et al. 1983). In an attempt to minimize cardiac stimulation, β_2-selective bronchodilators such as salbutamol, terbutaline and fenoterol were introduced (PATERSON et al. 1983, REED 1985).

Studies with β_2-selective agonists in isolated tissues from the guinea-pig, indicated that these drugs were highly selective for airway smooth muscle (BRITTAIN et al. 1976; PATERSON et al. 1983) In contrast, intravenous salbutamol in conscious humans was shown to be much less airway-selective than would be expected from the results of studies in guinea-pig isolated tissues (PATERSON et al. 1971, 1979). This may be partially explained by reflex activation of cardiac β_1-adrenoceptors following falls in peripheral vascular resistance (GIBSON and COLTART 1971). In addition, the presence of β_2-adrenoceptors in the human heart reduces the airway selectivity of such agents (ABLAD et al. 1974; HEITZ et al. 1983; ROBBERECHT et al. 1983; COREA et al. 1984).

The above data suggest that some cardiac stimulation should be expected even with very selective β_2-agonists. Furthermore, improving β_2 selectivity will not diminish the extent or frequency of adverse reactions to β_2-agonists such as skeletal muscle tremor, hyperglycaemia and hypokalaemia, since these effects are mediated via β_2-adrenoceptors (LULICH et al. 1986). However, in most cases, the severity of side effects from β_2-agonist therapy for asthma can be effectively minimized by using the inhalation route of administration (LULICH et al. 1986).

II. Efficacy

Efficacy is a measure of the capacity of a drug to initiate a response once it occupies the receptor (STEPHENSON 1956). Thus, a full agonist has a higher efficacy than a partial agonist. Furthermore, the higher the efficacy of a given full agonist, the fewer receptors that need to be occupied to produce a maximal response and thus the greater is the proportion of spare receptors.

In airway smooth muscle there is an inverse relationship between the level of tone induced by a spasmogen and the potency and maximal relaxant effect of β-agonists (VAN DEN BRINK 1973; BUCKNER and SAINI 1975; TORPHY et al. 1983). High concentrations of spasmogen produce functional antagonism by increasing the initial level of airway smooth muscle tone so that a greater fraction of the population of β-adrenoceptors needs to be activated to produce a relaxant response (BUCKNER and SANI 1975). Supramaximal concentrations of spasmogen signficantly reduce the maximal relaxation produced by β-agonists in human bronchial preparations (ADVENIER et al. 1988), guinea-pig isolated trachea, (BUCKNER and SAINI 1975), bovine trachea (VAN DEN BRINK 1973) and canine trachea (TORPHY et al. 1983).

In severe asthma, where there may be both marked bronchoconstriction and down-regulation of β-adrenoceptors, a β_2-selective agonist with high efficacy should be advantageous. Fenoterol has a greater bronchodilator efficacy than salbutamol (O'DONNELL and WANSTALL 1978), but it has the potential disadvantage of being a full agonist of cardiac β-adrenoceptors, while salbutamol is a weak partial agonist in the heart (BRITTAIN et al. 1976). HOCKLEY and JOHNSON (1983) reported that nebulised fenoterol (5 mg) was superior to nebulised salbutamol (5 mg) with respect to peak effect and duration of action. Nevertheless, there was little to choose between the two drugs clinically at these doses, since the additional level of bronchodilation produced by fenoterol was small and the incidence of side effects was similar. MADSEN et al. (1979) also reported that they could not discriminate between the clinical effectiveness of salbutamol and fenoterol. The clinical importance of high β_2-adrenoceptor efficacy in the treatment of asthma remains to be established.

G. Adverse Reactions to β-Adrenoceptor Agonists

At therapeutic doses, there are some clinically significant adverse effects of β_2-agonists which appear to be mediated via β-adrenoceptors. The most widely reported adverse effects are skeletal muscle tremor, cardiac effects, metabolic changes including hyperglycaemia and hypokalaemia and decrease in the partial pressure of arterial oxygen (PaO_2) (LARSSON and SVEDMYR 1977; PATERSON et al. 1979; SMITH and KENDALL 1984).

I. Skeletal Muscle Tremor

Skeletal muscle tremor is the most common adverse reaction to β_2-selective agonists (LARSSON 1977). The tremor is mediated via β_2-adrenoceptors in skeletal muscle and thus will be seen with all β_2-agonists (PATERSON et al. 1983). However, tolerance usually develops to the tremorogenic effects of β_2-agonists in patients receiving long-term treatment (PATERSON et al. 1979; SVEDMYR et al. 1976).

II. Cardiac Effects

Intravenous salbutamol may cause pronounced tachycardia in both asthmatics (PATERSON et al. 1971; SVEDMYR and THIRINGER 1971) and healthy volunteers (GUNDOGDU et al. 1979). Few patients complain of palpitations when β_2-agonists are administered by metered aerosol or in tablet form (FREEDMAN 1972; LARSSON 1977).

At present there is no evidence that recommended doses of metered aerosol exacerbate pre-existing cardiac arrhythmias. In such patients oral, nebulised or intravenous β_2-agonists do not generally produce serious disturbances in cardiac rhythm (LULICH et al. 1986). However, one should be aware that this may be a problem in patients with pre-existing serious cardiac arrhythmias. In these patients plasma PaO_2, plasma potassium and cardiac function should be carefully monitored.

III. Changes in Plasma Constituents

1. Potassium

β-Agonists can decrease plasma potassium by stimulation of β_2-adrenoceptors (COREA et al. 1981). The reduction in plasma potassium is produced by an increase in the transport of potassium ions into skeletal muscle (VICK et al. 1972; BUUR et al. 1982). Fortunately, marked tolerance develops to β-agonist-induced hypokalaemia with continued medication (BENGTSSON and FAGERSTROM 1982). In those individuals with healthy heart function, the extent of hypokalaemia produced by β_2-agonists should not have deleterious effects. However, plasma potassium changes in response to β-agonists should be monitored in patients with heart disease in view of the increased risk of cardiac arrhythmias (LULICH et al. 1986).

2. Glucose

β_2-Agonists increase plasma glucose via a β_2-adrenoceptor mediated increase in glycogenolysis. (SMITH and KENDALL 1984). Hyperglycaemic ketoacidosis may occur in some diabetic asthmatics who require β_2-agonist medication depending on the level of tolerance which has developed to the metabolic effects of these agents (LULICH et al. 1986).

3. Lipids

While β-agonist-induced lipolysis is mainly mediated via β_1-adrenoceptors (PATERSON et al. 1983), β_2-adrenoceptors appear to play a significant role in the degradation of stored triglycerides to fatty acids and glycerol (SMITH et al. 1984). The lipolytic effect of β_2-agonists may be a problem in diabetic asthmatics, but in most cases this effect would not be expected to be clinically significant (LULICH et al. 1986).

IV. PaO_2

Although β_2-agonists improve airway function (PATERSON et al. 1979), they may decrease PaO_2 by increasing blood flow through poorly ventilated areas of the lung, thereby increasing ventilation/perfusion mismatch (TAI and READ 1967). Hypoxia appears to be insignificant in stable asthmatics treated on a long term basis with β_2-agonists (LARSSON 1977). However, when the initial PaO_2 is below 60 mm Hg, if a decrease in PaO_2 is produced by a β_2-agonist, it may become clinically significant (PATERSON et al. 1971). Thus, PaO_2 should be carefully monitored when β_2-agonists are used for severe life-threatening asthma and supplemental oxygen should be administered routinely.

V. Neuropharmacological Effects

Salbutamol rapidly crosses the blood-brain barrier to reach a concentration in the brain approximately 5% of that in the plasma (CACCIA and FONG 1984). Salbutamol has been reported to cause appetite suppression, headache, nausea and sleep disturbances (PRATT 1982); central nervous system disturbances have also been reported with terbutaline and fenoterol (MILLER and RICE 1980). These symptoms tend not to be severe and abate with continued treatment (MILLER and RICE 1980). Cases of addiction to salbutamol from inhalers have been reported, but the incidence is extremely low (PRATT 1982; LULICH et al. 1986).

VI. Hypersensitivity

Sulphites present in β_2-agonist bronchodilator solutions can induce hypersensitivity responses, including bronchoconstriction, in some asthmatic subjects (TWAROQ and LEUNG 1982; KOEPKE et al. 1984). Hence, in individuals allergic to sulphite food preservatives, it is prudent to examine the composition of the bronchodilator preparation and to use one not containing sodium bisulphite or sodium metabisulphite. Furthermore, if a patient is suspected of exhibiting an adverse hypersensitivity reaction to a β-agonist preparation containing a bisulphite preservative, the bronchodilator should be replaced immediately with one not containing such substances.

H. Therapeutic Actions of β-Agonists

It is generally accepted that the major effect leading to clinical improvement and an increase in lung function tests is the production of bronchial smooth muscle relaxation. However, if a β-agonist is given prior to allergen challenge in humans, the early response is completely prevented even though there is obvious bronchial oedema in addition to bronchial smooth muscle constriction (METZGER et al. 1987). Because β-agonists are more effective when they are given prior to the induction of oedema in experimental models than in established oedema (PERSSON 1987), it is possible that in therapy the anti-oedema effect is of some importance in the prophylactic use of β-agonists. In vivo studies in humans have shown that the clearance of inhaled polystyrene particles radiolabeled with technetium-99 is reduced in subjects with asthma (PAVIA et al. 1985). This is thought to be due to defects in both ciliary and mucous function. Administration of β-agonists can improve the clearance of radiolabeled particles (PAVIA et al. 1987a,b), presumably by improving mucociliary function.

I. Adrenaline

Modern β-adrenoceptor agonists are related to the endogenous compound adrenaline. The structure of adrenaline is shown in Fig. 2. It is a

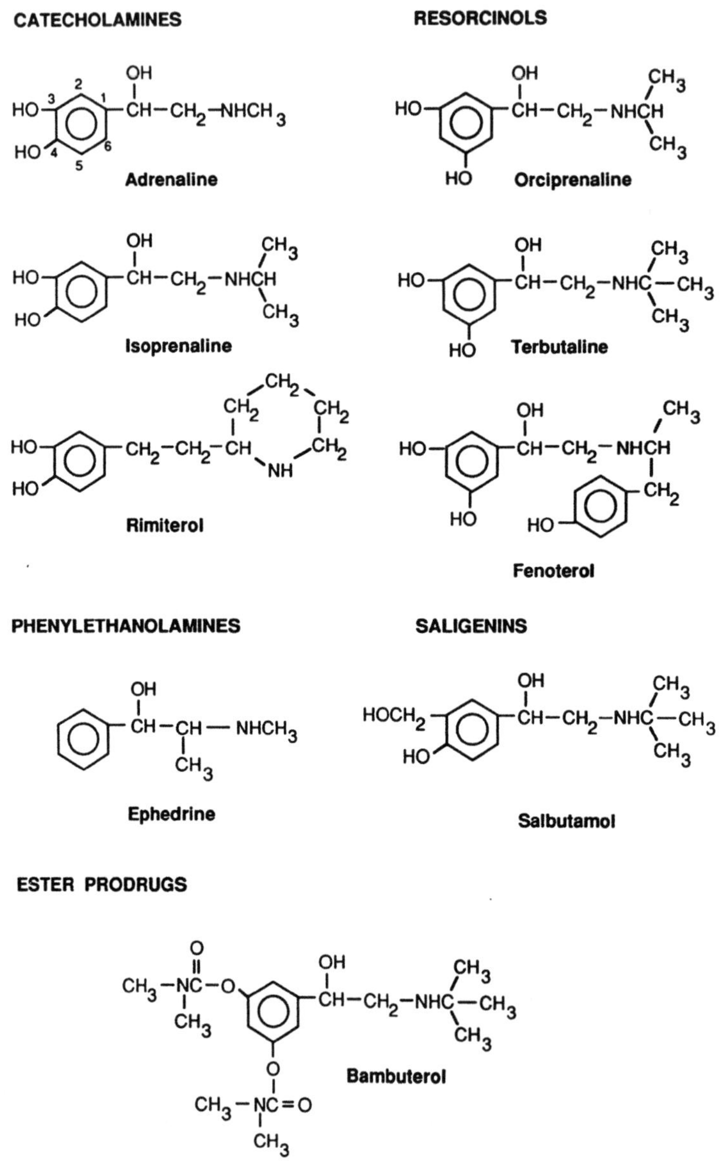

Fig. 2. Chemical structures of some β-adrenoceptor agonist bronchodilators

catecholamine that is, the benzene ring has hydroxyl groups at the 3,4 positions. Adrenaline stimulates both α- and β-adrenoceptors, although it is the powerful β-effects that are most important in asthma treatment. The major mechanism by which its pharmacologic effect is terminated is by tissue uptake. It is actively taken up into sympathetic nerve endings (neuronal uptake) (HERTTING et al. 1961; IVERSEN 1963) and into various other tissues including smooth muscle (extraneuronal uptake) (IVERSEN 1965). Following uptake into sympathetic nerve endings, the major route of metabolism is conversion of adrenaline by monoamine oxidase (MAO) to 3,4-hydroxymandelic acid (KOPIN and GORDON 1962).

Following extraneuronal uptake, the dominant enzyme metabolizing the drug is catechol-*O*-methyl transferase (COMT) (GULDBERG and MARSDEN 1975). This converts adrenaline to 3-*O*-methyl adrenaline. MAO and COMT are widely distributed in different organs including gut, liver, kidney and lung. The combined effect of these enzymes is to form the urinary excretion product 3-*O*-methyl-4-hydroxymandelic acid (vanillyl mandelic acid).

Adrenaline is ineffective when taken orally, as it is broken down by MAO in the gastrointestinal tract and liver. Thus, when used in the treatment of asthma, it is given by parenteral injection or by inhalation. The more selective effect is obtained by inhalation of either a wet aerosol or the commercially available multidose pressurised aerosol.

II. Ephedrine

This drug differs from adrenaline in the loss of the 3,4 hydroxyl groups. This prevents the action of COMT and also results in a loss of potency so that ephedrine has to be taken in higher doses. The loss of these groups also increases the effect of ephedrine on the central nervous system. An additional methyl group on the alpha carbon makes the drug resistant to MAO. Thus, in contrast to adrenaline, ephedrine is effective when taken orally. The major pharmacological effect is indirect, via release of noradrenaline from sympathetic nerve endings, although there is a small direct effect at higher doses. The major route of metabolism is by demethylation to norephedrine. This metabolite also has pharmacological properties similar to ephedrine and is marketed as phenylpropanolamine in a variety of decongestant preparations, many of which are avilable over the counter. As both ephedrine and phenylpropanolamine are indirectly acting amines, dangerous interactions may occur if they are taken with MAO inhibitors. Ephedrine and adrenaline are no longer widely used in present day asthma therapy.

III. Isoprenaline

Isoprenaline was the first β-adrenoceptor-selective agonist, since it apparently has no α-adrenoceptor stimulating effects in humans. The pharmacological effects of the circulating drug are terminated by extraneuronal uptake. Following oral administration, very large doses of the drug are necessary to produce pharmacological effects, as isoprenaline is significantly conjugated with sulphate in the gut wall and liver (GEORGE et al. 1974; DAVIES 1975). When this drug is given intravenously, a significant percentage is excreted as free isoprenaline, conjugated isoprenaline or free 3-*O*-methyl isoprenaline.

The drug has also been administered intrabronchially to both animals and humans (DAVIES 1975). A greater fraction is 3-*O*-methylated than when the drug is given intravenously. This suggests that 3-*O*-methylation may occur in the lung. The presence of COMT in the lung has been demonstrated and comparison between intravascular administration of isoprenaline and intrabronchial administration in isolated perfused lung has shown that a greater amount of 3-*O*-methyl isoprenaline is formed following intrabronchial administration (BLACKWELL et al. 1974). Thus, considerable O-methylation can occur during absorption from the airways. Kinetic studies and studies using radiolabeled tracers suggest that approximately 10% of the dose from a pressurised aerosol reaches the lung (DAVIES 1975). The majority of the dose is swallowed and in the case of isoprenaline will be inactivated, making the inhalation route using the pressurised aerosol airway-selective, while the intravenous route is not.

IV. Selective β_2-Agonists

The newer β-agonists used in the treatment of asthma are distinct from isoprenaline in that they show selectivity for the β_2-adrenoceptor, have pharmacological effects after oral administration and have a longer duration of effect. The first compound with these properties to be introduced was orciprenaline. The minor change in molecular structure from 3,4 to 3,5 dihydroxy, a resorcinol as opposed to a catecholamine, resulted in major pharmacokinetic changes. The resorcinol structure was no longer a substrate for either extraneuronal uptake or COMT. The drug is metabolized mainly by sulpho-conjugation, but a sufficient percentage of free drug is absorbed to make the drug an effective oral bronchodilator. Orciprenaline established a new class of orally active bronchodilators (CHU 1984). However, intravenous studies in humans suggested that selectivity for the β_2-adrenoceptor was only slightly improved over isoprenaline (McEVOY et al. 1973).

In 1966, two further sympathomimetics were developed which had much greater selectivity for the β_2-adrenoceptor. These were salbutamol (BRITTAIN et al. 1968) and terbutaline (BERGMAN et al. 1969) Two different approaches to overcome the effect of COMT were used. Terbutaline was based on the resorcinol moiety, whereas salbutamol had a methylene group inserted at the 3-hydroxyl position forming a saligen derivative. Both salbutamol and terbutaline, however, remain substrates for conjugating enzymes, and in humans mainly sulphate conjugates are formed (EVANS et al. 1973; NILSSON et al. 1972). It has been shown for both drugs that they are rapidly absorbed from airways and that no metabolism occurs in the lung (DAVIES 1975; SHENFIELD et al. 1976). The two drugs have virtually equal β_2 selectivity. It would appear that the relatively greater β_2 activity was due to substitution of the tertiary butyl group in the amine head. Both compounds are effective orally, by inhalation and by intravenous injection. Following the introduction of salbutamol and terbutaline, a variety of new β_2-agonists were developed.

Rimiterol retains a basic catecholamine nucleus and thus is not active orally due to sulpho-conjugation and 3-*O*-methylation as with isoprenaline. When introduced via the lung, it is also 3-*O*-methylated to a considerable extent (SHENFIELD et al. 1976). Thus the ratio of the dose of inhaled to intravenous rimiterol producing bronchodilation is higher than for salbutamol, a drug absorbed unchanged from the lung (TARALA et al. 1981). This drug is effective when given by inhalation or injection, but is relatively short acting compared with salbutamol and terbutaline (PATERSON et al. 1979). Fenoterol is based on the resorcinol nucleus and thus, like other compounds in this group, is metabolized by 5-sulpho-conjugation, sufficient free drug being absorbed that the drug is effective when given by the oral route. It can be given also by injection or inhalation and its β_2 selectivity depends on the large substituent group on the amine head.

All of the newer, orally active β_2-agonists are used in various inhalation formulations. These aim to give a small but highly effective dose to the lung. The small dose delivered to the lung by the pressurised multidose aerosol gives the desired therapeutic effect with a reasonably fast onset of action (5–10 min) and a duration of action of up to 6 h (CHU 1984; TATTERFIELD 1986). Following inhalation of such small doses, plasma concentrations of the active drug will be negligible.

V. Prodrugs

The most recent development in the β-agonist field has been the development of prodrugs (SVENSSON 1985). These drugs are obtained by the synthesis of esters of the corresponding β_2-agonist. Esters are an appropriate prodrug derivative since esterases are widely distributed in the human body and will therefore catalyse the hydrolysis of the ester back to the active parent drug. The most recent and promising of these drugs is bambuterol, the *bis-N, N*-dimethylcarbamate of terbutaline. Administration of this drug results in prolonged plasma levels of terbutaline. The drug displays high selectivity for pseudo-cholinesterase without affecting acetylcholinesterase. A prolonged duration of action has been reported in asthmatic patients at much lower terbutaline levels than would be required if terbutaline itself was given. Bambuterol also displays increased affinity for lung tissue (SVENSSON 1985). It is possible therefore that the drug is concentrated in the lung and then slowly released at the site of action. This would be an alternative method by which selectivity could be increased. Further clinical trials with this promising agent are in progress.

VI. Clinical Application

Although certain β-agonists have greater efficacy when studied in the laboratory, there is no convincing evidence in vivo to suggest differences between β-agonists in the magnitude of their maximal clinical bronchodilator effect (TATTERSFIELD 1986). Nonetheless, β_2 selectivity will be important in deciding which drug to use in humans. There are problems in translating in vitro studies to humans since a highly selective β_2-agonist may cause a reflex increase in heart rate as a consequence of a β_2-mediated fall in peripheral vascular resistance. In addition, it has been shown that there are β_2-adrenoceptors in the heart (HEITZ et al. 1983; ROBBERECHT et al. 1983). Studies in humans using intravenous drugs have suggested a seven- to tenfold selectivity for salbutamol compared to isoprenaline (PATERSON et al. 1971). Salbutamol and terbutaline appear to have similar selectivities (TATTERSFIELD 1986). Improved therapeutic effect has been achieved by the use of various spacer devices allowing evaporation of the fluorocarbons and reducing the need for exact timing of the release of a dose of pressurised aerosol (ELLUL-MICALLEF et al. 1980; GOMN et al. 1980; O'REILLY et al. 1986).

In addition to pressurised aerosol with various spacer devices, the drugs can be given by wet nebuliser. There are considerable discrepancies in the dose used with these two types of inhalation. Thus two puffs of a standard salbutamol aerosol contains 0.2 mg, but it is not unusual to see 5–10 mg administered from a wet nebuliser (SHENFIELD et al. 1973). The distribution of the drug leaving the nebuliser is of course different from that leaving the metered dose inhaler. Less of the nebuliser dose is swallowed although a great deal is left in the inhalation apparatus. A number of comparisons have been made to see whether there is a greater effect from the wet nebuliser than from the pressurised aerosol (COHEN and HALE 1965; CHOO-KANG and GRANT 1975; CAYTON et al. 1978; TARALA et al. 1980). If larger doses of pressurised aerosol are given there is little advantage in giving the drug by wet aerosol. It must be remembered that these studies are done in selected patients who, although their asthma is fairly severe, are trained and observed by operators.

Wet nebulisers are still of advantage in the domestic and early inpatient treatment of severe asthma. In very severe asthma there is a theoretical argument for parenteral injection of the β-agonists. Several studies have compared wet nebulised and intravenous salbutamol in the treatment of acute asthma. The majority have found the nebulised route to be at least as effective as the intravenous route (LAWFORD et al. 1978; BLOOMFIELD et al. 1979; WILLIAMS et al. 1981). In one study, patients responded poorly to nebulised salbutamol and subsequently responded well to intravenous salbutamol (WILLIAMS and SEATON 1977). On the basis of these studies it would seem sensible to start the patients on a nebulised β-agonist and if they are not improving or are unable to use the nebuliser, the intravenous route should be considered immediately. It is of interest that, recently in a group of severe brittle asthmatics, continuous subcutaneous or intermittent subcutaneous β_2-agonist administration caused considerable improvement in the control of these very sick subjects (O'DRISCOLL et al. 1988).

It has been postulated that the continuous use of β_2-agonist drugs could cause worsening of asthma by inducing "resistance" to their action and thus progressive worsening of the clinical condition. A number of studies in both animals and human subjects suggested that this was possible. SVEDMYR (1984) reviewed the available evidence from studies in asthmatic subjects and concluded that long-term administration of β_2-agonists seemed to cause no reduction in the airway response to these drugs. It seems therefore that prolonged treatment with a β_2-agonist is unlikely to cause adverse effects due to the development of tolerance or resistance.

β_2-Agonist drugs may be used acutely in the management of individual asthma attacks and their place in the management of the acute attack is still undisputed. Until recently regular β_2-agonist aerosol therapy alone was accepted as reasonable treatment for recurrent mild asthma (PATERSON et al. 1979; PATERSON and TARALA 1985). Now, however, persistent symptoms and/or lung function tests below predicted normal are thought to indicate continuing airway inflammation. β_2-Agonists alone are considered to be insufficient in this situation and should be combined with drugs that have anti-allergic or anti-inflammatory effects (PATERSON et al. 1988; WOOLCOCK 1986).

If it is accepted that reactivity is a measure of the degree of inflammation in the airways, then the recent review by Tattersfield (TATTERSFIELD 1987) demonstrated that in a large number of studies, whereas β_2-agonists will effectively decrease reactivity to histamine and other agents following acute administration, after prolonged treatment with the drugs, there is little change in reactivity from pretreatment levels. Thus, concern has been expressed that the β-agonists are disguising continuing inflammatory changes in the lung.

The most recent study from Tattersfield's group (VATHENIN et al. 1988) has suggested that prolonged treatment with an inhaled β_2-agonist may lead to a rebound increase in bronchial reactivity following cessation of treatment. The authors concluded that with patients on high doses of a β-agonist, sudden cessation may cause at least temporary sensitivity to provocative stimuli. Further studies are awaited, but there is no doubt that clinical practice is changing to that suggested in Table 1. Thus, in acute attacks which are mild to moderate, the primary emphasis in treatment is the use of inhaled β_2-agonists. Similarly, in the prophylaxis of acute attacks due to exercise or exposure to allergen, inhalation of a β_2-agonist would be the first treatment. In severe, life-threatening asthma with substantial additional pathology, although there is a greater emphasis on the need for corticosteroids and other modalities of therapy, β_2-agonists still have an important role. In the prophylactic treatment of mild to severe asthma, there is now an increase in emphasis towards combining inhaled β_2-agonists with drugs which have demonstrated anti-inflammatory properties, such as disodium cromoglycate, steroids and theophylline. The inhaled β_2-agonists have an important but supplementary role in this situation.

Table 1. The use of β-agonists for the treatment of asthma

Prophylactic Therapy	
Mild to severe	A primary emphasis on drugs with anti-inflammatory properties (corticosteroids, disodium cromoglycate, theophylline etc.); inhaled β_2-agonists should have an important supplementary role in further improving lung function
Acute Therapy	
Mild to moderate attacks	A primary emphasis on inhaled β_2-agonists
Severe, life-threatening attacks	Essential role for systemic corticosteroids, but inhaled and/or parenteral β_2-agonists still have an important role

I. Concluding Remarks

Despite the widespread distribution of β_2-adrenoceptors on non-airway smooth muscle in the lung, β_2-agonists reverse airway obstruction primarily by relaxing airway smooth muscle. The effects of β_2-agonists on mucus production and transport, tracheobronchial microvessels and cholinergic nerves may be therapeutically useful, but the clinical importance of these effects still remains to be established. β_2-Agonists appear to be ineffective in suppressing or controlling airway inflammation in asthmatics. They do not suppress late asthmatic reactions nor do they decrease bronchial hyperreactivity when taken on a long term basis. This brings into question the wisdom of exclusively relying on β_2-agonist bronchodilators for the prophylactic treatment of asthma, despite their potent bronchodilator action and relative lack of dangerous side effects. Indeed, their powerful bronchodilator action may mask the onset and/or exacerbation of airway inflammation in asthmatics, resulting in the possible underutilization of effective anti-inflammatory agents. Thus there should be a primary emphasis on the use of drugs with anti-inflammatory properties, such as corticosteroids, disodium cromoglycate and perhpas theophylline, for the prophylactic treatment of asthma. Nevertheless, it is important to stress that, if adequate treatment with anti-inflammatory prophylactics is maintained, it should be beneficial to also use long-acting inhaled β_2-agonists to produce supplementary bronchodilation. Inhaled β_2-agonists are and should be the drugs of choice for reversing acute attacks of asthma, since they not only rapidly produce powerful bronchodilation throughout the bronchial tree, but also are very safe to use. However, in severe asthmatic attacks, where obstruction may be mainly due to mucus plugging and bronchial wall oedema and where there may also be significant β-adrenoceptor dysfunction, inhaled β_2-agonists may be ineffective and a greater emphasis should be directed towards using alternative drugs including corticosteroids. Our recommendations concerning the use of β_2-agonists for the treatment of asthma are summarized in Table 1.

References

Ablad B, Carlsson B, Carlsson E et al. (1974) Cardiac effects of β-adrenergic receptor antagonists. Adv Cardiol 12:290–302

Advenier C, Naline E, Matran R et al. (1988) Interaction between fenoterol, ipratropium, and acetylcholine on human isolated bronchus. J Allergy Clin Immunol 82:40–46

Allegra L, Bossi R, Braga PC (1985) Influence of surfactant on mucociliary transport. Prog Respir Res 19:456–458

Arnoux B, Simoes-Caeioro MH, Landes A et al. (1982) Alveolar macrophages from asthmatics release PAF-acether and lyso-PAF-acether when stimulated with specific antigen. Am Rev Respir Dis 125:A70

Barnes PH, Basbaum CB, Nadel JA et al. (1982) Localization of beta-adrenoceptors in mammalian lung by light microscopic autoradiography. Nature 299:444–447

Barnes PJ (1984) Beta-adrenoceptors in lung tissue. In : Morley J (ed) Perspectives in asthma 2: beta-adrenoceptors in asthma. Academic, London, p 67

Barnes PJ (1986) Neural control of human airways in health and disease. Am Rev Respir Dis 134:1289–1314

Barnes PJ (1987) Asthma management – a new dimension. J Int Med Res 15:397–400

Barnes PJ, Basbaum CB (1983) Mapping of adrenergic receptors in the trachea by autoradiography. Exp Lung Res 5:183–192

Barnes PJ, Pride NB (1983) Dose-response curves to inhaled β-adrenoceptor agonists in normal and asthmatic subjects. Br J Clin Pharmacol 15:677–682

Barnes PJ, Karliner JS, Dollery CT (1980) Human lung adrenoceptors studied by radioligand binding. Clin Sci 58:457–461

Barnes PJ, Basbaum CB, Nadel JA (1983) Autoradiographic localization of autonomic receptors in airway smooth muscle: marked differences between large and small airways. Am Rev Respir Dis 127:758–762

Barnes PJ, Jacobs M, Roberts JM (1984) Glucocorticoids preferentially increase fetal alveolar beta-adrenoceptors: autoradiographic evidence. Pediatr Res 18: 1191–1194

Barnes PJ, Grandordy BM, Page CP et al. (1987) The effect of platelet activating factor on pulmonary β-adrenoceptors. Br J Pharmacol 90:709–715

Barnes PJ, Chung KF, Page CP (1988) Platelet-activating factor as a mediator of allergic disease. J Allergy Clin Immunol 81:919–934

Barnett DB, Rugg EL, Nahorski SR (1978) Direct evidence of two types of β-adrenoceptor binding site in lung tissue. Nature 273:166–168

Basbaum C, Carlson D, Davison E et al. (1988) Cellular mechanisms of airway secretion. Am Rev Respir Dis 137:479–485

Basbaum CB, Ueki I, Brezina L et al. (1981) Tracheal submucosal gland serous cells stimulated in vitro with adrenergic and cholinergic agonists. Cell Tissue Res 220:481–498

Bengtsson B, Fagerstrom P-O (1982) Extrapulmonary effect of terbutaline during prolonged administration. Clin Pharmacol Ther 31:726–732

Benovic JL, Stiles GL, Lefkowitz RJ et al. (1983) Photoaffinity labelling metal-dependent proteolysis explaining apparent heterogeneity. Biochem Biophys Res Commun 110:504–511

Benovic JL, Pikes LJ, Cerione RA et al. (1985) Phosphorylation of the mammalian β-adrenergic receptor by cyclic AMP-dependent protein kinase. J Biol Chem 260:7094–7101

Benovic JL, Strasser RH, Caron MG et al. (1986) β-Adrenergic receptor kinase: identification of a novel protein kinase that phosphorylates the agonist-occupied form of the receptor. Proc Natl Acad Sci USA 83:2797–2801

Benovic JL, Staniszewski C, Mayor F et al. (1988) β-Adrenergic receptor kinase. J Biol Chem 263:3893–3897

Bergman J, Persson H, Wetterlin K (1969) Two new groups of selective stimulants of adrenergic β-receptors. Experientia 25:899–901

Bertram JF, Goldie RG, Papadimitriou JM et al. (1983) Correlations between pharmacological responses and structure of human lung parenchyma strips. Br J Pharmacol 80:107–114
Blackwell EW, Briant RH, Conolly ME et al. (1974) Metabolism of isoprenaline after aerosol and direct intrabronchial administration in man and dog. Br J Pharmacol 50:587–591
Bloomfield P, Carmichael J, Petrie GR et al. (1979) Comparison of salbutamol given intravenously and by intermittent positive pressure breathing in life threatening asthma. Br Med J 1:848–850
Boat TF, Kleinerman JI (1975) Human respiratory tract secretions 2: effect of cholinergic and adrenergic agents on in vitro release of protein and mucous glycoprotein. Chest 67:32s–34s
Booij-Noord H, de Vries K, Sluiter HJ et al. (1972) Late bronchial obstructive reaction to experimental inhalation of house dust extract. Clin Allergy 2: 43–61
Bourne HR, Lichtenstein LM, Melmon KL et al. (1974) Modulation of inflammation and immunity by cyclic AMP. Science 184:19–28
Boushey HA, Holtzman MJ, Sheller JR et al. (1980) Bronchial hyperreactivity. Am Rev Respir Dis 121:389–413
Brittain RT, Farmer JB, Jack DJ et al. (1968) α-[(*t*-Butylamino)methyl]-4-hydroxy-*m*-xylene-α^1,α^3-diol (AH.3365) a selective β-adrenergic stimulant. Nature 219: 862–863
Brittain RT, Dean CM, Jack DJ (1976) Sympathomimetic bronchodilator drugs. Pharmacol Ther 2:423–462
Brodde O-E, Kuhloff F, Arroyo J et al. (1983) No evidence for temperature-dependent changes in pharmacological specificity of β_1- and β_2-adrenoceptors in rabbit lung membranes. Naunyn Schmiedebergs Arch Pharmacol 322:20–28
Brooks SM, McGowan K, Bernstein IL et al. (1979) Relationship between numbers of beta-adrenergic receptors in lymphocytes and disease severity in asthma. J Allergy Clin Immunol 63:401–406
Brown CM, Duncan GP, McKenniff MG et al. (1985) Characterization of β-adrenoceptors on cultured human embryonic lung cells. Br J Pharmacol 86:766P
Buckner CK, Saini RK (1975) On the use of functional antagonism to estimate dissociation constants for beta-adrenergic receptor agonists in isolated guinea-pig trachea. J Pharmacol Exp Ther 194:565–574
Buss WW, Sosman JM (1984) Isoproterenol inhibition of isolated human neutrophil function. J Allergy Clin Immunol 73:404–410
Busse WW, Bush RK, Cooper W (1979) Granulocyte response in vitro to isoproterenol, histamine, and prostaglandin E, during treatment with β-adrenergic aerosols in asthma. Am Rev Respir Dis 120:377–384
Butchers PR, Fullarton JR, Skidmore IF et al. (1979) A comparison of the anti-anaphylactic activities of salbutamol and disodium cromoglycate in the rat, the rat mast cell and in human lung tissue. Br J Pharmacol 67:23–32
Butchers PR, Skidmore IF, Vardey CJ et al. (1980) Characterization of the receptor mediating the anti-anaphylactic effects of β-adrenoceptor agonists in human lung tissue in vitro. Br J Pharmacol 71: 663–667
Buur T, Clausen T, Holmberg E et al. (1982) Desensitization by terbutaline of beta-adrenoceptors in the guinea-pig soleus muscle: biochemical alterations associated with functional changes. Br J Pharmacol 76:313–317
Caccia S, Fong MH (1984) Kinetics and distribution of the beta-adrenergic agonist salbutamol in rat brain. J Pharm Pharmacol 36:200–202
Caron MG, Lefkowitz RJ (1976) Solubilization and characterization of the β-adrenergic receptor binding sites of frog erythrocytes. J Biol Chem 251: 2374–2384
Caron MG, Cerione RA, Benovic JL et al. (1985) Biochemical characterization of the adrenergic receptors: affinity labelling, purification, and reconstitution studies.

In: Cooper DMF, Seamon KB (eds) Advances in cyclic nucleotide and protein phosphorylation research, vol 19. Raven, New York, p 1
Caron MG, Kobilka BK, Frielle T et al. (1988) Cloning of the cDNA and genes for the hamster and human β_2-adrenergic receptors. J Recept Res 8:7–21
Carstairs JR, Nimmo AJ, Barnes PJ (1984) Autoradiographic localization of beta-adrenoceptors in human lung. Eur J Pharmacol 103:189–190
Carstairs JR, Nimmo AJ, Barnes PJ (1985) Autoradiographic visualization of beta-adrenoceptor subtypes in human lung. Am Rev Respir Dis 132:541–457
Carswell H, Nahorski SR (1983) β-Adrenoceptors in guinea-pig airways: comparison of functional and receptor labelling studies. Br J Pharmacol 79:965–971
Cauvin C, Meisheri K, Mueller E et al. (1984) β-Adrenergic relaxation of smooth muscle and second messengers. In: Morley J (ed) Perspectives in asthma 2: beta-adrenoceptors in asthma. Academic, London, p 25
Cayton RM, Webber B, Paterson JW et al. (1978) A comparison of salbutamol given by pressure-pack aerosol or nebulization via IPPB in acute asthma. Br J Dis Chest 72:222–224
Cerrina J, Ladurie ML, Labat C et al. (1986) Comparison of human bronchial muscle responses to histamine in vivo with histamine and isoproterenol agonists in vitro. Am Rev Respir Dis 134:57–61
Choo-Kang YFJ, Grant IWB (1975) Comparison of two methods of administering bronchodilator aerosol to asthmatic patients. Br Med J 2:119–120
Chu SS (1984) Bronchodilators, part 1: adrenergic drugs. Drugs Today 20:439–464
Chung KF (1986) Role of inflammation in the hyperreactivity of the airways in asthma. Thorax 41:657–662
Church MK, Hiroi J (1987) Inhibition of IgE-dependent histamine release from human dispersed lung mast cells by anti-allergic drugs and salbutamol. Br J Pharmacol 90:421–429
Church MK, Young KD (1983) The characteristics of inhibition of histamine release from human lung fragments by sodium cromoglycate, salbutamol and chlorpromazine. Br J Pharmacol 78:671–679
Cockcroft DW, Murdoch KY (1987) Comparative effects of inhaled salbutamol, sodium cromoglycate and beclomethasone dipropionate on allergen-induced early asthmatic responses, late asthmatic responses, and increased bronchial responsiveness to histamine. J Allergy Clin Immunol 79:734–740
Cohen AA, Hale SC (1965) Comparative effects of isoproterenol aerosol on airway resistance in obstructive pulmonary diseases. Am J Med Sci 2449:309–315
Conner M, Reid LM (1984) Mapping of beta-adrenergic receptors in rat lung. Effect of isoproterenol. Exp Lung Res 6:91–101
Conolly ME, Greenacre JK (1976) The lymphocyte β-adrenoceptor in normal subjects and patients with bronchial asthma: the effect of different forms of treatment on receptor function. J Clin Invest 58:1307–1316
Cook N, Nahorski SR, Barnett DB (1987) Human platelet β_2-adrenoceptors: agonist-induced internalisation and down-regulation in intact cells. Br J Pharmacol 92: 587–596
Corea L, Bentivoglio M, Verdecchia P (1981) Hypokalaemia due to salbutamol overdosage. Br Med J 283:500
Corea L, Bentivoglio M, Verdecchia P et al. (1984) Noninvasive assessment of chronotropic and inotropic response to preferential beta-1 and beta-2 adrenoceptor stimulation. Clin Pharmacol Ther 35:776–781
Cuss FM, Dixon CMS, Barnes PJ (1986) Effects of inhaled platelet activating factor on pulmonary function and bronchial responsivenss in man. Lancet 2:189–192
Davies DS (1975) Pharmacokinetics of inhaled substances. Postgrad Med J 51 [Suppl 7]:69–75
Davis B, Marin MG, Yee JW et al. (1979) Effect of terbutaline on Cl^- and Na^+ fluxes across dog trachea in vitro. Am Rev Respir Dis 120:547–552

Davis PB, Simpson DM, Paget GL et al. (1986) β-adrenergic responses in drug-free subjects with asthma. J Allergy Clin Immunol 77:871–879
De Monchy JGR, Kauffman HF, Venge P et al. (1985) Bronchoalveolar eosinophilia during allergen-induced late asthmatic reactions. Am Rev Respir Dis 131: 373–376
De Troyer A, Yernault YC, Rodenstein D (1978) Influence of $beta_2$ agonist aerosols on pressure-volume characteristics of the lung. Am Rev Respir Dis 118:987–995
Dickinson K, Richardson A, Nahorski SR (1981) Homogeneity of $beta_2$-adrenoceptors on rat erythrocytes and reticulocytes. A comparison with heterogeneous rat lung beta-adrenoceptors. Mol Pharmacol 19:194–204
Dixon RAF, Kobilka BK, Strader DJ et al. (1986) Cloning of the gene and DNA for mammalian β-adrenergic receptor and homology with rhodopsin. Nature 321: 75–79
Dobbs LG, Mason RJ (1979) Pulmonary alveolar type II cells isolated from rats: Release of phosphatidylcholine in response to β-adrenergic stimulation. J Clin Invest 63:378–387
Ekelund L, Burgoyne R, Enhorning G (1983) Pulmonary surfactant release in fetal rabbits: Immediate and delayed response to terbutaline. Am J Obstet Gynecol 147:437–442
Ellul-Micallef R, Moren F, Wetterlin K et al. (1980) Use of a special inhaler attachment in asthmatic children. Thorax 35:620–623
Engel G (1980) Identification of different sub-groups of β-adrenoceptors by means of binding studies in guinea-pig and human lung. Triangle 19:69–76
Engel G, Hoyer D, Berthold R et al. (1981) (±)[^{125}Iodo]-cyanopindolol, a new ligand for β-adrenoceptors. Identification and quantitation of subclasses of β-adrenoceptors in guinea-pig. Naunyn Schmiedebergs Arch Pharmacol 317:277–285
Erjefalt IAL, Wagner ZG, Strand SE et al. (1985) A method for studies of tracheobronchial microvascular permeability to macromolecules. J Pharmacol Methods 14:275–283
Evans ME, Walker SR, Brittain RT et al. (1973) The metabolism of salbutamol in man. Xenobiotica 3:113–120
Fairshter RD, Wilson AF (1980) Relationship between the site of airflow limitation and localization of the bronchodilator response in asthma. Am Rev Respir Dis 122:27–32
Finkel MS, Quirion R, Pert C et al. (1984) Characterization and autoradiographic distribution of the β-adrenergic receptor in rat lung. Pharmacology 29:247–254
Freedman BJ (1972) Trial of a terbutaline aerosol in the treatment of asthma and a comparison of its effects with those of a salbutamol aerosol. Br J Dis Chest 66:222–229
Frigas E, Gleich GJ (1986) The eosinophil and the pathophysiology of asthma. J Allergy Clin Immunol 77:527–537
Fuller RW, O'Malley G, Baker AJ et al. (1988) Human alveolar macrophage activation: Inhibition by forskolin but not β-adrenoceptor stimulation or phosphodiesterase inhibition. Pulm Pharmacol 1:101–106
Galant SP, Underwood S, Duriseti L et al. (1978a) Characterization of high affinity $beta_2$-adrenergic receptor binding of (−)-[3H] dihydroalprenolol to human polymorphonuclear cell particulates. J Lab Clin Med 92:613–618
Galant SP, Dureseti L, Underwood S et al. (1978b) Decreased beta-adrenergic receptors on polymorphonuclear leukocytes after adrenergic therapy. N Engl J Med 299:933–936
Galant SP, Duriseti L, Underwood S et al. (1980) Beta-adrenergic receptors of polymorphonuclear particulates in bronchial asthma. J Clin Invest 65:577–585
George CF, Blackwell EW, Davies DS (1974) Metabolism of isoprenaline in the intestine. J Pharm Pharmacol 26:265–267
Gibson D, Coltart DJ (1971) Haemodynamic effects of intravenous salbutamol in patients with mitral valve disease: comparison with isoprenaline and atropine. Postgrad Med J 47 [Suppl]:40–44

Gillespie E, Valentine DD, Lichtenstein LN (1974) Cyclic AMP metabolism in asthma: studies with leukocytes and lymphocytes. J Allergy Clin Immunol 53:27–33
Gilman AG (1987) G proteins: transducers of receptor-generated signals. Annu Rev Biochem 56:615–649
Goldie RG, Paterson JW, Wale JL (1982a) Pharmacological responses of human and porcine lung parenchyma, bronchus and pulmonary artery. Br J Pharmacol 76:515–521
Goldie RG, Paterson JW, Wale JL (1982b) A comparative study of β-adrenoceptors in human and porcine lung parenchyma strip. Br J Pharmacol 76:523–526
Goldie RG, Paterson JW, Spina D et al. (1984) Classification of beta-adrenoceptors in human isolated bronchus. Br J Pharmacol 81:611–615
Goldie RG, Papadimitriou JM, Paterson JW et al. (1986a) Autoradiographic localization of β-adrenoceptors in pig lung using [^{125}I]-iodocyanopindolol. Br J Pharmacol 88:621–628
Goldie RG, Papadimitriou JM, Paterson JW et al. (1986b) Influence of the epithelium on responsiveness of guinea-pg isolated trachea to contractile and relaxant agonists. Br J Pharmacol 87:5–14
Goldie RG, Spina D, Henry P et al. (1986c) In vitro responsiveness of human asthmatic bronchus to carbachol, histamine, beta-adrenoceptor agonists and theophylline. Br J Clin Pharmacol 22:669–676
Goldie RG, Spina D, Lulich KM et al. (1987) The status of β-adrenoceptor function in asthma. Pharmacology: Excerpta Med Int Congr Series 750:465–468
Gomn SA, Keaney NP, Winsey NJP et al. (1980) Effect of an extension tube on the bronchodilator efficacy of terbutaline delivered from a metered dose inhaler. Thorax 35:552–556
Grieco MH, Pierson RN, Pi-Sunyer FX (1968) Comparison of the circulatory effects of isoproterenol, epinephrine and methoxamine in normal and asthmatic subjects. Am J Med 44:863–872
Guldberg HC, Marsden CA (1975) Catechol-*O*-methyltransferase. Pharmacological aspects and physiological role. Pharmacol Rev 27:135–206
Gundogdu AS, Brown PM, Juul S et al. (1979) Comparison of hormonal and metabolic effects of salbutamol infusion in normal subjects and insulin-requiring diabetics. Lancet 2:1317–1321
Harden TK (1983) Agonist-induced desensitization of beta-adrenergic receptor-linked adenylate cyclase. Pharmacol Rev 35:5–32
Heitz A, Schwartz J, Velly J (1983) β-Adrenoceptors of the human myocardium: determination of β_1 and β_2 subtypes by radioligand binding. Br J Pharmacol 80:711–717
Henry, PJ, Rigby, PJ, Goldie, RG (1990) Distribution of β_1- and β_2-adrenoceptors in mouse trachea and lung: a quantitative autoradiographic study. Br J Pharmacol 99:136–144
Hertting G, Axelrod J, Kopin IJ et al. (1961) Lack of uptake of catecholamines after chronic denervation of sympathetic nerves. Nature 189:66
Hockley B, Johnson N (1983) Fenoterol versus salbutamol nebulisation in asthma. Postgrad Med J 59:504–505
Hollingsworth M, Gilfillan AM (1984) The pharmacology of lung surfactant secretion. Pharmacol Rev 36:69–90
Howarth PH, Durham SR, Lee TH et al. (1985) Influence of albuterol, cromolyn sodium and ipratropium bromide on the airway and circulating mediator responses to allergen bronchial provocation in asthma. Am Rev Respir Dis 132:986–992
Hughes JM, Seale, JP, Temple DM (1983) Effect of fenoterol on immunological release of leukotrienes and histamine from human lung in vitro: selective antagonism by β-adrenoceptor antagonists. Eur J Pharmacol 95:239–245
Ito Y, Itoh T (1984) Effects of isoprenaline on the contraction-relaxation cycle in the cat trachea. Br J Pharmacol 83:677–686

Ito Y, Tajima K (1982) Dual effects of catecholamines on pre- and post-junctional membranes in the dog trachea. Br J Pharmacol 75:433–440

Iversen LL (1963) The uptake of noradrenaline by isolated perfused rat heart. Br J Pharmacol 21:523-537

Iversen LL (1965) The uptake of catecholamines at high perfusion concentrations in the rat isolated heart: a novel catecholamine uptake process. Br J Pharmacol 25:18–33

Kaliner M, Marom Z, Patow C et al. (1984) Human respiratory mucus. J Allergy Clin Immunol 73:318–323

Kariman K (1980) β-Adrenergic receptor binding in lymphocytes from patients with asthma. Lung 158:41–51

Kay AB (1986) The cells causing airway inflammation. Eur J Respir Dis 69 [Supp 147]:38–43

Kay AB, Walsh GM, Moqbel R et al. (1987) Disodium cromoglycate inhibits activation of human inflammatory cells in vitro. J Allergy Clin Immunol 80:1–8

Kerrebijn KF, van Essen-Zanduliet EEM, Neijens HJ (1987) Effect of long-term treatment with inhaled corticosteroids and beta-agonists on the bronchial responsiveness in children with asthma. J Allergy Clin Immunol 79:653–659

Koch-Weser J (1968) Beta-adrenergic blockade and circulating eosinophils. Arch Intern Med 121:255–258

Koepke JW, Christopher KL, Chai H et al. (1984) Dose-dependent bronchospasm from sulfites in isoetharine. J Am Med Assoc 251:2982–2983

Koeter GH, Meurs H, Kauffman HF et al. (1982) The role of the adrenergic system in allergy and bronchial hyperreactivity. Eur J Respir Dis 63 [Suppl 121]:72–78

Kopin IJ, Gordon EK (1962) Metabolism of norepinephrine-^{3}H released by tyramine and reserpine. J Pharmacol Exp Ther 138:351–359

Kraan J, Koeter GH, van der Mark THW et al. (1985) Changes in bronchial hyperreactivity induced by 4 weeks of treatment with antiasthmatic drugs in patients with allergic asthma: a comparison between budesonide and terbutaline. J Allergy Clin Immunol 76:628–636

Krzanowski JJ, Polson JB, Goldman AL et al. (1979) Reduced adenosine 3′5′-cyclic monophosphate levels in patients with reversible obstructive airways disease. Clin Exp Pharmacol Physiol 6:111–115

Kurland JI, Hadden JW, Moore MAS (1977) Role of cyclic nucleotides in the proliferation of committed granulocyte-macrophage progenitor cells. Cancer Res 37:4534–4538

Laitinen LA, Heino M, Laitinen A et al. (1985) Damage of the airway epithelium and bronchial reactivity in patients with asthma. Am Rev Respir Dis 131:599–606

Larsson S (1977) Long-term treatment with beta$_2$-adrenostimulants in asthma. Side effects, selectivity, tolerance and routes of administration. Acta Med Scand 202 [Suppl 608]:1–40

Larsson S, Svedmyr N (1977) Bronchodilating effect and side effects of beta$_2$-adrenoceptor stimulants by different modes of administration (tablets, metered aerosol, and combinations thereof). Am Rev Respir Dis 116:861–869

Lawford P, Jones BJN, Milledge JS (1978) Comparison of intravenous and nebulized salbutamol in initial treatment of severe asthma. Br Med J 1:84

Lefkowitz RJ, Benovic JL, Kobilka B et al. (1986) β-Adrenergic receptors and rhodopsin: shedding new light on an old subject. Trends Pharmacol Sci 7:444–448

Leikauf GD, Ueki IF, Nadel JA (1984) Autonomic regulation of viscoelasticity of cat tracheal gland secretions. J Appl Physiol 56:426–430

Lofdahl C-G, Svedmyr N (1982) Effects of prenalterol in asthmatic patients. Eur J Clin Pharmacol 23:297–302

Lohmann SM, Walter U (1984) Regulation of the cellular and subcellular concentrations and distribution of cyclic nucleotide-dependent protein

kinases. In: Greengard P (ed) Advances in cyclic nucleotide and protein phosphorylation research, vol 18. Raven, New York, p 63
Lopez-Vidriero MT, Jacobs M, Clarke SW (1985) The effect of isoprenaline on the ciliary activity of an in vitro preparation of rat trachea. Eur J Pharmacol 112:429–432
Lulich KM, Goldie RG, Ryan G et al. (1986) Adverse reactions to β_2-agonist bronchodilators. Med Toxicol 1:286–299
Lulich KM, Goldie RG, Paterson JW (1988) β-Adrenoceptor function in asthmatic bronchial smooth muscle. Gen Pharmacol 19:307–311
Mack JA, Nielson CP, Stevens CP et al. (1986) β-Adrenoceptor-mediated modulation of calcium ionophore activated polymorphonuclear leucocytes. Br J Pharmacol 88:417–423
Madsen BW, Tandon MK, Paterson JW (1979) Cross-over study of the effficacy of four β_2-sympathomimetic bronchodilator aerosols. Br J Clin Pharmacol 8:75–82
Marone G (1985) The role of basophils and mast cells in the pathogenesis of pulmonary diseases. Int Arch Allergy Appl Immunol 76 [Supp 1]:70–82
Massaro GD, Fischman CM, Chiang M-J et al. (1981) Regulation of secretion in Clara cells: studies using the isolated perfused rat lung. J Clin Invest 67:345–351
Mazzoni L, Morley J, Page CP et al. (1985) Prophylactic anti-asthma drugs impair the airway hyper-reactivity that follows exposure to platelet activating factor (PAF). Br J Pharmacol 86:571P
McEvoy JDS, Vall-Spinosa A, Paterson JW (1973) Assessement of orciprenaline and isoproterenol infusions in asthmatic patients. Am Rev Respir Dis 108:490–500
Melttler N, Gray ME, Schuffman S et al. (1981) β-Adrenergic induced synthesis and secretion of phosphatidylcholine by isolated pulmonary alveolar type II cells. Lab Invest 45:575–586
Metzger WJ, Zavala D, Richerson et al. (1987) Local allergen challenge and bronchoalveolar lavage of allergic asthmatic lungs. Am Rev Respir Dis 135:433–440
Meurs H, Koeter GH, De Vries K et al. (1982) The beta-adrenergic system and allergic bronchial asthma: changes in lymphocyte beta-adrenergic receptor number and adenylate cyclase activity after an allergen induced asthmatic attack. J Allergy Clin Immunol 70:272–280
Middleton E, Finke SR (1968) Metabolic responses to epinephrine in bronchial asthma. J Allergy 42:288–299
Miller WC, Rice DL (1980) A comparison of oral terbutaline and fenoterol in asthma. Ann Allergy 44:15–18
Minneman KP, Hegstrand LR, Molinoff PB (1979) Simultaneous determination of β_1 and β_2-adrenergic receptors in tissues containing both subtypes. Mol Pharmacol 16:34–46
Moqbel R, Walsh GM, MacDonald AJ et al. (1986) Effect of disodium cromoglycate on activation of human eosinophils and neutrophils following reversed (anti-IgE) anaphylaxis. Clin Allergy 16:73–83
Mossberg B, Strandberg K, Philipson K et al. (1976) Tracheobronchial clearance in bronchial asthma: response to beta-adrenoceptor stimulation. Scand J Respir Dis 57:119–128
Motojima S, Fukuda T, Makino S (1983) Measurement of β-adrenergic receptors on lymphocytes in normal subjects and asthmatics in relation to β-adrenergic hyperglycaemic response and bronchial responsiveness. Allergy 38:331–337
Nadel JA (1983) Neural control of airway submucosal gland secretion. Eur J Respir Dis 64 [Suppl 128]:322–326
Nielson CP (1987) β-Adrenergic modulation of the polymorphonuclear leukocyte respiratory burst is dependent upon the mechanism of cell activation. J Immunol 139:2392–2397

Nilsson HT, Persson K, Tegner K (1972) The metabolism of terbutaline in man. Xenobiotica 2:363–373
Norn S, Skov PS (1980) The pharmacological basis of drug treatment in bronchial asthma. Allergy 35:549–556
O'Bryne PM, Dolovich J, Hargreave FE (1987) Late asthmatic responses. Am Rev Respir Dis 136:740–751
O'Donnell SR, Wanstall JC (1978) Evidence that efficacy (intrinsic activity) of fenoterol is higher than that of salbutamol on β-adrenoceptors in guinea-pig trachea. Eur J Pharmacol 47:333–340
O'Driscoll BRC, Ruffles SP, Ayres JG et al. (1988) Long term treatment of severe asthma with subcutaneous terbutaline. Br J Dis Chest 82:360–367
Ohman JL, Lawrence M, Lowell FC (1972) Effect of propranolol on the isoproterenol, responses of cortisol, isoproterenol and aminophylline. J Allergy Clin Immunol 50:151–156
O'Reilly JF, Gold G, Kendrick AH et al. (1986) Domicilary comparison of terbutaline treatment by metered dose inhaler with and without conical spacer in severe and moderately severe chronic asthma. Thorax 41:766–770
Parker CW, Smith JW (1973) Alterations in cyclic adenosine monophosphate metabolism in human bronchial asthma. I. Leukocyte responsiveness to beta-adrenergic agonists. J Chem Invest 52:48–59
Paterson JW, Tarala RA (1985) The treatment of asthma, part II: practical aspects of management. Med J Aust 143:453–455
Paterson JW, Courtenay Evans RJ et al. (1971) Selectivity of bronchodilator action of salbutamol in asthmatic patients. Br J Dis Chest 65:21–38
Paterson JW, Woolcock AJ, Shenfield GM (1979) Bronchodilator drugs. Am Rev Respir Dis 120:1149–1188
Paterson JW, Lulich KM, Goldie RG (1983) A comment on beta$_2$-agonists and their use in asthma. Trends Pharmacol Sci 4:67–69
Paterson JW, Lulich KM, Goldie RG (1984) Drug effects on beta-adrenoceptor function in asthma. In: Morley J (ed) Perspectives in asthma, vol 2: beta-adrenoceptors in asthma. Academic, London, p 245
Paterson JW, Lulich KM, Goldie RG (1988) An overview of the current status of the drug therapy of asthma. Agents Actions 23:15–33
Pauwels R, Van Renterghem D, Van Der Straeten M et al. (1985) The effect of theophylline and enprofylline on allergen-induced bronchoconstriction. J Allergy Clin Immunol 76:583–590
Pavia D, Bateman JRN, Sheahan NF et al. (1985) Tracheobronchial mucociliary clearance in asthma: impairment during remission. Thorax 40:171–175
Pavia D, Agnew JE, Lopez-Vidriero MT et al. (1987a) General review of tracheobronchial clearance. Eur J Respir Dis 71 [Suppl 153]:123–129
Pavia D, Agnew JE, Sutton PP et al. (1987b) The effect of terbutaline administered from metered dose inhaler (2 mg) and subcutaneously (0.25 mg) on tracheobronchial clearance in mild asthma. Br J Dis Chest 81:361–370
Persson CGA (1986) Role of plasma exudation in asthmatic airways. Lancet 2:1126–1129
Persson CGA (1987) Leakage of macromolecules from the tracheobronchial microcirculation. Am Rev Respir Dis 135:S71–S75
Persson CGA, Erjefalt I, Andersson P (1986) Leakage of macromolecules from guinea-pig tracheobronchial microcirculation. Effects of allergen, leukotrienes, tachykinins, and anti-asthma drugs. Acta Physiol Scand 127:95–105
Peters SP, Schulman ES, Schleimer RP et al. (1982) Dispersed human lung mast cells: pharmacologic aspects and comparison with human lung tissue fragments. Am Rev Respir Dis 126:1034–1039
Phipps RJ, Nadel JA, Davis B (1980) Effect of alpha-adrenergic stimulation on mucus secretion and on ion transport in cat trachea in vitro. Am Rev Respir Dis 121:359–365

Phipps RJ, Williams IP, Richardson PS et al. (1982) Sympathomimetic drugs stimulate the output of secretory glycoproteins from human bronchi in vitro. Clin Sci 63:23–28

Pratt HF (1982) Abuse of salbutamol inhalers in young people. Clin Allergy 12:203–208

Raaijmakers JAM, van Rosen AJ, Terpstra GK et al. (1985) Autonomic receptor deficiences in the pathogenesis of chronic obstructive lung disease. Prog Respir Dis 19:152–158

Reed CE (1985) Adrenergic bronchodilators: pharmacology and toxicology. J Allergy Clin Immunol 76:335–341

Reed CE (1986) New therapeutic approaches in asthma. J Allergy Clin Immunol 77:537–543

Reed CE, Cohen M, Enta T (1970) Reduced effect of epinephrine on circulating eosinophils in asthma and after beta-adrenergic blockade or bordetella pertussis vaccine. J Allergy 46:90–102

Rhoden KJ, Meldrum LA, Barnes PJ (1987) β-Adrenoceptor modulation of cholinergic neurotransmission in human airways. Am Rev Respir Dis 135: A91

Richardson PS, Sterling GM (1969) Effects of β-adrenergic receptor blockade on airway conductance and lung volume in normal and asthmatic subjects. Br Med J 3:143–145

Robberecht P, Delhaye M, Taton G et al. (1983) The human heart beta-adrenergic receptors. 1. Heterogeneity of the binding sites: presence of 50% $beta_1$- and 50% $beta_2$-adrenergic receptors. Mol Pharmacol 24:169–173

Rugg EL, Barnett DB, Nahorski SR (1978) Coexistence of $beta_1$ and $beta_2$ adrenoceptors in mammalian lung: Evidence from direct binding studies. Mol Pharmacol 14:996–1005

Sano Y, Watt G, Townley RG (1983) Decreased monouclear cell beta-adrenergic receptors in bronchial asthma: parallel studies of lymphocyte and granulocyte desensitization. J Allergy Clin Immunol 72:495–503

Santa Cruz R, Landa J, Hirsch J et al. (1974) Tracheal mucous velocity in normal man and patients with obstructive lung disease: effects of terbutaline. Am Rev Respir Dis 109:458–463

Schenkelaars EJ, Bonta I (1985) β_2-Adrenoceptor agonists reverse the leukotriene C_4-induced release response of macrophages. Eur J Pharmacol 107:65–70

Schlimmer P, Austgen M, Ferber E (1983) Classification and possible function of phospholipids obtained from central airways. Eur J Respir Dis 64 [Suppl 128]:318–321

Schultz RM, Pavlidis NA, Stoychkov JN et al. (1979) Prevention of macrophage tumoricidal activity by agents known to increase cellular cyclic AMP. Cell Immunol 42:71–78

Schwartz S, Davies S, Jeur JA (1980) Life threatening cold- and exercise-induced asthma potentiated by administration of propranolol. Chest 78:100–101

Seamon KB, Padgett W, Daly JW (1981) Forskolin: unique diterpene activation of adenylate cyclase in membranes and in intact cells. Proc Natl Acad Sci USA 78:3363–3367

Shelhamer JH, Marom Z, Kaliner M (1980) Immunologic and neuropharmacologic stimulation of mucous glycoprotein release from human airways in vitro. J Clin Invest 66:1400–1408

Shenfield GM, Evans ME, Walker SR et al. (1973) The fate of nebulized salbutamol (albuterol) administered by intermittent positive pressure respiration to asthmatic patients. Am Rev Respir Dis 108:501–505

Shenfield GM, Evans ME, Paterson JW (1976) Absorption of drugs by the lung. Br J Pharmacol 3:583–589

Silver PJ, Stull JT (1982) Regulation of myosin light chain and phosphorylase phosphorylation in tracheal smooth muscle. J Biol Chem 257:6145–6150

Sleigh MA (1983) Ciliary function in transport of mucus. Eur J Respir Dis 64 [Suppl 128]:287–292
Sly RM, Anderson JA, Bierman CW et al. (1985) Adverse effects and complications of treatment with beta-adrenergic agonist drugs. J Allergy Clin Immunol 75:443–449
Smith SR, Kendall MJ (1984) Metabolic responses to beta$_2$ stimulants. J R Coll Physicians Lond 18:190–194
Smith SR, Ryder C, Kendall MJ et al. (1984) Cardiovascular and biochemical responses to nebulised salbutamol in normal subjects. Br J Clin Pharmacol 18:641–644
Sorokin SP (1970) The cells of the lung. In: Nettesheim P, Hanna MG, Deatherage SW (eds) Morphology of experimental respiratory carcinogensis. Proceedings of the Biology Division, US Atomic Energy Commission, Oak Ridge, Tennessee, pp 3–41
Stephenson RP (1956) A modification of receptor theory. Br J Pharmacol 11:379–393
Stiles GL, Caron MG, Lefkowitz RJ (1984) β-Adrenergic receptors: biochemical mechanisms of physiological regulation. Physiol Rev 64:661–743
Strader CD, Candelore MG, Rands E et al. (1987) β-Adrenergic receptor subtype is an intrinsic property of the receptor gene product. Mol Pharmacol 32:179–183
Strasser RH, Benovic JL, Caron MG, et al. (1986) β-Agonist- and prostaglandin E$_1$-induced translocation of the β-adrenergic receptor kinase: evidence that the kinase may act on multiple adenylate cyclase-coupled receptors. Proc Natl Acad Sci USA 83:6362–6366
Svedmyr N (1984) Is beta-adrenoceptor sensitivity a limiting factor in asthma therapy? In: J Morley (ed) beta-adrenoceptors in asthma. Academic, London, pp 181–210
Svedmyr N, Thiringer G (1971) The effects of salbutamol and isoprenaline on beta-receptors in patients with chronic lung disease. Postgrad Med J 47 [Suppl]:44–46
Svedmyr N, Larsson SA, Thiringer GK (1976) Development of "resistance" in beta-adrenergic receptors of asthmatic patients. Chest 69:479–483
Svensson LA (1985) Sympathomimetic bronchodilators: increased selectivity with lung-specific prodrugs. Pharm Res 4:151–194
Szentivanyi A (1968) The beta-adrenergic theory of the atopic abnormality in bronchial asthma. J Allergy 42:203–232
Szentivanyi A (1979) The conformational flexibility of adrenoceptors and the constitional basis of atopy. Triangle 18:109–115
Taki F, Takagi K, Satake T et al. (1986) The role of phospholipase in reduced beta-adrenergic responsiveness in experimental asthma. Am Rev Respir Dis 133: 362–366
Tai E, Read J (1967) Response of blood gas tensions to aminophylline and isoprenaline in patients with asthma. Thorax 22:543–549
Tarala RA, Madsen BW, Paterson JW (1980) Comparative efficacy of salbutamol by pressurized aerosol and wet nebulizer in acute asthma. Br J Clin Pharmacol 10:393–397
Tarala RA, Martin V, Paterson JW (1981) Effect of intravenous injection of rimiterol in asthma. Br J Clin Pharmacol 12:333–340
Tashkin DP, Conolly ME, Deutsh RI et al. (1982) Subsensitization of β-adrenoceptors in airways and lymphocytes of healthy and asthmatic subjects. Am Rev Respir Dis 125:185–193
Tattersfield AE (1986) Clinical applications of beta-agonists. Asthma: clinical pharmacology and therapeutic progress. In: Kay AB (ed) Asthma chemotherapy. Blackwell, Oxford, pp 161–170
Tattersfield AE (1987) Effect of beta-agonists and anti-cholinergic drugs on bronchial reactivity. Am Rev Respir Dis 136:S64–S68
Tattersfield AE, Holgate ST, Harvey JE et al. (1983) Is asthma due to partial beta-bockade of airways? Agents Actions [Suppl] 13:265–271

Torphy TJ, Rinard GA, Rietow MG et al. (1983) Functional antagonism in canine tracheal smooth muscle: inhibition by methacholine of the mechanical and biochemical responses to isoproterenol. J Pharmacol Exp Ther 227:694–699

Twaroq FJ, Leung DYM (1982) Anaphylaxis to a component of isoetharine (sodium bisulfite). J Am Med Assoc 248:2030–2031

Van Den Brink FG (1973) The model of functional interaction. II. Experimental verification of a new model: the antagonism of β-adrenoceptor stimulants and other agonists. Eur J Pharmacol 22:279–286

Vathein AS, Knox AJ, Higgins BG et al. (1988) Rebound increase in bronchial responsiveness after treatment with inhaled terbutaline. Lancet 1:554–557

Verdugo P, Johnson NT, Tam PY (1980) β-Adrenergic stimulation of respiratory ciliary activity. J Appl Physiol 48:868–871

Vermeire PA, Vanhoutte PM (1979) Inhibitory effects of catecholamines in isolated canine bronchial smooth muscle. J Appl Physiol 46:787–791

Vick RL, Todd EP, Luedke DW (1972) Epinephrine-induced hypokalaemia: relation to liver and skeletal muscle. J Pharmacol Exp Ther 181:139–146

Wanner A (1979) The role of mucociliary dysfuntion in bronchial asthma. Am J Med 67:477–485

Wardlaw AJ, Kay AB (1987) The role of the eosinophil in the pathogenesis of asthma. Allergy 42:321–335

Whicker SD, Armour CL, Black JL (1988) Responsiveness of bronchial smooth muscle from asthmatic patients to relaxant and contractile agonists. Pulm Pharmacol 1:25–31

Williams LT, Snyderman R, Lefkowitz RJ (1976) Identification of β-adrenergic receptors in human lymphocytes by (−) [^{3}H]-alprenolol binding. J Clin Invest 57:149–155

Williams SJ, Seaton A (1977) Intravenous or inhaled salbutamol in severe acute asthma. Thorax 32:555–558

Williams SJ, Winner SJ, Clark TJH (1981) Comparisons of inhaled and intravenous terbutaline in acute severe asthma. Thorax 36:629–631

Woolcock AJ (1986) Therapies to control airway inflammation of asthma. Eur J Respir Dis 69 [Suppl 147]:166–174

Young SL, Silbajoris RA (1985) Type II cell response to chronic β-adrenergic agonist and antagonist infusions. J Pharmacol Exp Ther 233:271–276

Young WS, Kuhar MJ (1979) A new method for receptor autoradiography: [^{3}H]opioid receptors in rat brain. Brain Res 179:255–270

Zaagsma J, van der Heijden PJ, van der Schaar MW et al. (1984) Differentiation of functional adrenoceptors in human and guinea-pig airways. Eur J Respir Dis 65 [Suppl 135]:16–33

Zaid GM, Beall GN (1966) Bronchial response to beta-adrenergic blockade. New Engl J Med 75:580–584

Ziment I (1987) Mucokinetic agents. In: Hollinger MA (ed) Current topics in pulmonary pharmacology and toxicology, vol 3. Elsevier, New York, p 122

CHAPTER 7

Pharmacology of Anti-Asthma Xanthines

C.G.A. PERSSON and R. PAUWELS

A. Introduction

The xanthine drugs are clinically effective in asthma. In select patients they produce therapeutc effects which cannot be accomplished by other drugs, singly or in combination, however high the dosages. Hence, if we can learn more about the mode of action of theophylline, the major anti-asthma xanthine, we may gain new insights into the mechanisms of asthma. The pharmacology of anti-asthma xanthines remains a topical field for experimental and clinical research.

I. Developmental Aspects

We will probably never learn the details of the very earliest uses of xanthines in asthma. A two-page letter written by William Withering in 1786 had "coffee made very strong" as the number one reliever of asthmatic symptoms. Withering gives no clue as to the source of this information. Books on asthma remedies in seventeenth and eithteenth centuries mention coffee (containing caffeine), and one asthmatic patient in Dover's "The ancient physician's legacy to his country" (from 1742) reported that drinks of chocolate (containing theobromine) did her well (see PERSSON 1990a). Though thus not the very first, Henry Hyde Salter must be considered the pioneer in the field of treatment of asthma with xanthines (PERSSON 1985a). Salter was an astute observer whose descriptions in the nineteenth century of asthma and its treatments are outstanding. From the complex therapeutic armamentarium of that time, he selected strong coffee as the best remedy for this own asthma as well as for that of his many patients. He prescribed coffee in amounts which should have produced therapeutic blood levels of caffeine.

Coffee must have been an essential anti-asthma drug in the 'prepharmacological' days. It is therefore surprising that there is rarely any mention of this xanthine remedy in the asthma literature of the late 1800s and early 1900s (PERSSON 1985a). Indeed, when caffeine was purified and synthetized, it did not become a drug for asthma, nor did theophylline when it became available a few years later. Around 1890 theophylline was promoted as a diuretic agent. High dosages were often employed in attempts to compensate for the tachyphylactic nature of its natriuretic action. Soon serious side effects were encountered and during the first decade of this century, there was an interesting debate as to whether or not the occurrence of seizures could be blamed on the use of this drug (ALLARD 1904; SCHMIEDEBERG 1905).

In the early 1920s a mixture of theophylline and theobromine was successfully used in asthma and also evaluated in vitro on airway preparations. The breakthrough for a widespread use of xanthines came towards the end of the 1930s when these drugs were reported to have a unique efficacy. Thus parenteral theophylline produced dramatic effects in severe asthma where adrenaline was no longer effective. The interest in i.v. and oral theophylline grew steadily. For a number of decades the use of this drug, its salts and weakly active 7-derivatives (Fig. 1) rested a great deal on clinical impression and an evolving therapeutic tradition. At the beginning of the 1970s when a range of new anti-asthma drugs were being introduced (β_2-agonists, cromoglycate, and inhaled glucocorticoids), the clinical scientific documentation of theophylline was not advanced and its role in therapy was questioned (in particular in Europe). The time lag between Europe and USA concerning the introduction of the novel drugs probably contributed to the fact that the pharmacokinetics and efficacy of theophylline became well studied in North America. The advent of long-acting oral formulations allowing twice daily medications was an important development that contributed to make theophylline the most frequently prescribed anti-asthma drug worldwide.

In 1957, the same year as Ther et al. first demonstrated an "antagonism between adenosine and methylxanthine", Berthet and coworkers reported on the first steps on the road to the discovery of the enzyme cyclic 3',5'-nucleotide-phosphodiesterase and its inhibition by xanthines. Drug-oriented

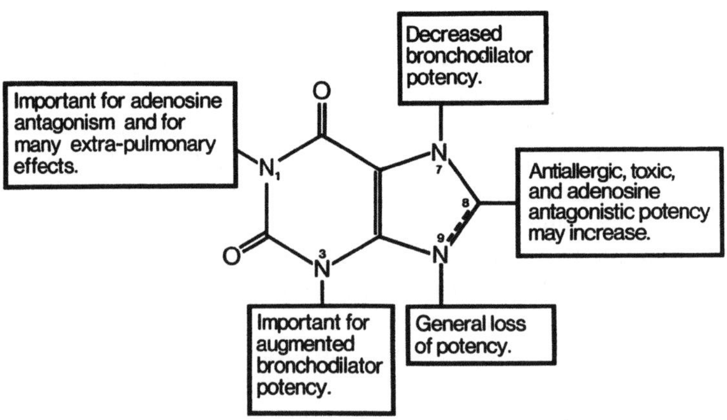

Fig. 1. Summary of some of the pharmacological effects that are achieved by substitution with alkyls at various positions of the xanthine molecule. "Bronchodilator" denotes both relaxant and anti-inflammatory actions that seem to covary with many xanthine derivatives

research in the field of phosphodiesterase inhibition has since been going on in several laboratories but still no novel phosphodiesterase-inhibiting anti-asthma drugs have resulted. Old and new compounds with potent enzyme-inhibitory effects have been examined but with little success. Bergstrand and coworkers (BERGSTRAND 1985) identified a number of phosphodiesterase isoenzymes in human lung and bronchi, but this advancement appeared to be little helpful in explaining the airway actions of xanthines (PERSSON 1985b). There is currently a renewed interest in the role of phosphodiesterase isoenzymes. Xanthines as well as other types of compounds aimed at asthma treatment continue to be studied as to their ability to inhibit these enzymes (TORPHY 1988).

More recently, adenosine antagonism by xanthines has been taken as a ground for developing novel drugs for the treatment of asthma. Indeed, some characteristics of adenosine are compatible with a role as a mediator of asthma. It may increase the allergen-induced release of mediators (MARQUARDT et al. 1978). As in other organs, the lung tissue level of adenosine is increased by stresses such as hypoxia (MENTZER et al. 1975). The intriguing observation has been made that atopic as well as nonatopic asthmatic subjects, but not normal subjects, respond with a marked bronchoconstriction to inhaled adenosine (CUSHLEY et al. 1983). Adenosine antagonism was thus an attractive explanation of the anti-asthma effects of xanthines (MARQUART et al. 1978; FREDHOLM 1980; WELTON 1980; SNYDER 1981; CUSHLEY et al. 1983). Considering the clinical effectiveness of xanthines, this possibility, if true, would make adenosine a significant single agent among the numerous proposed mediators of asthma. Novel adenosine antagonists have also been synthetized in attempts to find improved asthma remedies. However, the advent of enprofylline has reduced the interest in this particular approach (PERSSON 1982a, 1983). Enprofylline is a xanthine derivative without adenosine antagonism, yet it is more potent in asthma than theophylline.

Besides adenosine antagonism and phosphodiesterase inhibition, a large number of subcellular mechanisms have been proposed for the xanthines, including the release of catecholamines, 5′-nucleotidase inhibition, prostaglandin antagonism, and effects on calcium metabolism. These compounds are also included in increasingly detailed studies of cell membrane mechanisms involved in the regulation of cytosolic calcium concentration and other pathways of importance in the control of cell function. It is a matter of speculation whether the current lines of research will reveal the crucial anti-asthmatic mechanisms of the xanthines.

II. Chemistry and Subdivision of Xanthines

Xanthines (and adenosine) belong to the group of purine compounds which have a fascinating chemical history (KJELLIN and PERSSON 1985). Purines were first isolated by von Scheele in 1776. In 1820, Runge managed to isolate caffeine (so named a few years later by Pelletiér) from coffee. There was then some confusion as to the purine content of tea leaves. The leaves contain a considerable amount of caffeine, which was isolated and for some years called "theine", until its identity was clarified by Berzelius and others. In 1888, Kossel successfully isolated a small amount of a dimethylxanthine from tea and named it theophylline. Emil Fischer pioneered work on the synthesis and determination of the chemical structure of caffeine and theophylline. A significant further step was taken by Traube, who devised a versatile method of synthesis of different xanthine derivatives. His method is used by chemists today.

The xanthine molecule can be substituted at at least four different positions to produce xanthine derivatives (Fig. 1). The relationships between structure and activity are complex (PERSSON 1985b). A few points related to anti-asthma xanthines may be mentioned. Substitutions at the 7 (or 9) position result in low potency, whereas substitutions at the 1 and 8 positions may be associated with increased potency, possibly at the expense of an out of proportion increased toxicity. Substitutions at the 1 and 8 positions are critical for potent adenosine antagonism. The 8 position may also be used to produce compounds which selectively inhibit cyclic GMP phosphodiesterase. By proper substitution at the 3 position, anti-asthma potency is increased. Furthermore, compounds having an alkyl only in the 3 position may lack adenosine antagonism and, probably because of this, have a reduced number of pharmacological actions. By increasing the bulkiness in the 3 position, airway potency is increased with an optimum for 3-propyl or 3-butyl derivatives (PERSSON and KJELLIN 1981; PERSSON, 1985b; PERSSON, 1986). The 3-propyl-xanthine, enprofylline, is also more potent than theophylline (1,3-dimethylxanthine) as an anti-asthma agent (LUNELL et al. 1983, 1984; PAUWELS et al. 1985; CHAPMAN et al. 1989; LJUNGHOLM et al. 1990).

It should be remembered that xanthine is the name of a group of chemicals and may not be used as a term indicating a special range of activities. Indeed, xanthine derivatives have a great variety of effects, including diuretic, CNS-excifatory, smooth muscle relaxant, anti-inflammatory, and anti-allergic properties. Xanthine derivatives may also be useful as anti-thrombotic and anti-cancer agents, as sperm motility stimulants, and as anti-viral and anti-mycotic compounds. They may even be employed in the future on the basis of their pesticidal actions (see PERSSON 1985b). Also, by proper substitution, xanthines may be anti-muscarinics, anti-histamines, and even β_2-agonists. However, in these latter cases the xanthine moiety is merely a chemical substitution and, at the dose levels employed, adds no xanthine-like action to the pharmacological property of these specific and, usually, potent drugs (PERSSON 1985b). Among xanthine derivatives which exert therapeutic effects in obstructive airway disease there appear to be two major pharmacological groups, i.e. xanthines which are adenosine antagonists, such as theophylline, and xanthines which are not adenosine antagonists, such as the experimental drug enprofylline.

B. Clinical Use

I. Acute Severe Asthma

Both enprofylline and theophylline have been shown to cause bronchodilation in patients with reversible airway obstruction and have therefore been used in the treatment of acute severe asthma. The bronchodilating effect is related to the serum concentration of the drug (RACINEUX et al. 1981; LUNELL et al. 1983) Although the bronchodilating effect is observed shortly after intravenous administration, the maximal bronchodilating effect may be seen later than the serum level would indicate (ISHIZAKI et al. 1988). Perhaps actions other than the smooth muscle relaxant effect have contributed to this finding.

Sympathomimetic β_2-receptor agonists, in an inhaled or intravenous form, are now the agents of choice in the treatment of acute severe asthma. The concomitant use of intravenous xanthines has been questioned in view of the risk of serious side effects linked to an overdose of theophylline. A recent meta-analysis of 13 double blind controlled studies on the treatment of acute asthma came to the conclusion that none of the studies published up to now had sufficient power to give a final answer as to the benefit-risk ratio of theophylline added to sympathomimetics (LITTENBERG 1988).

Enprofylline, a potent xanthine molecule without the risk for serious central nervous side effects (PERSSON and ERJEFÄLT 1982; VILSVIK et. al 1990) has been compared to high dose inhaled terbutaline in a multicenter study in Australia involving 69 patients (RUFFIN et al. 1988). The bronchodilating effect of intravenous enprofylline (2 mg/kg) was comparable to that of nebulized terbutaline (10 mg). The major side effects reported were tremor in the terbutaline-treated patients and nausea in the enprofylline-treated group. In another multicenter study carried out in Sweden and involving 176 patients, i.v. theophylline was given 1 h after administration of albuterol, either 5 μg/kg b.w. i.v. or 0.15 mg/kg b.w. by inhalation of nebulized drug. Theophylline produced a significant ($P<0.001$) further increase in peak expiratory flow values (BOE 1988). The possibility for this positive interaction is supported by in vitro findings where xanthines produced larger effects than β-agonists (KARLSSON and PERSSON 1981) and where an additive or slightly supraadditive interaction between xanthines and β-agonists has been recorded (PERSSON and GUSTAFSSON 1986; PERSSON and KARLSSON 1987).

II. Chronic Asthma

Chronic treatment with oral sustained-release theophylline or enprofylline significantly improves symptoms and lung function in patients with chronic asthma (WEINBERGER 1984; LJUNGHOLM et al. 1990). Chronic treatment with sustained-release enprofylline has thus been shown to produce a dose-dependent improvement of symptoms and lung function in asthmatics who have a baseline therapy with inhaled β-agonists (95%) plus inhaled glucocorticoids (77%) (Chapman et al. 1989). However, the establishment of anti-asthmatic effects of xanthines does not per se determine the place of theophylline with respect to other treatment forms with long-acting oral and inhaled drugs.

The development of topically active inhaled steroids has significantly altered the therapeutic approach to asthma in large parts of the world. They are safe to use and frequently they are more effective than oral theophylline in the treatment of chronic asthma. There also have been further developments in the field of β_2-agonists, raising questions about the role of theophylline in the treatment of obstructive airway diseases and more specifically, its place relative to treatments with β_2-agonists and inhaled steroids. As will be outlined below, it appears that the anti-asthma pharmacology of xanthines makes these latter drugs essential, at least in subgroups of patients.

Ten years ago NASSIF et al. (1981) showed in a small group of patients that maintenance treatment with theophylline was beneficial also for those patients who required small to moderate dosages of inhaled glucocorticoids. More recent large scale clinical trials with theophylline and enprofylline, comprising over 1000 patients, further demonstrate that the clinical effect of maintenance treatment was the same whether or not patients also received inhaled glucocorticoids. Indeed, the same xanthines-induced improvement was recorded also in those patients who were on regular treatment with oral and inhaled glucocorticoids as in those who were without glucocorticoid treatment (LJUNGHOLM et al. 1990).

BRENNER et al. (1988) carried out perhaps the most significant study so far to address the possibility that xanthines are essential drugs in asthma. Concern about potential side effects of theophylline prompted these workers to investigate whether this drug could be eliminated from the multi-medication regimen of a group adolescents with very severe asthma. They all received alternate day prednisone (10–30 mg) orally, inhaled beclomethasone dipropionate 100–200 μg qid, inhalations of nebulized β_2-agonists and atropine qid, and inhaled cromoglycate 20 mg qid. The design of the trial specifically instructed that non-theophylline drug therapy should be increased if the asthma got worse. The authors (and probably many others) were surprised to learn that elimination of theophylline was not possible in these patients. Thus, despite the fact that the doses of oral glucocorticoid and inhaled bronchodilators were significantly increased, the asthmatic condition underwent severe deterioration when theophylline was replaced by placebo. None of the patients completed the intended study period of 4 weeks without theophylline. The observed deterioration also stopped recruitment of further patients of this trial that was carried out with only five subjects. It seems clear that in patients with severe asthma, theophylline has to play a role supplementing that of high dose inhaled steroids. At what does of inhaled steroids xanthines should be added is, in our view, dependent on the individual sensitivity for systemic effects of inhaled steroids and requires further investigation.

III. Chronic Obstructive Pulmonary Disorders

The role of theophylline in the treatment of chronic obstructive pulmonary disorders (COPD) is yet another point of discussion. Most studies have looked at short-term symptomatic relief and very little attention has been given to any long-term effects including the prognosis of these diseases and the further progression of airflow obstruction. Several studies demonstrate that theophylline may have favourable effects on spirometry, exercise performance, and the sensation of dyspnea in COPD patients (ALEXANDER et al. 1980; JENNE et al. 1984; MURCIANO et al. 1984; TAYLOR et al. 1985). A recently published study by CHRYSTYN et al. (1988) suggests that the major effect of theophylline on the pulmonary function of COPD patients may be a decrease of the trapped gas volume. They compared in a double blind, randomized, cross over study in COPD patients the effect of placebo and three dose levels of theophylline. The effect on FEV_1 and FVC was minimal. A dose dependent reduction of the trapped gas volume (the difference between total lung capacity measured by whole body plethysmography and that measured by helium dilution) and an increase of the slow vital capacity was observed together with a decrease in dyspnea-score and the use of rescue-inhaler of sympathomimetics. These data would suggest that in COPD the major pharmacological effect of theophylline may be situated at the level of the small airways. It could in part be a bronchodilator action because xanthines relax human small airways (FINNEY et al. 1985). It is also possible that theophylline inhibits the airway inflammation present in the small airways of patients with COPD.

Numerous studies have compared β-agonists, anticholinergics and theophylline in COPD (BARCLAY et al. 1982; FILUK et al. 1985; PASSAMONTE and MARTINEZ 1984). It is difficult to conclude from the mean data in these studies that theophylline can have additional effects beyond the results obtained with optimized doses of inhaled β-agonists and anti-cholinergics. However, it seems that those patients who benefit most from theophylline are not the same as those who respond well to the inhaled bronchodilators (GUYATT et al. 1987). This clinical observation agrees with the fact that the airway pharmacology of xanthines is distinct from that of the β-agonists.

The bronchodilating effects and protective effect of theophylline on nocturnal asthma are sufficiently known (BARNES et al. 1982). The protective effect of theophylline on bronchial responsiveness is dependent both on the type of stimulus used in the assessment of the bronchial responsiveness and the severity of the asthma in the patients studied. Theophylline has been shown to inhibit in mild asthmatics the airway responsiveness to histamine (CARTIER et al. 1986), methacholine (MAGNUSSEN et al. 1987), exercise (POLLOCK et al. 1977), distilled water (FABBRI et al. 1986) and adenosine

(MANN and HOLGATE 1985). As would be expected from the distinct pharmacology of enprofylline, it does not specifically antagonise adenosine-induced bronchoconstriction (CLARK et al. 1989), but in other respects it seems to act similarly to theophylline (PAUWELS et al. 1985). A study by DUTOIT et al. (1987) failed to show any significant effect of 10-week treatment with theophylline on the histamine responsiveness in a group of severe asthmatics.

The effect of theophylline on allergen-induced bronchoconstriction had not been extensively studied until quite recently and only the effect on the immediate bronchoconstriction had been investigated. PAUWELS et al. (1985) studied the effect of theophylline and enprofylline on the allergen-induced immediate and late bronchoconstriction in nine asthmatics. The patients were challenged three times at weekly intervals with the same does of allergen. This dose had previously been chosen as causing an immediate bronchoconstriction with an FEV_1 decrease between 20% and 50% of the prechallenge value. On the three days FEV_1 and sGaw were followed up for 6 h after challenge. The drugs were given intravenously. Placebo was given on the first occasion. Theophylline and enprofylline were administered on test day 2 and 3 with a double blind randomized cross over technique. One hour before the allergen challenge, a loading dose was given for 60 min followed by a constant infusion for 6 h. The loading infusion was 7.2 mg/kg of theophylline or 2.7 mg/kg of enprofylline. The maintenance dose was 74 mg/h and 71 mg/h, respectively. Theoplylline and enprofylline both caused a minor initial bronchodilation. Theophylline and enprofylline slightly but significantly inhibited the immediate bronchoconstrictor reaction after allergen inhalation. Both drugs had a significant inhibitory effect on the late reaction. The mean plasma level of theophylline was 0, 10.8, 10.5 and 10.5 mg/l at 0, 1, 4 and 7 h after the start of the loading infusion. The corresponding mean plasma levels of enprofylline were 0, 2.6, 2.7 and 2.7 mg/l. The data thus showed that theophylline, at plasma levels considered therapeutic, inhibits both the immediate and the late bronchoconstrictor reaction after allergen challenge. MAPP et al. (1987) investigated the effect of different anti-asthmatic drugs on the immediate and late bronchoconstrictor reaction following challenge with toluene diisocyanate (TDI) in TDI sensitive asthmatics. The drugs were administered for 1 week before the challenge. The last dose was administered a short time before the TDI exposure. Theophylline was administered as a dose of 6.5 mg/kg of a slow release preparation twice daily. The last dose of 10 mg/kg was given 2 h before challenge. The bronchial response to TDI was followed and the bronchial responsiveness to methacholine was studied before and 8 h after the TDI challenge. Theophylline significantly inhibited the late reaction following TDI challenge but did not have a significant effect on the increase in methacholine responsiveness. The mean theophylline serum concentration in the six patients was 18 mg/l before and 20 mg/l 8 h after the TDI challenge.

The same group of investigators (CRESCIOLI et al. 1988) has recently confirmed the observation that theophylline inhibits both the immediate and late reaction following allergen challenge but could again not demonstrate a significant inhibitory effect on the increase in methacholine responsiveness measured 8 h after the allergen challenge.

The result of animal studies agree with clinical observations. In a guinea-pig model of allergen induced bronchial reactions ANDERSSON et al. (1985) demonstrated that the intravenous injection of a low dose of theophylline immediately before the antigen challenge significantly inhibited the immediate bronchoconstriction and the late reaction. The late reaction in the guinea-pig is characterized by the influx of neutrophils, eosinophils and lymphocytes as well as by plasma exudation (ANDERSSON et al. 1985; ANDERSSON and PERSSON 1988). Pretreatment with theophylline significantly reduced the airway inflammation. The intravenous injection of theophylline 90 min after the antigen-induced immediate bronchoconstriction also had an inhibitory effect on the late reaction. Similar findings were obtained in an allergic sheep model (PERRUCHOUD et al. 1984). Theophylline at a serum concentration of 10 mg/l inhibited the late reaction after allergen challenge both when given prior to and follwoing the allergen challenge. Also, enprofylline was demonstrated to be as effective as the PAF-antagonist WEB 2086 in preventing the late phase inflammatory exudative response occurring in guinea-pig tracheobronchial airways after topical application of PAF (O'DONNELL et al. 1990).

C. Mode of Action of Xanthines

I. Smooth Muscle Relaxation or Inhibition?

Smooth muscle relaxation was the first airway action of the xanthines to be demonstrated. Partly because of this priority in time and partly because of a long-standing emphasis on the role of bronchoconstriction in asthma, the airway relaxant effect of xanthines is generally considered the major therapeutic action. Xanthines do have interesting tracheobronchial relaxant characteristics, but these may not suffice to explain their clinical efficacy.

Xanthines relax airway smooth muscle in a manner that is relatively independent of the contractile agent that has contributed to an increased tone. As demonstrated in animal and human airways in vitro, this general relaxation is independent of age, is without tachyphylaxis, and may be more effective (at large concentrations) than that induced by other bronchodilator drugs (PERSSON and KARLSSON 1987). The relaxant action of xanthines is also independent of the presence of an intact eptiheilial lining in the airways (LUNDBLAD and PERSSON 1988).

Another question which has been addressed in isolated airway experiments is whether or not the relaxant action of xanthines protects against the effect of a contractile mediator. Thus concentration response to carbachol was evaluated in the presence and absence of relaxant concentrations of theophylline (40 and 80 μg/ml) and enprofylline (10 and 20 μg/ml) (PERSSON et al. 1988). The only interaction recorded was that the induced relaxation allowed carbachol to contract over a wider tension range. The concentration-response curves to carbachol were not shifted. Even the same absolute maximum tension as in controls was produced in the presence of the xanthines. The xanthines thus did not attenuate the initiation and development of a contractile effect. In asthmatic subjects protection by xanthines against challenge with bronchoconstrictors has been recorded but this protection has not corresponded to the degree of the initial xanthine-induced bronchodilation. Taken together, in vitro and in vivo data suggest that the protective action of xanthines in asthma cannot be explained only on basis of the airway smooth muscle relaxant effects of these drugs. The possibility remains that bronchodilatation is partly responsible for the acute symptomatic effects in asthma.

II. Airway Anti-Inflammatory Actions

A consistent sign of ongoing inflammatory processes in asthma and rhinitis is plasma exudation. The extravasated plasma distributes not only in the mucosa/submucosa but it appears to always enter the airway lumen, too (ERJEFÄLT and PERSSON 1989). Hence, measurement of plasma proteins and plasma-derived mediators in mucosal surface liquid may enable us to quantify the inflammation in airway tissue (PERSSON 1990b). Interestingly, in patients with allergic rhinitis theophylline has been demonstrated to reduce nasal plasma exudation (NACLERIO et al. 1986; PERSSON 1988). Similar studies in asthmatics are now lacking but there are other indices of the tracheobronchial anti-inflammatory effects of theophylline and enprofylline in these patients (see above). Data obtained in guinea-pig tracheobronchial mucosa demonstrate that xanthines are effective locally in the airways in reducing inflammatory stimulus-induced plasma exudation responses (PERSSON et al. 1986a). Furthermore, xanthines such as enprofylline and theophylline dose-dependently reduce the plasma exudation, and the data suggest that this action is in part directly on the endothelial barrier cells of the airway microvascular wall (PERSSON et al. 1988). Since plasma exudation is a cardinal sign of inflammation, an anti-exudative action could result from an effect at any important level of the inflammatory process.

Several types or stationary and migrating cells of the airway may contribute to inflammation in asthma. Xanthines have been demonstrated to reduce the activity of basophils, macrophages, mast cells, platelets, and polymorphonuclear (PMN) leukocytes. In cooperation with other anti-inflammatory factors, xanthines exert particularly potent inhibitory effect. In 1971, ORANGE et al. demonstrated that 10^{-5} M of theophylline, if given together with isoprenaline, had pronounced inhibitory effects on the immunologic release of mediators from human lung in vitro. Recent in vitro work by NIELSON et al. (1986) also demonstrates that theophylline and enprofylline themselves may be potent inhibitors of the activation of human PMN leukocytes, suggesting that synergistic interactions with other autacoids may not always be required for in vivo effects of xanthines on inflammatory cells.

In a rat model of endotoxin-induced airway hyperresponsiveness theophylline reduced both the influx of neutrophils and the associated responsiveness (KIPS et al. 1989). Sensitized guinea-pigs challenged with particulate antigen developed a late phase reaction. Theophylline and enprofylline inhibited this response (ANDERSSON et al. 1985) as well as the assoicated plasma leakage. These animal observations are consistent with the view that the functional airway anti-inflammatory effects of xanthines are involved in the therapeutic response to these drugs in asthma. Xanthines may increase the mucociliary transport rate and thus improve the airway function (SUTTON et al. 1981). This action may be caused by anti-inflammatory actions as well as by direct stimulation of ciliary activity.

Inhibition of inflammatory cells may contribute to pulmonary effects of xanthines other than the anti-asthma actions. Thus theophylline significantly reduces bleomycin-induced airway inflammation and fibrosis in experimental animals (LINDENSCHMIDT and WITSCHI 1985). Theophylline and enprofylline also act prophylactically to reduce inflammatory stimulus-induced pulmonary edema (PERSSON et al. 1979; PERSSON et al. 1988). Another xanthine derivative, pentoxifylline, developed for the treatment of intermittent claudication, has now been demonstrated to decrease endotoxin-induced neutrophil sequestration and protein-rich edema in canine lungs (WELSCH et al. 1988). Both theophylline and pentoxifylline have been demonstrated to reduce multiple organ damage after *Escherichia coli* sepsis in guinea-pigs (HARADA et al. 1989).

Theophylline has been reported to increase the number and activity of "suppressor" T-cells in asthma (FINK et al. 1987). Studies addressing the possibility that immunosuppressive effects of xanthines may play a role seem warranted.

III. Anti-Asthma Actions Outside the Airways?

A variety of actions xanthines outside the airways have been considered important for the clinical efficacy in asthma. In the nineteenth century Salter rather convincingly emphasized the CNS-stimulant action of coffee. He referred to the fact that another excitatory influence, sudden alarm, could instantaneously abolish asthmatic symptoms and that the opposite, a few hours of sleep, induces asthma. More recently it has been speculated that an urge for vigilant actions may explain the popularity that theophylline has with many patients. Evidence supporting a CNS locus is now lacking. Indeed, a xanthine derivative without CNS-stimulant behavioural effects, enprofylline, has been demonstrated to be a more potent anti-asthma drug and with no less efficacy than theophylline (PERSSON 1983; LUNELL et al. 1983; PAUWELS et al. 1985; CHAPMAN et al. 1989).

Since the 1950s several attempts have been made to administer xanthines via inhalation. The lack of success in these trials, experienced also with more potent xanthines than theophylline, apparently favours an extra-airway site of action. However, there are alternative explanations for the failure of topical theophylline in asthma. Studies involving guinea-pig airways and lungs indicate that topically applied theophylline distributes promptly, as sucrose does, in extracellular water and disappears very rapidly from airway tissue into pulmonary veins (KRÖLL et al. 1990). Hence, the high local concentration reached immediately after inhalation will not remain in the airway tissue for a sufficiently long period of time to reduce an asthmatic obstruction. A further complicating factor is the low potency of xanthines. By comparison, inhaled β-agonists bind to airway tissue and are about a thousand times more potent than the xanthines. Taken together, these facts indicate that the absence of acceptable effects via inhalation with the current xanthines does not contradict the hypothesis that the airway is the site of the clinically important anti-asthma action.

A catecholamine releasing action has been suggested to be important. However, actual observations in patients receiving theophylline are not consistent on this point. Any rise in plasma adrenaline is not related in magnitude or time course to the anti-asthma effect of theophylline (ANDERSSON et al. 1984; ISHIZAKI et al. 1988). As expected from its lack of adenosine antagonism enprofylline is without catecholamine releasing effects (ANDERSSON et al. 1984). Enprofylline is nevertheless as efficacious as theophylline in acute and long term treatment of asthma.

Theophylline is known to stimulate the heart and reduce the pulmonary artery pressure but the clinical relevance of these effects in asthma is unknown. A few years ago it was suggested that xanthines produced a highly important stimulant effect on diaphragmatic contractions. Although it was reported in patients too (MURCIANO et al. 1984), it has not been confirmed in several recent studies involving asthmatic subjects. Enprofylline seems to be completely devoid of this action in both animals and humans. Again, the equal clinical efficacy of enprofylline and theophylline suggests that the importance of any respiratory muscle action of theophylline would only be marginal.

IV. Subcellular Mechanisms of Action

One obvious problem in a discussion about the subcellular mode of action of anti-asthma xanthines is that the important target cells have not been finally identified. Even if the locus of action can be said to the airways, the number of possible target cells is great. Another problem is that a biochemical mechanism is difficult to put into the context of cell function in vivo without making several undue simplifications of the complex cellular and intercellular chemistry that exist in healthy and diseased living tissue.

For a long time xanthines have been popular compounds in biological-biochemical research, which has been successful in attaching interesting mechanisms of action to them (see Sect. A. I.) This development has, in turn, increased the usefulness of xanthines as tools, making them some of the most frequently studied compounds in biomedical research. However, few studies have critically addressed the issue of whether the proposed mechanisms are critical for therapeutic response in asthma. Further studies are obviously warranted to support, or refute, many of the proposed modes of actions of xanthines in asthma, be it specific effects on calcium metatolism or select actions on an isoenzyme of phosphodiesterase. One mechanism that has recently attracted interest, the adenosine antagonism, may now be considered of little importance for the anti-asthma action. The work with two xanthines, however, has given important insights into other roles of adenosine.

V. Role of Endogenous Adenosine

Methods used in adenosine research have undergone impressive developments in recent years. However, a number of factors may still make it difficult to delineate specific roles for adenosine, in particular in humans.

- Adenosine receptors are ubiquitous and adenosine may well have different actions even in a single cell type.
- Administration of the nucleoside itself may not always yield valid data due to its rapid breakdown and limited distribution.
- When adenosine is given intravascularly, its potent vascular and reflex action will dominate and adequate doses for effects on a variey of tissues may not readily be given by this route.
- Even when topical application is employed, exogenous adenosine may not reach important biophases in the same way as the endogenously released nucleoside would.
- The action of adenosine analogues may not be entirely identical to that of adenosine itself and few, if any, analogues can be used in human experiments.
- Caffeine and theophylline can readily be given to man in adenosine antagonsitic doses, but an unknown variety of adenosine-independent actions are produced by these xanthines.

Theophylline and caffeine are both useful as adenosine antagonists because they are excellently distributed and equally effective as A_1 and A_2 adenosine receptor antagonists. However, these blocking drugs lack specificity. Enprofylline therefore has an outstanding value, because it indicates the actions of theophylline (and caffeine) where adenosine most likley is and is not involved (Fig 2).

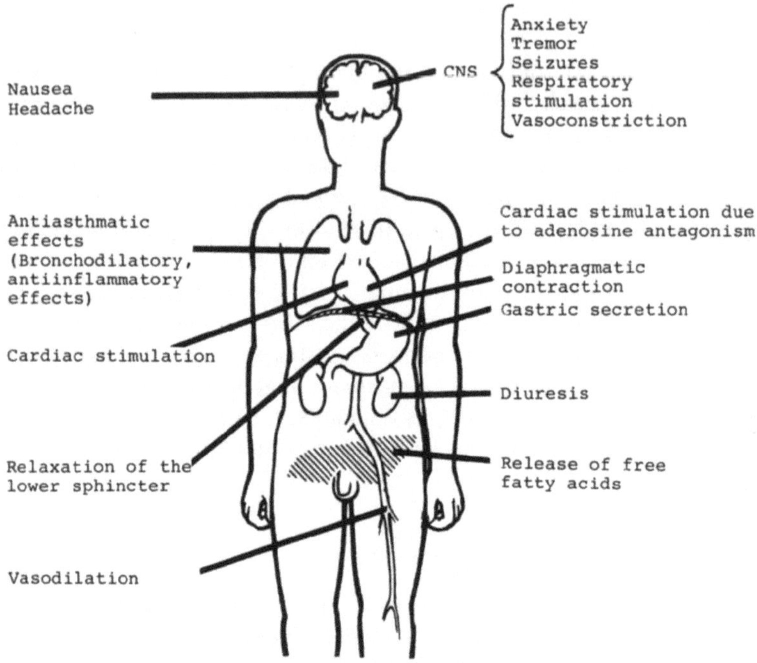

Fig. 2. Select actions of xanthines which have been demonstrated in man. *Right*, A number of excitatory actions of theophylline which are not showed by enprofylline, indicating that they reflect adenosine antagonism. *Left*, Such action which do not depend on adenosine, including the anti-asthma effects

As a result of the lack of theophylline-like diuretic effects, fee fatty acid releasing effects, gastric secretory effects and CNS-stimulant behavioural effects, enprofylline has demonstrated that adenosine is likely to have physiologically important inhibitory effects in these systems in animals and humans. Indeed the employment of enprofylline has led to the discovery that adenosine plays novel important roles in man even before their possible existence had been suggested by in vitro and animal research (PERSSON et al. 1986b).

D. Conclusion

Progress in research, partly sparked by the advent of enprofylline, has helped to focus on the importance of airway anti-inflammatory actions taking some of the attention away from the bronchodilator and extrapulmonary aspects of the anti-asthma efficacy of xanthines. However, it is still not possible to grade the importance of individual target cells for xanthines in asthamtic airways and, despite the enormous amount of data, the subcellular mechanism(s) of action of the anti-asthma xanthines are still a matter of speculation. One popular mechanism may now be excluded: enprofylline data have provided evidence against the importance of adenosine antagonism as a mode of action of anti-asthma xanthines. Further pharmacological work is required to positively reveal the important xanthines drug actions and the particular interactions that may occur between these compound and disease processes in asthmatic airways. Perhaps the advent of novel types of specific anti-asthma xanthines, even for topical use, may be required for the further assessment of which cellular and subcellular mechanisms are important for the clinical efficacy of this class of drugs.

References

Alexander MR, Dull WL, Kasik JE (1980) Treatment of chronic obstructive pulmonary disease with orally administered theophylline. JAMA 244:2286–2290

Allard E (1904) Über Theocinvergiftung. Arch Klin Med 80:510–519

Andersson PT, Persson CGA (1988) Developments in anti-asthma glucocorticoids. In: O'Donnell SR, Persson CGA (eds). Directions for New Antiasthma Drugs. Birkhäuser, Basel; 239–260

Andersson KE, Johannesson N, Karlberg B, Persson CGA (1984) Increase in plasma free fatty acids and natriuresis by xanthines may reflect adenosine antagonism. Eur J Clin Pharmacol 26:33–38

Andersson P, Brange C, Sonmark B, Stahre G, Erjefält I, Wieslander E, Persson CGA (1985) Anti-anaphylactic and anti-inflammatory effects of xanthines in the lung. In: Andersson KE, Persson CGA (eds) Anti-asthma xanthines and adenosine. Amsterdam, Excerpta, pp 187–192

Barclay J, Whiting B, Addis GJ (1982) The influence of theophylline on maximal response to salbutamol in severe chronic obstructive pulmonary disease. Eur J Clin Pharmacol 32:389–393

Barnes PJ, Greening AP, Neville L, Timmers J, Poole GW (1982) Single-dose slow release aminophylline at night prevents nocturnal asthma. Lancet 1:299–301

Bergstrand H (1985) Xanthines as phosphodiesterase inhibitors. In: Andersson KE, Persson CGA (eds) Antiasthma xanthines and adenosine. Amsterdam, Excerpta, pp 16–22

Berthet J, Sutherland EN, Rall TW (1957) The assay of glucagon and epinephrine with use of liver homogenates. J Biol Chem 229:351–357

Boe J, Swedish Socity of Chest Medicine (1988) Salbutamoel in acute asthma – a multicenter study. Am Rev Respir Dis 137:36A

Brenner M, Berkowitz R, Marshall N, Strunk RC (1988) Need for theophylline in severe steroid-requiring asthmatics. Clin Allergy 18:143–150

Cartier A, Lemaire I, L'Archeveque J, Ghezzo H, Martin RR, Malo JL (1986) Theophylline partially inhibits bronchoconstriction caused by inhaled histamine in subjects with asthma. J Allergy Clin Immunol 77:570–575

Chapman KR, Bryant D, Marlin GE, Mitchell C, Ruffin R, Inouye T, Pedersen B, Koskinen S, Osen SS, Ringdal N, Willey RF, Formgren H, Persson G, Källén A, Ljungholm K (1989) A placebo-controlled dose-response study of enprofylline in the maintenance therapy of asthma. Am Rev Respir Dis 139:688–693

Chrystyn H, Mulley BA, Peake MD (1988) Dose response relation to oral theophylline in severe chronic obstructive airways disease. Br Med J 297:1506–1510

Clark H, Cushley MJ, Persson CGA, Holgate ST (1989) The protective effects of intravenous theophylline and enprofylline against histamine-and adenosine-5'-monophosphate-provoked bronchoconstriction: implications for the mechanisms of action of xanthine derivatives in asthma. Pulm Pharmacol 2:147–154

Crescioli S, Spinazzi A, Paleari D, Pozzan M, Mapp CE, Fabbri LM (1988)

Theophylline inhibits early and late asthmatic reactions induced by allergens in atopic subjects with asthma. Am Rev Respir Dis 137:A35
Cushley MJ, Tattersfield AE, Holgate ST (1983) Inhaled adenosine and guanosine on airway resistance in normal and asthmatic subjects. Br J Clin Pharmacol 15:161–166
Dutoit J, Salome CM, Woolcock AJ (1987) Inhaled corticosteroids reduce the severity of bronchial hyperresponsiveness in asthma but oral theophylline does not. Am Rev Respir Dis 136:1174–1178
Erjefält I, Persson CGA (1989) Inflammatory passage of plasma macromolecules into airway wall and lumen. Pulm Pharmacol 2:93–102
Fabbri LM, Allessandri MV, De Marzo N, Zocca E, Paleari D (1986). Long-lasting protective effect of slow-release theophylline on asthma induced by ultrasonically nebulized distilled water. Ann Allergy 56:171–175
Fink G, Mittelman M, Shohat B, Spitzer SA (1987) Theophylline-induced alterations in cellular immunity in asthmatic patients. Clin Allergy 17:316–321
Finney M, Karlsson JA, Persson CGA (1985) Effects of bronchoconstrictors on a novel human small airway preparation. Br J Pharmacol 85:29–37
Fliuk RB, Easton PA, Anthonisen NR (1985) Responses to large doses of salbutamol and theophylline in patients with chronic obstructive pulmonary disease. Am Rev Respir Dis 132:871–874
Fredholm BB (1980) Are the actions of methylxanthines due to antagonism of adenosine? Trends Pharmacol Sci 1:129–132
Guyatt GH, Townsend M, Pugsley O (1987) Bronchodilators in chrònic airflow limitation. Am Rev Respir Dis 135:1069–1074
Harada H, Ishizaka A, Yonemaru M, Mallick AA, Hatherill JR, Zheng H, Lilly CM, O'Hanley PT (1989) The effects of aminophylline and pentoxifylline on multiple organ damage after escherichia coli sepsis. Am Rev Respir Dis 140:974–980
Ishizaki T, Minegishi A, Morishita M, Odaijma Y, kanagawa S, Nagai T, Yamaguchi M (1988) Plasma catecholamine concentrations during a 72-hour aminophylline infusion in children with acute asthma. J Allergy Clin Immunology 82:146–154
Jenne JW, Siever JR, Druz WS, Solano JV, Cohen SM, Sharp JT (1984) The effect of maintenance theophylline therapy on lung work in severe chronic obstructive pulmonary disease while standing and walking. Am Rev Respir Dis 130:600–605
Karlsson JA, Persson CGA (1981) Influence of tracheal contraction on relaxant effects in vitro of theophylline and isoprenaline. Br J Pharmacol 74:73–79
Kips J, Pauwels R, Van Der Straeten M (1989) Effect of theophylline on the endotoxin-induced airway inflammation and bronchial hyperresponsiveness. J Allergy Clin Immunol
Kjellin G, Persson CGA (1985) Xanthine derivatives. In:Andersson K-E, Persson CGA (eds) Anti-asthma xanthines and adenosine. Excerpts, Amsterdam, pp 223–229
Kröll F, Karlsson JA, Nilsson E, Ryrfeldt Å, Persson CGA (1990) Rapid clearance of xanthines from airway and pulmonary tissues. Am Rev Respir Dis 141:1167–1171
Lindenschmidt RC, Witschi HP (1985) Attenuation of pulmonary fibrosis in mice by aminophylline. Biochem Pharmacol 34:4269-4273
Littenberg B (1988) Aminophylline treatment in severe, acute asthma. JAMA 259:1678–1684
Ljungholm K, Kallen A, Persson CGA, Persson G (1990) The anti-asthma efficacy of xanthines added to regimens of non-glucocorticoid and glucocorticoid therapies. Eur Respir J (in press)
Lundblad KA, Persson CGA (1988) The epithelium and the pharmacology of guinea-pig tracheal tone in vitro. Br J Pharmacol 93:909–917
Lunell E, Svedmyr N, Andersson KE, Persson CGA (1983) A novel broncholdilator xanthine apparently without antagonism and tremorogenic effect. Eur J Resp Dis 64:333–339

Lunell E, Andersson KE, Persson CGA, Svedmyr N (1984) Intravenous enprofylline in asthma patients. Eur J Resp Dis 65:28–34

Magnussen H, Reuss G, Jorres R (1987) Theophylline has a dose related effect on the airway response to inhaled histamine and methacholine in asthmatics. Am Rev Respir Dis 136:1163–1167

Mann JS, Holgate ST (1985) Specific antagonism of adenosine induced bronchoconstriction in asthma by oral theophylline. Br J Clin Pharmacol 19:85–92

Mapp C, Boshetto P, Dal Veccho L, Crescioil S, De Marzo N, Paleari D, Fabbri LM (1987) Protective effect of antiasthma drugs on late asthmatic reactions and increased airway responsiveness induced by toluene diisocyanate in sensitized subjects. Am Rev Respir Dis 136:1403–1407

Marquardt DL, Parker CN, Sullivan TJ (1978) Potentiation of mast-cell mediator release by adenosine. J Immunol 120:871–878

Mentzer RM, R, Berne RM (1975) Release of adenosine by hypoxic canine lung tissue and its possible role in pulmonary circulation. Am J Physiol 229:1625–1631

Murciano D, Aubier M, Lecocguic Y, Pariente R (1984) Effect of theophylline on diaphragmatic strength and fatigue in patients with chronic obstructive pulmonary disease. N Engl J Med 311:349–353

Naclerio RM, Bartenfelder D, Proud D, Togias AG, Meyers DA, Kagey Sobotka A, Norman PS, Lichtenstein LM (1986) Theophylline reduces histamine release during pollen-induced rhinitis. J Allergy Clin Immunol 78:874–876

Nassif EG, Weinberger M, Thompson R, Huntley W (1981) The value of maintenance theophylline in steroid-dependent asthma. N Engl J 304:71–75

Nielson CP, Crowley JJ, Cusak BJ, Vestal RE (1986) Therapeutic concentrations of theophylline and enprofylline potentiate catecholamine effects and inhibit leukocyte activation. J Allergy Clim Immunol 78:660–667

O'Donnell SR, Erjefält I, Persson CGA (1990) Early and late tracheobronchial plasma exudation by platelet activating factor administered to the airway mucosal surface in guinea-pigs: effect of WEB 2086 and enprofylline. J Pharmacol Exp Ther 254:65–70

Orange PP, Kaliner MA, Laraia PJ, Austen KF (1971) Immunological release of histamine and slow reacting substance of anaphylaxis from human lung II. Influence of cellular levels of cyclic AMP. Fed Proc 30:1725–1729

Passamonte PM, Martinez AJ (1984) Effect of inhaled atropine or metaproterenol in patients with chronic airway obstruction and therapeutic serum theophylline levels. Chest 85:610–615

Pauwels R, Van Renterghem D, Van Der Straeten M, Johannesson N, Persson CGA (1985) The effect of theophylline and enprofylline on allergen-induced bronchoconstriction. J Allergy Clin Immunol 76:583–590

Perruchoud AP, Yerger L, Abraham W (1984) Differential effects of aminophylline on the early and late antigen-induced bronchial obstruction in allergic sheep. Respiration 46:44

Persson CGA (1982) Universal adenosine receptor antagonism is neither necessary nor desirable with xanthine antiasthmatics. Med Hypoth 8:515–526

Persson CGA (1983) The profile of action of enprofylline or why adenosine antagonism seems less desirable with xanthine antiasthmatics. In: Morley J, Rainsford KD (eds) Pharmacology of asthma. Birkhäuser, Basel, pp 115–129

Persson CGA (1985a) On the medical history of xanthines and other remedies for asthma: a tribute to HH Salter. Thorax 40:881–886

Persson CGA (1985b) Subdivision of xanthines. In: Andersson KE, Persson CGA (eds) Anti-asthma xanthines and adenosine. Excerpta, Amsterdam, pp 23–40

Persson CGA (1986) Development of safer xanthine drugs for treatment of obstructive airway disease. J Allergy Clin Immunol 78:817–824

Persson CGA (1988) Xanthines as airway antiinflammatory drugs. J Allergy Clin Immunol 81:615–617

Persson CGA (1990a) On the medical history of asthma and rhinitis. In: Mygind N, Pipkorn U, Dahl R (eds) Rhinitis and asthma. Simliarities and differences. Munksgaard, Copenhagen, pp 9–21
Persson CGA (1990b) Exudative indices in airways inflammation. In: Persson CGA, Brattsand R, Laitinen L, Venge P (eds) Inflammatory indices in chronic bronchitis. Birkhäuser, Basel 243–256
Persson CGA, Erjefält I (1982) Seizure activity in animals given enprofylline and theophylline, two xanthines with partly different mechanisms of action. Arch Int Pharmacodyn Ther 258:267–282
Persson CGA, Gustafsson B (1986) Tracheal relaxation of combinations between xanthines and between a β_2-receptor agonist and xanthines. Lung 164:33–40
Persson CGA, Karlsson JA (1987) In vitro responses to bronchodilator durgs. In: Jenne JW, Murphy S (eds) Drug therapy for asthma. Research and clinical practice. Dekker, New York, pp 129–176
Persson CGA, Kjellin G (1981) Enprofylline, a principally new antiasthmatic xanthine. Acta Pharmacol Toxicol 49:313–316
Persson CGA, Ekman M, Erjefält I (1979) Vascular antipermeability effects of β-receptor antagonist and theophylline in the lung. Acta Pharmacol Toxicol 44:216–220
Persson CGA, Erjefält I, Andersson P (1986a) Leakage of macromoleculs from guinea-pig tracheobronchial microcirculation. Effects of allergen, leukotrienes, tachykinins, and antiasthma durgs. Acta Physiol Scand 127:95–104
Persson CGA, Andersson KE, Kjellin G (1986b) Effects of enprofylline and theophylline may show the role of adenosine. Life Sci 38:1057–1072
Persson CGA, Erjefält I, Gustafasson B (1988) Xanthines – symptomatic or prophylactic in asthma? In: O'Donnel SR, Persson CGA (eds) Direction for new anti-asthma drugs. Birkhäuser, Basel, pp 137–155
Pollock J, Kiechel F, Cooper D, Weinberger M (1977) Relationship of serum theophylline concentration to inhibition of exercise-induced bronchospasm and comparison with cromolyn. Pediatrics 60:840–844
Racineaux JL, Troussier J, Turcant J, Tuchais E, Allain P (1981) Comparison of bronchodilation effects of sablutanol and theophylline. Bull Eur Physiopathol Respir 17:799–806
Ruffin R, Bryant D, Burdon J, Marlin G, Mitchell C, O'Hehir R, Wilson J, Woolocock A, Webb S (1988) Comarison of the effects of nebuilzed terbutaline with intravenous enprofylline in patients with acute asthma. Chest 93:510–514
Schmedeberg O (1905) Über die Anwendung des Theophyllins als Diuretikum. Dtsch Arch Klin Med 82:395–408
Snyder SH (1981) Adenosine receptors and the actions of methylxanhines. Trends Neurosci 4:242–244
Sutton PP, Pavia D, Bateman JRM, Clarke SW (1981) The effect of oral aminophylline on lung mucociliary clearance in man. Chest 80 (suppl):889–892
Taylor DR, Buick B, Kinney C, Lowry RC, McDevitt DG (1985) The efficacy of orally administered theophylline, inhaled salbutamol, and a combinaion of the two as chronic therapy in the management of chronic bronchitis with reversible air-flow obstruction. Am Rev Respir Dis 131:747–751
Ther L, Maschaweck R, Hergott J (1957) Antagonismus zwischen Adenosine und Methylxanthinen am Reizleitungssystem des Herzens. Arch Exp Pathol Pharmacol 231:586–601
Torphy T (1988) Action of mediators on airway smooth muscle: Functional antagonism as a mechanism for bronchodilator drugs. In: O'Donnell SR, Persson CGA (eds) Directions for new antiasthma drugs. Birkhauser, Basel, pp 37–54
Vilsvik JS, Persson CGA, Amundsen T et al. (1990) Comparison between theophylline and an adenosine non-blocking xanthine in acute asthma. Eur Respir J 3:27–32
Weinberger M (1984) Pharmacology and therapeutic use of theophylline. J Allergy Clin Immunol 73:525–540

Welsh CH, Lien D, Worthen GS, Veil JV (1988) Pentoxifylline decreases endotoxin-induced pulmonary neutrophil sequestration and extravascular protein accumulation if the dog. Am Rev Respir Dis 138:1106–1114
Welton AF (1980) Regulatory role of adenosine in antigen-induced histamine-release from the lung tissue of actively sensitized guinea-pigs. Biochem Pharmacol 29:1085–1092

CHAPTER 8

Glucocorticosteroids in Asthma

S.F. SMITH, C.P. PAGE, P.J. BARNES, and R.J. FLOWER

A. Mechanism of Action of Glucocorticosteroids

The mechanisms by which glucocorticosteroids act to control asthma are unclear, partly because the underlying pathology causing this condition remains to be fully elucidated, and also because many of the actions of glucocorticosteroids are still unexplained. In particular, their anti-inflammatory effects are the most frequently harnessed by clinicians, but the least well understood. Glucocorticosteroids can cause certain cellular responses by direct interaction with cell membranes or, more commonly, they complex with specific receptors to modulate gene expression and protein synthesis. In the context of asthma therapy, steroid-induced protein synthesis may occur either in the respiratory tract, or elsewhere in the body, with secondary effects occurring in the airways. For example, stimulation by glucocorticosteroids of α_1-proteinase inhibitor synthesis in the liver increases circulating levels of this protein and thus may modify the proteinase-antiproteinase balance of the airways and lungs.

Likewise down-regulation of interleukin-1 (IL-1) synthesis may also be important in resolving the inflammatory response in the airways (KNUDSEN et al. 1987).

One thing that can be stated with certainty is that steroids reduce practically every facet of airway inflammation and this makes a description of the totality of steroid actions very difficult indeed.

I. Induction of Protein Synthesis

Glucocorticosteroid receptors are found in most cell types (BALLARD et al. 1974). The interaction of glucocorticosteroids with the specific receptor induces a conformational change in the latter, which in turn permits it to interact with a specific region of the nuclear DNA, thereby regulating the production of complementary mRNAs and the proteins which they encode (see RINGOLD 1985 for a recent review). It is important to remember that not all genes which are modulated by steroids are up-regulated: many, for example the IL-1 gene, are down-regulated.

Glucocorticosteroids have been shown to induce the synthesis of a number of proteins. However, most of these regulate the metabolic effects of glucocorticosteroids (SCHUTZ et al. 1975; IYNEDIJIAN and HANSON 1977; HAYNES and MURAD 1985) rather than the anti-inflammatory actions which are believed to be important in the control of asthma. The action of some potentially anti-inflammatory proteins, the synthesis of which is thought to be glucocorticosteroid-modulated, is discussed below.

1. Lipocortins

The lipocortins are a family currently consisting of six member proteins which have differing molecular weights and N-terminal sequences (PEPINSKY et al. 1988; CROMPTON et al. 1988). They share considerable sequence homology, their basic structure consisting of four or more repeats of a single sequence of about 70 amino acids. Interestingly, this primary structure occurs in a number of other proteins, known collectively as annexins (GEISOW and WALKER 1986) including the calpactins, calcimedins, calelectins, endonexins and chromobindins, all of which share the common property of binding calcium and phospholipid (SARIS et al. 1986; HUANG et al. 1986; GEISOW 1986). Therefore, it has been suggested that this primary protein structure may facilitate the association of the proteins with membranes. Calcium ions are essential for this interaction and may form a "bridge" between phospholipid and protein moieties. However, there is recent evidence to suggest that there are a discrete number of saturable binding sites for at least one lipocortin (lipocortin 1) on the surface of some cell types but not others (GOULDING et al., in press) and it is unlikely that this particular protein associates in a non-specific way with plasma membranes.

Since only one member of the lipocortin subfamily has been studied in depth worth respect to steroid-inducibility, the remainder of this section will focus on this protein, lipocortin 1. Lipocortin 1 has a molecular weight of 37 kDa and the gene encoding the human protein has been cloned (WALLNER et al. 1986). The protein consists of 348 amino acids of which roughly 30% are charged, thus rendering it a highly polar molecule (WALLNER et al. 1986). It has one potential glycosylation site, and several potential sites for phosphorylation. The significance of the latter will be discussed below. Interestingly, the protein has no leader peptide, even though it is found in extracellular fluids (BLACKWELL et al. 1982; SMITH et al. 1990) and thus its mechanism of release from cells is currently unknown. Beyond these basic facts, almost every role or action suggested for this enigmatic protein by one research team has been disputed and rejected by another.

In this chapter, we suggest how lipocortin 1 may have some therapeutic role in asthma, and discuss our hypothesis in the light of the evidence to date. It is proposed that lipocortin 1 is glucocorticosteroid-regulated and that it is anti-inflammatory, suppressing inflammation possibly by down-regulating the activity of phospholipase A_2 (PLA_2), the rate-limiting enzyme for the production of the eicosanoid inflammatory mediators and platelet activating factor (PAF). PLA_2 is also an important enzyme in the control of other inflammatory processes, such as neutrophil migration and some mechanisms of oxidant formation (MARIDONNEAU-PARINI et al. 1989).

a) Evidence that Lipocortin Synthesis and Release is Modulated by Glucocorticosteroids

An early study on rat isolated peritoneal macrophages showed that glucocorticosteroids initially caused release of a stored PLA_2 inhibitor, now known to be one of the lipocortin family, from these cells and subsequently induced de novo synthesis of further protein (BLACKWELL et al. 1982). A later study by the same authors showed that only the binding of an agonist to the glucocorticosteroid receptor would stimulate increased lipocortin release from these cells (BLACKWELL 1983). However, it seems that not all cells contain lipocortin stores, and moreover some may not synthesize lipocortin in response to glucocorticosteroid exposure. For example, in a recent publication by BRONNEGARD et al. (1988) several cell types including human alveolar macrophages were maintained in culture. In none of them was mRNA coding for lipocortin increased following glucocorticosteroid treatment, and protein levels were similarly unchanged. Similarly, BIENKOWSKI et al. (1989) found no evidence for dexamethasone-induced lipocortin 1 synthesis by U937 cells in culture – both mRNA and protein levels of lipocortin were unchanged – whilst in a study of human endothelial cells in culture, HULLIN and co-workers (1989) found no evidence for induction of lipocortin 1 synthesis by dexamethasone. In contrast, AMBROSE and HUNNINGHAKE (1989) found that the lipocortin content of human alveolar macrophages in culture was significantly increased on exposure to glucocorticosteroids, but not in response to other steroids. Furthermore, in a study on feline airways, tracheal explants cultured in the presence of 10^{-5} *M* dexamethasone showed an increase in lipocortin content of about 60% within 8 h and 220% over 24 h (JD LUNDGREN et al. 1988).

In contrast to the variation in the results of in vitro studies, data from in vivo studies on the glucocorticosteroid inducibility of lipocortin 1 seem more coherent. Early studies showed that glucocorticosteroid treatment of rats increased the PLA_2 inhibitory activity (subsequently shown to be lipocortin) released from their peritoneal macrophages compared to control animals (reviewed by FLOWER 1988). Later when lipocortin 1 was cloned and sequenced in 1986, it was shown by means of Northern blotting that the lipocortin mRNA transcription was increased sixfold in rat resident peritoneal cells (mainly macrophages) 2 h after intraperitoneal injection of these animals with dexamethasone (WALLNER et al. 1986). The increase was less in elicited exudate cells, and interestingly, protein levels were unchanged in either group.

SMILLIE et al. (1989) were able to demonstrate rapid (within 40 min) induction of lipocortin 1 and lipocortin 2 synthesis in peritoneal lavage macrophages after administration to rats of dexamethasone or hydrocortisone. Furthermore, the administration of the glucocorticosteroid receptor antagonist RU486 prevented this response and also lowered the resting levels of the protein, suggesting that the endogenous steroid hormones had a tonic influence upon levels of the protein.

In a recent publication (BROWNING et al. 1990) the generation of lipocortin 1 by human peripheral monocytes in primary culture was shown to be increased substantially following exposure to dexamethasone, also suggesting that the production of the protein can be glucocorticosteroid driven. A significant fraction of the protein in both these studies could be recovered from the cell surface, once again pointing to some translocation mechanism which enables the protein to move from the site of generation to its pericellular location.

Recently, studies of lipocortin levels have been facilitated by the development of an accurate enzyme-linked immunosorbent assay for estimation of lipocortin 1. In a recent investigation on human subjects, a positive correlation was observed between serum cortisol levels and the concentrations of lipocortin 1 present in bronchoalveolar lavage (BAL) fluid, giving support to the hypothesis that lipocortin may be glucocorticosteroid inducible in humans (GOULDING et al. 1989). In a second, retrospective study on patients undergoing glucocorticosteroid therapy for a variety of disorders (none for asthma), lipocortin concentrations were increased in BAL fluid from patients receiving oral dexamethasone, but not in that from patients inhaling beclamethasone (SMITH et al. 1989a). These data suggest that some exogenous glucocorticosteroids increase lipocortin production, at least in the lung.

The most convincing evidence that lipocortin is induced in humans comes from a study by GOULDING et al. (1990). In this study volunteers were given 100 mg hydrocortisone intravenously and blood samples were taken before and at regular time intervals following the injection. Cells were separated from the blood samples and the amount of intracellular and cell surface lipocortin was measured in the various subpopulations. No free lipocortin was detectable in plasma at any time point, although there was a significant amount associated with mononuclear and polymorphonuclear cell surfaces. Within 120 min concentrations of lipocortin had risen in the monocytes and remained elevated for the duration of the experiment (4 h). No equivalent change was seen in polymorphonuclear cells at this time, although this does not exclude the possibility that induction may incur at a later time point. The mononuclear and polymorphonuclear cells were the richest sources of lipocortin in human blood with much less being associated with the lymphocyte fraction and only tiny amounts detectable in the platelets and red blood cell fraction. Placebo injections had no effect on the generation of lipocortin by human mononuclear cells.

Inductions were also seen in mononuclear cells taken from volunteers 30 min after they had received the steroid, but not if the cells were first prepared and isolated and then incubated with the steroid. These data point to a fact often seen, that there is the difficulty of obtaining inductions in vitro whereas the induction process in vivo seems to proceed readily. Presumably the environment of the cells during the critical induction phase is of great importance.

Thus, although there are contradications in the data, the consensus of the in vivo studies is that lipocortin 1 synthesis (and, in some cells, release) can be modulated by endogenous and exogenous glucocorticosteroids. The inconsistencies in the results of in vitro studies may well reflect differences between the cell types investigated, between laboratories in methods of cell isolation and culture, or in techniques of lipocortin estimation. Alternatively, some as yet unidentified co-factor may be necessary for the glucocorticosteroids to be effective inducers of lipocortin synthesis in vitro.

b) Evidence that Lipocortin is Present in the Lung

When the distribution of lipocortin was investigated in the rat by analysis of tissue homogenates, the lung was found to be a rich source of this protein (PEPINSKY et al. 1986). In addition, bovine (BOUSTEAD et al. 1988), murine (ERRASFA and RUSSO-MARIE 1988) and porcine lungs have been used as a source for the isolation several members of the annexin family including lipocortin 1. Since macrophages contain abundant lipocortin 1, the high levels of pulmonary lipocortin were originally attributed to the presence of large numbers of macrophages in the lung (reviewed in FLOWER 1988). Western blotting of BAL fluid collected from healthy volunteers and patients with lung disease has shown that lipocortin is always present in human lung lavage fluids, irrespective of the age, sex, smoking history or clinical status of the subject (SMITH et al. 1990). Interestingly, however, although human macrophages in culture certainly contain and release lipocortin 1 (AMBROSE and HUNNINGHAKE 1989; SMITH et al, unpublished observation), no relationship has yet been demonstrated between lipocortin and alveolar macrophage recovery in BAL fluid. In contrast, recent immunocytochemical studies of ovine foetal and human foetal and neonatal lung show most of the detectable lipocortin to be in the airway epithelial cells, ducts of the submucosal glands and mucous acinal cells, (JOHNSON et al. 1989; JOHNSON et al. 1990) which suggests that lipocortin detected in lung lavage fluids may be produced, at least in part, within the lung itself.

Thus, if the distribution found by JOHNSON and colleagues (1989; 1990) is observed in the mature, human airways, lipocortin could be in regions of the respiratory tract which is important in the pathogenesis of asthma.

c) Evidence that Lipocortin is Anti-Inflammatory

The majority of early studies in which the existence of glucocorticosteroid-inducible PLA_2 inhibitors was elucidated were on isolated, perfused, guinea pig lung models. These important experiments were instrumental in demonstrating the existence of the protein mediators, now called lipocortins, in the lung, and these studies have been comprehensively reviewed by

FLOWER (1988). Early investigations of the anti-inflammatory actions of lipocortin were performed using partially purified lipocortin and showed that this protein could mimick the effects of exogenous glucocorticosteroids against inflammation induced in the rat paw (HIRATA 1983; PARENTE et al. 1984) or the rat pleura (BLACKWELL et al. 1982) by carrageenan. These papers have also been reviewed by FLOWER (1988). More recently, pure protein has been used in in vivo investigations. CIRINO et al. (1989) have shown that recombinant human lipocortin 1 reduces carrageenan-induced inflammation of the rat paw when given simultaneously with the inflammatory stimulus or during ensuing inflammation. In their study, the reduction in swelling was dose related and was lost when the protein was boiled before administration.

Interestingly, lipocortin was only effective in models of inflammation which directly depend upon eicosanoids for their generation. Thus while the protein was active against carrageenan-induced oedema and against oedema generated by the injection of PLA_2, and partially effective against oedema induced with the compound 48/80, it was ineffective against stimuli which acted in a different manner. For example, it failed to control the oedema induced by the injection of PAF or by the injection of dextran. In the case of the PLA_2 oedema, which is also mast cell dependent, a direct interaction of lipocortin with the enzyme seemed to be responsible for the inhibition. Furthermore, lipocortin was effective in controlling the oedema in rats in which mast cell amines, complement or kininogen had been depleted by appropriate treatment, but was ineffective in animals pretreated with the dual lipoxygenase and cyclooxygenase inhibitor BW755C (CIRINO et al. 1989). All this evidence suggests that lipocortin exerts its anti-inflammatory effect by controlling eicosanoid generation.

The data indicating an anti-inflammatory role for lipocortin are in direct contrast to a study by NORTHUP et al. (1988), in which the authors showed that lipocortin purified from human placenta caused no reduction in carrageenan-induced oedema – indeed if anything it exacerbated the inflammation. In another very recent in vivo investigation, it was observed that lipocortin isolated from the murine lung had a potent anti-inflammatory action in a murine model of subcutaneous inflammation (ERRASFA and RUSSO-MARIE 1989). Clearly there are inconsistencies in the data which are not easy to resolve. However, it is possible that the purification processes used by different authors and the different sources of the material used result in the production of proteins with varying degrees of activity.

This idea is given credence by the recent work of BROWNING et al. (1990) who found that lipocortin 1 was easily denatured at gas-water interphases, losing all its biological activity, and that the recombinant material displayed marked heterogeneity at the N-terminus and in its folding properties. When these factors were taken into account and corrected for, biologically active material was invariably obtained. These studies enabled Browning and his colleagues to demonstrate that whilst the intact, correctly refolded lipocortin

1 was very effective when given intravenously as an anti inflammatory agent in the rat, the non-refolded material was totally inactive.

To date, there have been very few studies demonstrating that highly purified lipocortin 1 has any anti-inflammatory actions in the lung or airways. The study by JD LUNDGREN et al. (1988) mentioned above showed that dexamethasone-induced inhibition of respiratory glycoconjugate synthesis in feline tracheal rings is mediated via lipocortin, since it could be prevented by the use of neutralizing monoclonal antilipocortin antibodies. These data support the hypothesis that lipocortin could be anti-inflammatory in the airways.

Thus, although the evidence is not yet entirely unanimous, the consensus appears to be in favour of lipocortin possessing anti-inflammatory activity, and of induction of its synthesis being one of the mechanisms by which glucocorticosteroids (endogenous or exogenous) suppress inflammation.

d) How Does Lipocortin Suppress Inflammation?

There is now abundant evidence that the eicosanoids (prostaglandins, leukotrienes and thromboxanes), PAF and the cells which release them play an important role in airway inflammation (BARNES et al. 1988). One enzyme, PLA_2, is thought to play a key role in the biosynthesis of many of these lipid mediators. PLA_2 releases arachidonic acid, the precursor of all eicosanoids from membrane phospholipids. Furthermore, following removal of the fatty acid at the 2 position by PLA_2, certain lysophospholipids can be further metabolized to PAF. Clearly, then, PLA_2 is likely to be an important enzyme in the regulation of the inflammatory process. Thus, retrospectively, it seems unsurprising that anti-inflammatory steroids were shown in 1976 to prevent the release of arachidonic acid from the 2′ position of phospholipids in fibroblasts in vitro, that is, to apparently down-regulate PLA_2 activity (HONG and LEVINE 1976). Furthermore, it was found that the same occurred in the isolated, perfused guinea pig lung (BLACKWELL et al. 1978). Subsequent experiments in the isolated lung and in a variety of cell types in vitro (DANON and ASSOULINE 1978; RUSSO-MARIE et al. 1979; DI ROSA and PERSICO 1979; FLOWER and BLACKWELL 1979) showed that this effect could be prevented by inhibitors of protein or mRNA synthesis.

The next step was to demonstrate that the protein synthesized in response to glucocorticosteroids did inhibit PLA_2. Attempts to demonstrate this apparently simple point have created confusion and acrimony. Early studies were carried out on isolated guinea pig lungs arranged in such a way that effluent from one lung (the generator) perfused a second lung (the target). PLA_2 activity was assessed in the target which was treated with cycloheximide to prevent any independent response to glucocorticosteroids. Infusion of dexamethasone into the generator lung was followed by a fall in PLA_2 activity in the target. The factor was shown to be soluble and protease and heat sensitive, strongly suggesting that it was a protein. Several groups working independently on different PLA_2-regulated functions came to the same conclusion, namely, that glucocorticosteroids induce that synthesis of a second messenger protein which inhibits PLA_2 activity; comparison of data showed that the investigators were all dealing with the same protein, now known as lipocortin (DIROSA et al. 1984). Again, this has been fully reviewed by FLOWER (1988).

Interestingly, phosphorylation of lipocortin 1 has been shown to reduce its PLA_2-inhibiting capacity (HIRATA 1981), so it is plausible that a change in phosphorylation state could facilitate the development of chronic inflammation. The mechanism by which phosphorylation might modify PLA_2-inhibitory capacity remains to be elucidated, although two recent studies have shown that phosphorylated lipocortin 1 may associate more readily than its phosphate-free counterpart with phospholipid (SCHLAEPFER and HAIGLER 1987; POWELL and GLENNEY 1987). It now appears that lipocortin possesses a number of potential phosphorylation sites, different sites being prefered by different kinases (VARTICOVSKI et al. 1988; SCHLAEPFER and HAIGLER 1988). Lipocortin 1 can act as a substrate for both epidermal growth factor (EGF) (PEPINSKY and SINCLAIR 1986) and insulin receptor kinases (KARASIK et al. 1988), which suggests that lipocortin may have a role in signal transduction although this remains to be fully elucidated. It has been suggested that variation in the position of phosphorylation sites in the N-terminal sequence of different lipocortins may create variation in their potential targets, or their control mechanisms (VARTICOVSKI et al. 1988).

Since the identification and purification of lipocortin, a controversy concerning its PLA_2-inhibitory capacity has developed. This is probably due, at least in part, to the problems involved in the assay of PLA_2 activity. First, which is the target enzyme in vivo? Many different PLA_2s have been identified to date (although a good number of these are non-mammalian), (CHANG et al. 1987). The soluble pancreatic enzyme is usually used in in vitro experiments because it is readily available and relatively easy to work with, but in vivo the target may be a membrane-bound enzyme. Furthermore, even the pancreatic enzyme fails to obey basic Michaelis-Menton Kinetics (CHANG et al. 1987). Selection of phospholipid substrate may also be important. It is perhaps unsurprising that whilst lipocortin 1 and lipocortin peptides have been shown to inhibit PLA_2 in some studies (HIRATA 1983; PEPINSKY et al. 1986; HUANG et al. 1986; TOUQUI et al. 1986; MIELE et al. 1988) other authors contend that lipocortin 1 inhibits PLA_2 only when substrate is rate limiting, and that it does so by substrate sequestration, that is, it is not a true enzyme inhibitor (DAVIDSON et al. 1987; AARSMAN et al. 1987; HAIGLER et al. 1987; LENTING et al. 1988).

An alternative method of determining the PLA_2 inhibitory capacity of lipocortin 1, perhaps more relevant to the study of inflammation, is to assess its effect on a PLA_2-mediated process in a cellular system. Again, the results of such studies are conflicting. ERRASFA et al. (1985) showed that a monoclonal antibody to renocortin (closely related to the lipocortin family) could prevent the inhibition by dexamethasone of prostaglandin production by cultured human embryonic skin fibroblasts. More relevant to asthma, a recent study by the same authors (ERRASFA et al. 1988) showed that lipocortin purified from mouse lung inhibited PLA_2 activity in alveolar macrophages from the guinea pig. In addition, preincubation of cells with mouse lung lipocortin inhibits the release of prostaglandin E_2, leukotriene B_4 (LTB_4) and PAF from rat pleural neutrophils in vitro – an effect which is prevented by the use of antilipocortin antisera (FRADIN et al. 1988). These observations are important, since these cells participate early in the inflammatory response. Some studies have shown that lipocortin blocks eicosanoid and PAF release from rat leucocytes in a dose-dependent way (PARENTE et al. 1984; PARENTE and FLOWER 1985). Lipocortin 1 reduced release of prostacyclin from human umbilical artery rings and blocked thromboxane release from isolated, perfused guinea pig lung (CIRINO et al.). In contrast, NORTHUP et al. (1988) found that lipocortin 1 had no effect on arachidonate release from zymosan-stimulated murine resident peritoneal macrophages.

Possibly results would be easier to interpret if cell culture methods were standardized between laboratories – for example, some authors routinely add foetal calf serum to culture media, thereby also adding unknown amounts of glucocorticosteroids and growth factors. Furthermore, it is quite possible that lipocortin 1 may be rapidly inactivated on addition to living cells either reversibly by phosphorylation (see above) or irreversibly by proteolysis. Incubation of human recombinant lipocortin to medium conditioned by exposure to mixed populations of BAL cells from human subjects results in proteolytic cleavage of the N-terminal which is believed to be essential for PLA_2-inhibitory activity (SMITH et al. 1989b, 1990). In addition, media exposed to neutrophils degrade lipocortin with greater avidity than those conditioned by alveolar macrophages alone, and neutrophil elastase is at least one neutrophil-derived enzyme which causes this specific cleavage in vitro (SMITH et al. 1989b). Cathepsin D and calpain I can also cleave lipocortin 1 at its N-terminal end (ANDO et al. 1989). Interestingly, the proteolysed protein has a greater sensitivity to calcium and enhanced phospholipid binding, although the significance of this remains to be fully established.

In addition, calcium status is important. Lipocortin has four calcium binding sites, and the calcium may be an important component in the binding of lipocortin to phospholipid (SCHLAEPFER and HAIGLER 1987). Calcium is certainly necessary to hold lipocortin to cell membranes, since washing the cells in a calcium-free medium or adding chelating agents followed by centrifugation leads to the repartitioning of lipocortin from the cell pellet to the acellular supernatant (CIRINO and FLOWER 1987; WILLIAM et al. 1988). In contrast, the addition of calcium ionophore to HL-60 cells causes lipocortin to associate with the particulate fraction (WILLIAM et al. 1988).

In summary, it seems likely that under some conditions, and in some cell types, lipocortin does reduce the activity of PLA_2 either directly or by interaction. With the substrate, however, under other conditions, or in the presence of different cell types (or different subpopulations of cells), it is equally possible that lipocortin itself is inactivated, reversibly or irreversibly, and this may account for some of the differences between studies.

Inhibition of PLA_2 activity by whatever mechanism would explain how lipocortin 1 can suppress inflammation by the down-regulation of eicosanoid inflammatory mediators, PAF and other PLA_2-dependent processes such as cell migration. Interestingly there appear to be no differences in basal or A-23187-induced arachidonic acid metabolism between alveolar macrophages collected from control atopic and asthmatic subjects, or in their regulation by methylprednisone, suggesting that there is no intrinsic difference in arachidonic acid metabolism in asthmatics (BALTER et al. 1988).

e) Does Lipocortin Have Other Potentially Anti-Inflammatory Actions Which Might Contribute to the Treatment of Asthma?

A number of other functions have been proposed for lipocortin 1. As mentioned above, it has been suggested that it plays a role in cell-cell signalling (important in immunoregulation) and that the phosphorylation state is probably important in signal transduction. For example phosphorylation of lipocortin is though to participate in glucocorticoid-regulated cell differentiation (HATTORI et al. 1983). In addition, it has been suggested that it is a cytoskeletal linkage protein, that it modifies arachidonic acid metabolism in platelet activation (TOUQUI et al. 1986), that it possesses anticoagulant activity and that it has effects on oxidant production. Another possibly anti-inflammatory activity of lipocortin is the inhibition of phagocytosis by macrophages. BECKER and GRASSO (1988) have shown that treatment of murine peritoneal macrophages with dexamethasone in vitro leads to the synthesis of a protein with phagocytosis-inhibitory activity. This action could be prevented by heat or trypsin treatment or by treating the cells with a neutralizing antibody to lipocortin, so supporting the hypothesis that this protein can down-regulate phagocytosis.

An elegant study by MARIDONNEAU-PARINI et al. (1989) has recently shown that lipocortin (from mouse lung, human peripheral monocytes or recombinant human lipocortin 1) decreases the generation of oxygen free radicals by guinea pig alveolar macrophages in a dose-dependent way by down-regulation of a PLA_2-dependent pathway of generation.

Cyclo-oxygenase, a key enzyme in prostanoid synthesis can be inactivated irreversibly by free radical intermediates of its metabolites. Aspirin also irreversibly inactivates this enzyme, recovery of enzyme activity depending on de novo protein synthesis (PASH and BAILEY 1988). Recovery of enzyme activity can be enhanced by EGF (PASH and BAILEY 1988) whilst preincubation with lipocortin can inhibit EGF-induced recovery of cyclo-oxygenase activity by vascular smooth muscle in culture (PASH and BAILEY 1988). As lipocortin is a target of the EGF receptor tyrosine kinase, it is possible that in vivo EGF may inactivate endogenous lipocortin by phosphorylation, thus permitting cyclo-oxygenase resynthesis (PASH and BAILEY 1988) Such an action might be overcome by a rise in lipocortin levels suggesting that this may be another mechanism by which lipocortin can down-regulate the activity of prostaglandins in vivo.

f) Lipocortin Summary

Although the role of lipocortin 1 remains to be fully elucidated, there seem to be several possible mechanisms by which its induction by glucocorticosteroids could suppress the inflammation which is thought to be the fundamental process underlying the bronchospasm of asthma. A recent review of the possible roles of lipocortin in the regulation of inflammation in another organ, the skin, has been compiled by GREAVES and CAMP (1988).

2. Other Proteins Modulated by Glucocorticosteroids

a) Recently Described Anti-inflammatory Proteins

It appears that apart from lipocortin, glucocorticosteroids induce the synthesis and/or release of another PLA_2 inhibitor found in human serum (SORENSON et al. 1988). The importance of this effect, and its relevance to the treatment of asthma, if any, remains to be established. However, its large size, >300 kDa, suggests that it may not readily cross the epithelial barrier onto the lung surface although it could play a role in the bronchial interstitium.

In addition, it is possible that glucocorticosteroids act on other enzymes involved in arachidonic acid metabolism. For example, a study by DUVAL et al. (1986) showed that dexamethasone inhibited acyltransferase activity in U937 cells in culture and that this inhibition was associated with an increase in the release of unmetabolized arachidonic acid from the cells. The authors suggested that the reduction in acyltransferase activity was due to steroid-induced changes in the levels of co-factors such as ATP. This study demonstrated that glucocorticosteroids can modulate arachidonic acid metabolism by at least one mechanism other than PLA_2 inhibition.

Vasocortin is a recently described, glucocorticosteroid-inducible, anti-inflammatory protein, distinct from lipocortin (CARNUCCIO et al. 1987; Table 1). To date, it has been identified only in the rat. Interestingly, like glucocorticosteroids, but unlike lipocortin, it can down-regulate histamine release from rat peritoneal cells in response to dextran or conconavalin A (CARNUCCIO et al. 1989) and thus, glucocorticosteroid-induced vasocortin and lipocortin may prove to play complementary roles in the control of the inflammatory response.

Table 1. Early nomenclature of lipocortin 1 and its relationship to vasocortin[a]

Original name	Original source	Reference	Current name
Macrocortin	Peritoneal and lung macrophages	BLACKWELL et al. 1980	Lipocortin 1
Lipomodulin	Neutrophils	HIRATA et al. 1980	Lipocortin 1
Renocortin	Renomedullary interstitial cells	RUSSO-MARIE et al. 1979	Lipocortin 1
Vasocortin	Peritoneal lavage fluid	CARNUCCIO et al. 1987	Vasocortin

[a] For the sake of simplicity, the relationship of lipocortin 1 to other members of the annexin family and other current names of lipocortin 1 have been omitted. The interested reader is referred to SARIS et al. (1986).

b) Acute Phase Reactant Proteins

A number of the acute phase reactant proteins have been shown to be steroid inducible. Plasma levels of these proteins can increase rapidly up to 100-fold in response to an appropriate stimulus, such as bacterial infection, inflammation, trauma and emotional stress.

These proteins form a diverse family including α_1-proteinase inhibitor (PI), α_1-antichymotrypsin (protease inhibitors), ceruloplasmin (antioxidant), fibrinogen, C-reactive protein (binds pneumococci, promoting their opsonization) and acid glycoprotein (function unclear). These proteins are synthesized mainly in the liver and there is no evidence that modified levels of any of them are directly implicated in the pathogenesis of asthma, although several of them may play roles in the control of inflammation in general [e.g. PI can limit proteolytic activity of many serine proteases, being particularly active against the elastase produced by the polymorphonuclear neutrophil (TRAVIS and SALVESEN 1983) whilst ceruloplasmin can scavenge and inactivate oxygen free radicals released from the same cells (GUTTERIDGE and STOCKS 1981)]. These actions are well known and therefore will not be described at length here.

Since cortisol levels and circulating levels of acute phase reactant proteins tend to rise in response to the same stimuli, it is difficult to tell whether the protein rise is driven directly by the stimulus or indirectly in response to the steroid. However, there is evidence from animals that the synthesis of at least some of the acute phase proteins is under the control of glucocorticosteroids. For example, in an early study by GRIENINGER et al. (1978), it was demonstrated that a variety of glucocorticosteroids increase fibrinogen synthesis by chick hepatocytes in culture. A later report by BAUMANN et al. (1983) showed a rise in concentrations of the mRNAs coding for acid glycoprotein and haptoglobin in the rat following treatment of the animal with dexamethasone. Subsequently, it was found by VANNICE et al. (1984)

that for acid glycoprotein at least the mechanism was an indirect one – the glucocorticosteroid receptor complex activated a gene, the product of which was necessary to stabilize acid glycoprotein mRNA transcrips.

It has been shown that the presence of both glucocorticosteroids and cytokines, specifically interleukins 1 and 6 and tumour necrosis factor (TNF) are required to achieve optimal induction of many of the acute phase reactant proteins (BAUMANN et al. 1987; MARINKOVIC et al. 1989). Exactly how this interaction occurs is unclear, since 1L-1 and perhaps other cytokines can increase glucocorticosteroid release via the hypothalamo-pituitary-adrenal axis (SAPOLSKY et al. 1987).

Finally, as well as up-regulating the synthesis of some proteins, glucocorticosteroids are believed to down-regulate the synthesis and/or release of others. As discussed above, neutrophils, eosinophils and monocytic cells can all release proteases at sites of inflammation. Down-regulation of protease activity will reduce tissue damage and lessen the chances of inflammation becoming chronic. There is evidence that glucocorticosteroids can reduce the release of a number of neutral proteases from macrophages in culture (WERB 1979). If this occurs in vivo it may well contribute to their anti-inflammatory effect in man.

II. Effects on Cell Activation and Recruitment

It has now been established that the infiltration of inflammatory cells is a central feature of the asthmatic lung and that such cells may contribute to the tissue damage and bronchial hyperresponsiveness characterizing this disease. It has also been suggested that part of the therapeutic efficacy of glucocorticosteroids in asthma may be via an effect on the recruitment or activation of inflammatory cells (BARNES et al. 1988). Certainly the beneficial effect of glucocorticosteroids in asthma is not due to an effect on airway smooth muscle since steroids do not affect this tissue directly (PERSSON 1989).

Eosinopenia was one of the earliest recognized effects of glucocorticosteroids. This may be the result of a combination of factors including a rapid tissue sequestration of eosinophils (ELLUL-MICALLEF 1987) and inhibition of mature cells leaving the bone marrow (HUDSON 1966). It has also additionally been reported that glucocorticosteroids will diminish the recruitment of eosinophils into tissues (ZWIEMAN et al. 1976), possibly via an effect on eosinophil adherence to vascular endothelium (ALTMAN et al. 1981). Such clinical observations have been confirmed and extended by laboratory investigations which have reported that glucocorticosteroids are capable of reducing eosinophil infiltration into lung tissues following treatment of normal animals with the phospholipid PAF (SANJAR et al. 1990) or allergic animals with allergen (BRATTSAND 1989). Glucocorticosteroids have also been shown to inhibit the production of eosinophil activating factor (EAF) by peripheral blood monocytes obtained from patients with moderate eosinophilia (THORNE et al. 1988) suggesting an additional mode of action for the glucocorticosteroids in reducing eosinophil recruitment. However, it must be pointed out that whilst chronic treatment with glucocorticsoteroids has been shown to reduce the number of eosinophils within pulmonary tissues of asthmatics (R LUNDGREN et al. 1988), other investigators have reported a marked depostion of eosinophil-derived mediators in pulmonary tissue despite chronic glucocorticosteroid treatment. These results suggest that even steroids may not be able to reduce all the eosinophil activation that accompanies bronchial asthma.

In addition to eosinophils, neutrophils have been shown to be activated in asthma and experimentally have been shown to contribute to airway hyperresponsivenss (AHR) induced by a variety of inflammatory stimuli (BARNES et al. 1988). The administration of glucocorticosteroids to humans results in neutrophil leucocytopenia, probably via release of these cells from the bone marrow and their reduced migration from the intravascular space (BISHOP et al. 1968). Glucocorticosteroids have been shown to inhibit neutrophil adherence both in vivo and in vitro (CLARK et al. 1979; VENEZIO et al. 1982). Furthermore, glucocorticosteroids have been shown to inhibit neutrophil influx into the lungs of both patients and control subjects following inflammatory insults such as allergen or toluene diisocyanate (BARNES 1990). Whilst the mechanism by which glucocorticosteroids inhibit neutrophil influx into tissues is not understood, steroids have been shown to inhibit degranulation, reduce phagocytosis and reduce O_2-generation in neutrophils (KLEBANOFF and CLARK 1978), albeit at high concentrations.

Perhaps the most important effect of glucocorticosteroids in this regard is their ability to inhibit the adherence of neutrophils to endothelial cells following exposure to stimuli such as PAF and *N*-formyl-methuiony-leucyl-phenylalanine (FMLP) or activators of endothelial cells such as IL-1 and lipopolysaccharide (SCHLEIMER et al. 1989). Furthermore, neutrophil activation in the presence of glucocorticosteroids causes a modest reduction in *CR*3, the C3bi receptor protein present on neutrophils (PETRONI et al. 1988) and a member of the leucocyte adherence glycoprotein family (CD18) which is essential for binding to endothelium and subsequent migration of neutrophils into epithelial tissues. Such changes in adherence properties may explain the decreased spontaneous adherence of neutrophils to endothelial cells or foreign surfaces such as nylon fibres (SCHLEIMER er al. 1989) and the demargination that occurs in vivo after treatment with glucocorticosteroids. Interferon-γ will induce a substantial increase in the expression of CR3 on neutrophils, which interestingly is susceptible to inhibition by glucocorticosteroids (PETRONI et al. 1988). This is potentially relevant to asthma since viral infections are known to be important trigger factors in the exacerbation of asthma.

Mast cells and basophils have been suggested for some time to have an important role in the aetiology of allergic asthma. However, concentrations of glucocorticosteroids that are in the normal therapeutic range are singularly ineffective at inhibiting the release of inflammatory mediators derived from arachidonic acid metabolism from these cells (COHAN et al. 1989). Leukotriene release from human lung tissue or purified human lung mast cells cannot be inhibited by therapeutically relevant concentrations of glucocorticosteroids (COHAN et al. 1989). Similar data have been obtained with basophils (DUNSKY et al. 1979) despite the fact that one of the effects of glucocorticosteroid treatment is a reduction of circulating basophils (SAAVEDRA-DELGADO et al. 1980). However, it has also been reported that prolonged culture of peripheral blood basophils with glucocorticosteroids inhibits immunoglobulin E- (IgE)-dependent release of histamine and leukotriene C_4 (LTC_4) (SCHLEIMER et al. 1982), although circulating basophils obtained from patients receiving chronic glucocorticosteroid treatment release histamine normally in response to stimulation with anti-IgE (LAMPL et al. 1985).

Macrophages are the most abundant resident inflammatory cell type found in the airways, and there are now numerous demonstrations that alveolar macrophages can be activated following stimulation of allergic asthmatics with allergen based on both the morphological appearance of the macrophages in BAL fluid or the appearance of glycosidases in the lavage fluid (BARNES et al. 1988). In vitro studies have also revealed that alveolar macrophages possess low affinity IgE receptors (FcϵRII) which when activated can lead to the release of a variety of preformed products (e.g. IL-1) and newly synthesized mediators (e.g. PAF) (FULLER et al. 1986). Administration of glucocortiosteroids markedly reduces the number of circulating monocytes, especially those bearing FcϵRII (SCHLEIMER 1990), and similarly causes a reduction in tissue macrophages (Langerhans cells) in skin (BELSITO et al. 1982). Certainly macrophages are very sensitive to the actions of glucocorticosteroids and the release of a variety of mediators such as arachidonic acid metabolites, elastase, collagenase and cytokines are all markedly reduced by treatment with glucocorticosteroids (GUYRE and MUNCK 1988).

There is now a growing awareness of the role of lymphocytes in the pathogenesis of allergic inflammatory conditions such as asthma, both as inflammatory cells and because of the important role they play in the production of IgE. It is therefore of interest that glucocorticosteroids are effective inhibitors of T-lymphocyte proliferation and cytokine production. Glucocorticosteroids achieve this effect in part by reducing the generation of interleukin-2 (IL-2) (SCHLEIMER 1990) and inhibiting the expression of IL-2 receptors on T-lymphocytes (REED et al. 1986). Furthermore, glucocorticosteroids have also been shown to inhibit the production of IL-1 by macrophages, dendritic cells and Langerhans cells, an event important to IL-2 secretion (reviewed by SCHLEIMER 1990). Furthermore, hydrocortisone has been demonstrated to inhibit the release of an interleukin-5 (IL-5)-like factor from T-cells cloned from hypereosinophilic patients (RAGHAVACHER et al. 1987).

III. Other Effects

1. Vascular Permeability and Effects on Vascular Endothelium

Submucosal oedema is now recognized as an important clinical feature contributing to the airway obstruction characterizing bronchial asthma and has been demonstrated in the lungs of allergic asthmatics following allergen exposure (BARNES et al. 1988). The oedema is thought to be secondary to the induction of vascular permeability within the bronchial microcirculation, and there are now a range of inflammatory mediators that have been demonstrated to induce microvascular leakage within this vascular bed (BARNES et al. 1988). In addition to the effect of glucocorticosteroids on the generation of inflammatory mediators by various inflammatory cells, it is of interest that glucocorticosteroids have been shown to inhibit protein leakage by an effect on the responses of endothelial cells to permeability-increasing mediators. Oedema formation induced by a range of inflammatory mediators such as bradykinin, serotonin (5-HT), PAF and histamine has been shown to be reduced by glucocorticosteroids using animal models such as the rat paw and the hamster cheek pouch preparation (TSURUFUJI et al. 1989, 1990; BJORK et al. 1985; SVENSJO and ROEMPKE 1985). The inhibition of rat paw oedema was time dependent and was blocked by inhibitors of protein synthesis such as cyclocheximide and actinomycin D. These observations suggest that glucocorticosteroids induce the synthesis of a protein that inhibits the response of the vascular endothelium and, at least in the rat paw, this protein has been suggested to be distinct from lipocortin (OYANAGUI and SUTUKI 1985). The permeability inhibitory activity referred to as vasocortin was approximately 100 kDa, whereas the phospholipase inhibitory activity appeared at 40 kDa (KOLTAI et al. 1987). Furthermore, the lipocortin fractions had little activity in inhibiting dextran-induced oedema in the rat paw, and vasocortin was unstable to heat whereas lipocortin fractions were stable (WILLIAMS and YARWOOD 1990; YARWOOD et al. 1988). Corticosteroids have also been shown to inhibit neutrophil-dependent oedema induced by C5a, FMLP and LTB_4 in rabbit skin, an effect that correlates with the ability of these drugs to inhibit the neutrophil accumulation induced by these chemoattractant agents. It is not known whether this effect is due to an action on the vascular endothelium and/or the neutrophils themselves, although there is evidence that glucocorticosteroids can affect both sites.

2. Effects on β-Adrenoceptors

Stimulation of β-adrenoceptors by the administration of agonists is an important therapeutic strategy in the management of asthma, but interestingly an inappropriate balance between endogenous β-adrenergic and α-adrenergic or cholinergic mechanisms may be responsible for at least part of the asthmatic reaction.

The synthesis as well as the release of many inflammatory and anaphylactic mediators are inhibited by β-adrenoceptor agonists, or by β-adrenergic stimulation, presumbaly acting through the adenyl cyclase system. Conversely, α-adrenergic or cholinergic stimulation increases the release of such mediators (LAMANSKE et al. 1985). Atopic individuals may exhibit β-adrenergic hyporeactivity (KALINER et al. 1982).

The responsiveness of cells to endogenous and exogenous β-agonists is governed by the number of β-receptors on the cell surface. These can be down-regulated following chronic treatment with β-agonists leading to tolerance to the drug and the possibility of rebound effects if the drug is suddenly discontinued.

Steroids act to prevent this receptor desensitization (DAVIES and LEFKOWITZ 1981; SAMUELSON and DAVIES 1984) probably by preventing, or reversing, the mechanism responsible for the down-regulation of the receptors. Interestingly, lipocortin has been implicated in this response. In a series of experiments relevant to the lung, KUNOS et al. (1985) examined the regulation of β-adrenergic receptors in two cell lines including A-549 cells (an type II alveolar cell line). Incubation of these cells with hydrocortisone doubled the number of β-adrenergic receptors (as estimated by the binding of labelled agonists to the membranes) with a maximum effect attained after 24 h. Apparently the affinity of the receptors for the ligand, as estimated by Scatchard analysis, was not changed. If a neutralizing antibody to lipocortin was included in the culture medium, the increase in the number of β-adrenergic receptors was completely blocked. Similar effects have been seen on other cell lines.

The exact mechanism by which steroids or lipocortin achieve this result is not clear, although the effect can be mimicked by other putative inhibitors of PLA_2 in addition to lipocortin (HIRATA 1984). This action of steroids in preventing the desensitization of β-receptors may be therapeutically important and the use of these drugs to restore β-receptor numbers in desensitized ariways should be borne in mind.

3. Effects on Airway Function

Corticosteroids were introduced for the treatment of asthma shortly after their discovery in the 1950s and remain the most effective therapy currently available. However, side effects and fear of adverse effects have limited their use and there has therefore been considerable research into discovering new or related agents which retain the beneficial action on the airways without the unwanted effects. The introduction of inhaled steroids has been a major advance in the treatment of chronic asthma (KERREBIJN 1990; REED 1990). Now that asthma is viewed as a chronic inflammatory disease, inhaled steroids may even be considered as first-line therapy (BARNES 1989).

Steroids have no direct effect on airway smooth muscle and improvement in lung function is probably due to an effect on the chronic airway inflammation and bronchial reactivity (BARNES 1990). Steroids therefore do not have a bronchodilator effect, although airway obstruction may improve over hours after intravenous steroids (ELLUL-MICALEF and FENECH 1975).

a) Protection Studies

Given as a single dose, both oral and inhaled corticosteroids prevent the late response to allergen (which may reflect inflammatory changes in the airway wall) (BOOIJ-NOORD et al. 1971; PEPYS et al. 1974; COCKCROFT and MURDOCK 1987) and the increased AHR which follows allergen challenge (COCKCROFT and MUDOCK 1987) while having no effect on the immediate bronchoconstrictor response. More prolonged therapy does protect against the immediate response to allergen, however (see later). Similarly, both oral (prednisone 50 mg daily for 3 days) and inhaled corticosteroids (beclomethasone dipropionate, BDP, 1 mg bid for 7 days) inhibit the late response and the associated increase in AHR induced by exposure to toluene diisocyanate (FABBRI et al. 1985; MAPP et al. 1987). The effect of inhaled BDP is dose related, with complete protection at a dose of 2 mg daily and less protection at a dose of 0.4 mg daily (DE MARGO et al. 1988). An inhaled steroid (budesonide 0.8 mg for 5 days) also reduces the maximal (plateau) bronchoconstriction, induced by inhaled leukotriene D_4 (LTD_4) in normal subjects, although the sensitivity is not changed, possibly indicating an effect on submucosal tissue swelling (BEL et al. 1988).

b) Effects on AHR

Corticosteroids also reduce AHR in asthmatic subjects. An early study of oral hydrocortisone in asthmatic subjects failed to demonstrate any effect on AHR induced by a cholinergic agonist (TIFFENEAU and DUNOYER 1956) and this was supported by a subsequent uncontrolled study using oral prednisone (WOLF et al. 1979), although the doses used were probably too low. In a more recent study, a short course of high-dose oral steroids (prednisone 60 mg daily × 1 week) caused a significant reduction in methacholine responsiveness and an increase in spirometric values in asthmatic children (BHAGAT and GRUNSTEIN 1985).

Oral methylprednisolone given as a divided dose 12 h prior to challenge in adult asthmatics provides significant protection against cholinergic challenge (ISRAEL et al. 1984), and there is no difference between low dose (32 mg) or high dose (128 mg). The patients with the mildest asthma appear to respond to the greatest extent. More prolonged treatment with oral methylprednisolone (16 mg daily for 7 days) in pollen-sensitive asthmatics reduced AHR to the levels measured before the pollen season (SOTOMAYOR et al. 1984). This suggests that the increase in AHR (and symptoms) observed during the period of allergen exposure may be reversed or prevented by oral steroid therapy.

Inhaled corticosteroids have, in general, proved even more effective in reducing AHR, although it is now becoming clear that it may take several weeks for the maximal effects to be established. In a comparative study, while inhaled BDP (1.2 mg daily) reduced AHR, there was no effect of oral prednisolone (12.5 mg daily) over the same 3-week period (JENKINS and WOOLCOCK 1988). In an early study, no significant reduction in methacholine responsiveness was seen after 4 months of low-dose BDP (0.4 mg daily) treatment, although patients tended to improve (EASTON 1981). Although this study was placebo controlled, the control group was not matched with the treatment group and PC_{20} values (index of brondrial hyperresponsiveness) were not log transformed before analysis. In a longer-term (up to 30 months) study of AHR in asthmatic patients, there was improvement in reactivity to histamine after prolonged treatment with inhaled BDP compared with a non-steroid treated group; this study was not however placebo controlled (JUNIPER et al. 1982). In a double-blind crossover study, inhaled BDP (0.4 mg daily over 4 weeks) was found to decrease AHR to histamine in mild asthmatics (RYAN et al. 1985), although the protection was small. Interestingly, the reduction in AHR was independent of any improvement in spirometry. In a similar study, inhaled budesonide (0.4 mg daily) given over 4 weeks gave a statistically significant reduction in histamine responsiveness and the effect at 4 weeks was greater than at 2 weeks (KRAAN et al. 1985). The reduction in responsiveness was again rather small (PC_{20} values increasing from 4.0 to 9.5 mg/ml), although in patients with only a mild increase in AHR larger changes might be difficult to demonstrate. A more recent study in more reactive asthmatics demonstrated an increase in mean PC_{20} from 0.9 to 2.7 mg/ml 8 weeks after treatment with budesonide (0.8 mg daily) (KRAAN et al. 1988). A greater effect was found with a dose of 0.8 mg daily compared with 0.2 mg daily. The decrease in reactivity was progressive over an 8-week period, although forced expiratory volume in 1 s (FEV_1) only improved over the first 2 weeks of treatment, again emphasizing a dissociation between AHR and spirometric improvement. A similar study in 38 moderate asthmatics showed a significant reduction in AHR after BDP (0.4 mg daily) over 8 weeks, and there was no correlation between changes in AHR and improvement in spirometry (SVENDSEN et al. 1987). A study in asthmatic children also showed a progressive reduction in AHR after inhaled budesonide (0.2 mg daily), which was maximal 3 months after starting treatment and was maintained at 6 months (KERREBIJN et al. 1987). similarly, in more severe adult patients, inhaled BDP (0.8 mg daily) gave a marked reduction in AHR to histamine from a mean PC_{20} of 0.09 to 0.74 μmol after 10 weeks of treatment. This reverted to pretreatment values after 4 weeks of stopping the inhaled steroids (and changing to theophylline) (DUTOIT et al. 1987).

Nocturnal asthma may be a reflection of AHR and airway inflammation (BARNES 1988), and inhaled steroids are effective in reducing morning wheeze in some patients (HORN et al. 1984). In a double-blind crossover study, inhaled budesonide (0.8 mg daily over 3 weeks) significantly reduced nocturnal symptoms and early morning peak expiratory flow volumes (Dahl et al. 1989).

c) Challenge Studies After Prolonged Therapy

Although neither oral nor inhaled corticosteroids are effective in preventing the immediate response to allergen after single doses (while effective in blocking the late response; (Booij-Noord et al. 1971; Pepys et al. 1974; Cockrcroft and Murdock 1987), more prolonged therapy with steroids reduces even the early response to allergen. After administration of oral prednisolone (40 mg daily) for 1 week, there was significant protection against the immediate response to allergen (Martin et al. 1980). Similarly, after inhaled BDP (0.8 mg daily) given for 1 week there was significant protection against the immediate bronchoconstrictor response to allergen, and this was probably greater for the dose of 0.8 mg than for 0.4 mg daily (Burge 1982). The protective effect was even greater with more prolonged inhaled therapy, since although significant protection against the immediate response to allergen was seen after treatment with budesonide (1 mg daily) for 1 week there was almost complete protection after treatment for 4 weeks (Dahl and Johansson 1982). The late response to treatment, by contrast, was inhibited after only a single dose of budesonide. Whether this effect of prolonged treatment with inhaled steroids on the early response is due to a reduction in AHR to the mediators released in the immediate response (including histamine, prostaglandins and, possibly, leukotrienes) or whether it is due to a more fundamental effect on the release of mediators is not yet certain.

A similar effect of prolonged steroid administration is also seen in exercise-induced asthma (Barnes 1990). However, inhaled budesonide given over 4 weeks significantly reduced the fall in FEV_1 after exercise in asthmatic children (Henriksen and Dahl 1983; Henriksen 1985) and was found to be more effective than treatment with sodium cromoglycate (Ostergaard and Pedersen 1987).

d) Clinical Implications

The fact that corticosteroids have an anti-inflammatory effect in asthma is now established and, although their precise mode of action has not yet been elucidated, it seems clear that steroids must play an important role in the long-term managment of asthma. It is now apparent that inflammatory changes are present, even in the mildest of asthmatics, and that this may lead to AHR and symptoms of asthma which are readily suppressed by a bronchodilator, such as a β_2-agonist or theophylline. Yet β-agonists do not modify the chronic inflammation in the airways, although the symptoms of asthma may disappear. By contrast, corticosteroids have no immediate bronchodilator action, so do not lead to immediate symptom relief. This means that patients have a preference for β-agonists and are less inclined to take inhaled steroids regularly. Indeed, it is possible that an over-reliance on β-agonists develops, yet these drugs mask the airway inflammation. Current clinical practice is usually to introduce a bronchodilator first and only add prophylactic therapy such as steroids later if symptoms are not adequately controlled. It now seems more logical to introduce inhaled steroids as *first-line* therapy for asthma, with β-agonists as required for symptom control (BARNES 1989). The effects of inhaled steroids are dose related and it is also likely that inflammation may be variable over time, so that the dose of inhaled steroids necessary to reduce AHR will change from time to time. The need for extra inhaled β-agonists, therefore, will indicate the dose of inhaled steroids required.

In several of the clinical studies discussed above, although AHR was reduced by inhaled steroids, the airway responsiveness never returned to the normal range, in the same way that lung function never normalized in chronic asthmatics (BROWN et al. 1984). It is possible that chronic inflammation eventually leads to structural changes which are irreversible, even with high-dose inhaled steroids for prolonged periods (R. LUNDGREN et al. 1988). For this reason it seems important that asthmatic inflammation is effectively suppressed over long periods of time and that effective anti-inflammatory treatment is used in children with asthma in order to prevent the decline in lung function and the persistent AHR seen in the poorly controlled adult asthmatic.

B. Clinical Use

I. Pharmacokinetics of Steroids

Prednisolone is readily and consistently absorbed after oral administration with little inter-individual variation. Enteric coatings to reduce the incidence of dyspepsia delay absorption but not the total amount of drug absorbed. Prednisolone is metabolized in the liver, and drugs such as rifampicin, phenobarbitone or phenytoin which induce hepatic enzymes lower the plasma half-life of prednisolone (GAMBERTOGLIO et al. 1980). Although the biological half-life of prednisolone is approximately 24 h, its plasma half-life is 2–3 h so that it is suitable for daily dosing. There is no evidence that previous exposure to steroids changes their subsequent metabolism. Prednisolone is approximately 95% protein bound. the majority to a specific protein transcortin and the remainder to albumin; it is the unbound fraction which is biologically active.

Some patients, usually those with severe asthma, apparently fail to respond to corticosteroids. "Steroid-resistant" asthma is not due to an impaired absorption or metabolism of steroids and may be assoicated with a defect in responsiveness of certain cells to steroids (CARMICHAEL et al. 1981).

II. Routes of Administration

1. Oral

Prednisolone or prednisone are the most commonly used steroids. Prednisone is converted to prednisolone in the liver. Clinical improvement with oral steroids may take several days and the maximal beneficial effect is usually achieved with 30 mg prednisone daily, although a few patients may need 60 mg daily to achieve control of symptoms. The usual maintenance dose is in the order of 10 mg/day.

Oral steroids should be given in a single dose in the morning, since this coincides with the normal increase in plasma cortisol and therefore there is less adrenal suppression that if given in divided doses or at night. Furthermore, the amount of steroid bound to transcortin is less during the day resulting in higher free concentrations, and this might contribute to the greater functional effect (REINBERG et al. 1983). Alternate day treatment has advantages, since there is less adrenal suppression of other side effects with similar control of asthma, although in some patients control is not optimal on this regime.

2. Intravenous

Parenteral steroids are indicated in acute severe exacerbations of asthma. Hydrocortisone is the steroid of choice as it has the most rapid onset (5–6 h after administration) (ELLUL-MICALLEF and FENECH 1975), being more rapid than prednisolone (8 h). The dose required is still uncertain, but it is common to give hydrocortisone 4 mg/kg initially followed by a maintenance dose of 3 mg/kg every 6 h. These doses are based on the argument that it is necessary to maintain "stress" levels of plasma cortisol (COLLINS et al. 1975).

3. Inhaled

Inhaled topical steroids have been a great advance in the management of chronic asthma as it may be possible to control symptoms without adrenal suppression or side effects and their administration allows a reduction in the dose of oral maintenance steroids (TOOGOOD et al. 1985; ELLUL-MICALLEF 1988; REED 1990). The high topical activity of inhaled steroids means that only small doses are required and any swallowed drug is immediately metabolized by the liver. Only when much larger doses are inhaled is sufficient steroid absorbed to cause adrenal suppression. Most patients get a maximal response at a dose of 400μg BDP per day, but some patients may benefit from higher doses (up to 1.5 mg/day) and high-dose inhalers have therefore been introduced in Europe. Traditionally steroid inhalers have been used four times daily, but twice daily administration is usually effective and compliance is better (MELTZET et al. 1985). Several inhaled preparations are available, including BDP, triamcinolone, flunisolide and budesonide (the latter having the highest topical potency).

III. Side Effects

1. Adrenal Suppression

Steroids inhibit adrenocorticotrophic hormone (ACTH) and cortisol secretion by a negative feedback effect on the pituitary gland and hypothalamus. This suppression is dependent on dose, and usually only occurs when a dose of prednisolone greater than 7.5-10 mg daily is used. Significant suppression after short courses of steroid therapy is not usually a problem, but prolonged suppression may occur after several months or years. Steroid doses after prolonged oral therapy must therefore be reduced slowly. Symptoms of "steroid withdrawal syndrome" include lassitude, musculoskeletal pains and occasionally fever.

The advantage of inhaled steroids is that airway inflammation may be suppressed by the local action of the steroids with less risk of systemic effects and adrenal suppression. For BDP and budesonide, systemic effects are not usually seen until doses of 1.5 mg daily are used. However, there is wide inter-individual variation and it is therefore important to monitor for systemic effects of inhaled steroids. In children even low doses of inhaled steroids have been found to inhibit the spontaneous "spikes" of cortisol release at night, although the functional significance of this is not yet apparent (LAW et al. 1986).

2. Systemic Side Effects

Side effects of long-term corticosteroid therapy are well described and include fluid retention, increased appetite, weight gain, osteoporsis, capillary fragility, hypertension, peptic ulceration, diabetes, cataracts and psychosis. Their frequency tends to increase with age. Very occasionally adverse reactions (such as anaphylaxis) to intravenous hydrocortisone have been describted, particularly in aspirin-sensitive asthmatics (DAJANI et al. 1981).

High-dose inhaled steroids may also have systemic effects, as discussed above. However, in a recent study even high-dose inhaled BDP had no effect on calcium excretion (TOOGOOD et al. 1988). Systemic effects of inhaled steroids may be reduced by the use of large volume spacers, since this reduces the oropharyngeal deposition and thus the dose of steroid absorbed (PRAHL and JENSEN 1987).

3. Local Side Effects

Side effects of inhaled steroids are few. The most common problem is oropharyngeal candidiasis (which may occur in 5% of patients). Hoarseness and weakness of the voice (dysphonia) may also occur and may be due to atrophy of the vocal chords. The incidence of these side effects may be related to the local concerntrations of the steroid deposited and may be reduced by the use of various spacing devices which reduce oropharyngeal deposition (TOOGOOD et al. 1984). There is no evidence to date for atrophy of the lining of the airways, or of an increase in lung infections after inhaled steroids.

IV. Use in Clinical Practice

1. Acute Asthma

Hydrocortisone is given intravenously in acute asthma. While the value of corticosteroids in acute severe asthma has been questioned by some investigators, others have found that they speed the resolution of attacks (ELLUL-MICALLEF 1988). There is no apparent advantage in giving very high doses of intravenous steroids (such as methylprednisolone 1 g). Intravenous steroids are indicated in acute asthma if lung function is less than 30% predicted and for patients in whom there is no significant improvement with a nebulized β-agonist. Intravenous therapy is usually given until a satisfactory response is obtained and then oral prednisolone (30–60 mg) has a similar effect to intravenous hydrocortisone and is easier to administer (HARRISON et al. 1986). Inhaled steroids have no proven effect in acute asthma.

2. Chronic Asthma

Corticosteroids are indicated if asthma is not adequately controlled with broncodilators alone, although increasing usage of inhaled steroids are indicated as first-line therapy for chronic asthma (BARNES 1989). Inhaled steroids are the treatment of choice and oral steroids are reserved for patients who cannot be controlled on other therapy, the dose being titrated to the lowest which provides acceptable control of symptoms. For any patient taking regular oral steroids objective evidence of steroid responsiveness should be obtained before maintenance therapy is instituted. Short courses of oral steroids (such as 30 mg prednisolone daily for 1–2 weeks) are indicated for exacerbations of asthma.

Inhaled steroids should be used twice daily to improve compliance once control of asthma has been achieved (which may require four times daily dosing initially). If a dose of more than 500 μg daily is used, a spacer device should be considered as this reduces the risk of oropharyngeal side effects. Inhaled steroids may also be used in children but sodium cromoglycale is the initial preferred anti-inflammatory treatment. In children, the dose should be kept under 500 μg daily, if possible, to reduce the risk of inhibitory effects on growth (KONIG 1988). Chronic bronchitis patients occasionally respond to steroids and these patients are likely to be undiagnosed asthmatics. Steroids have no objective benefit on airway function in patients with chronic bronchitis, although they may often produce subjective benefit because of their euphoric effect.

C. Future Developments

There is little doubt that corticosteroids are the most effective antiasthma drugs in current use in clinical practice since almost every patient with asthma responds and, when given by inhalation, they are relatively free of side effects.

As discussed above, steroids have many effects on the inflammatory response which may be relevant to their antiasthma action. Indeed, it may be the multiplicity of effects which make steroids so effective. An important aim of research is to discover which particular actions of steroids are critical for their antiasthma effects. This knowledge will be of great value in understanding the pathophysiology of asthma, but may also aid the development of new agents with fewer side effects.

Improvement in inhaled steroids would be desirable. Inhaled steroids, which have an even greater topical effect, may allow higher doses to be given without systemic effects (BRATTSAND 1989). First pass metabolism of steroids is important in reducing systemic concentrations. Effective first pass metabolism of budesonide accounts for the fact that higher doses of this steroid may be given than BDP before systemic effects are seen. Inhaled steroids which are metabolized locally in the airways may also be useful in the future. Targetting of inhaled steroids to specific cells in the ariways such as macrophages and the use of liposomes to administer steroids may also be a useful new approach for the future.

In the long-term future it is possible that non-steroidal drugs which mimic the critical action of steroids may be developed. These are likely to arise from a further understanding of the molecular mechanisms of steroids.

References

Aarsman AJ, Mynbeek G, van den Bosch H, Rothhut B, Prieur B, Comera C, Jordan L, Russo-Marie F (1987) Lipocortin inhibition of extracellular and intracellular phospholipases A_2 is substrate concentration dependent. FEBS Lett 219:176–180

Altman LC, Hill JS, Harfield WM et al. (1981) Effects of corticosteroids on eosinophil chemotaxis and adherance. J Clin Invest 67:28–36

Ambrose M, Hunninghake GW (1989) Human alveolar macrophages produce immunoreactive lipocortin. Am Rev Respir Dis 139:A309

Ando Y, Imamura S, Hong Y-M, Owada MK, Kakunaga T, Kannagi R (1989) Enhancement of calcium sensitivity of lipocortin 1 in phospholipid binding induced by limited proteolysis and phosphorylation at the amino-terminus as analyzed by phospholipid affinity column chromatography. J Biol Chem 264:6948–6955

Ballard PL, Baxter JD, Higgins SJ, Roussen GC, Tomkins GM (1974) General presence of glucocorticoid receptors in mammalian tissues. Endocrinology 94:998–1002

Balter MS, Eschenbacher WL, Peters-Golden M (1988) Arachidonic acid metabolism in cultured alveolar macrophages from normal, atopic and asthmatic subjects. Am Rev Respir Dis 138:1134–1142

Barnes PJ (1988) Inflammatory mechanisms and nocturnal asthma. Am J Med 85:64–70

Barnes PJ (1989) A new approach to asthma therapy. N Engl J Med 321:1517–1427

Barnes PJ (1990) Effect of corticosteroids on airway hyperresponsiveness. Am Rev Respir Dis 137 [Suppl] S70–S76

Barnes PJ, Chung KF, Page CP (1988) Inflammatory mediators and asthma. Pharmacol Rev 40:49–84

Baumann H, Firestone GL, Burgess TL, Gross KW, Kamamoto KR, Held WA (1983) Dexamethasone regulation of α-1-acid-glycoprotein and other acute phase reactants in rat liver and hepatic cells. J Biol Chem 258:563–570

Baumann H, Richards C, Gauldie J (1987) Interaction among hepatocyte stimulating factors, interleukin-1 and glucocorticoids for regulation of acute phase plasma proteins in human hepatoma (hep G2) cells. J Immunol 139:4122–4128

Becker J, Grasso RJ (1988) Suppression of yeast injestion by dexamethasone in macrophage cultures: evidence for a steroid-induced phagocytosis inhibitory protein. Int J Immunopharmacol 10: 325–338

Bel EH, van der Veen H, Dijkman JH, Sterk RJ (1988) Budesonide reduces the maximal degree of airway narrowing to leukotriene D4 in normal subjects. Am Rev Respir Dis 137 [Suppl] 97

Belsito DV, Flotke TJ, Lim HW, Baer RC, Thornbecke GJ, Gigli I (1982) Effects of glucocorticoids on epidermal Langerhan cells. J Exp Med 155:291–302

Bhagat RG, Grunstein MM (1985) Effect of corticosteroids on bronchial responsiveness to methancholine in asthmatic children. Am Rev Respir Dis 131:902–906

Bienkowski MJ, Petro MA, Robinson LJ (1989) Inhibition of thromboxane A synthesis in U937 cells by glucocorticoids. Lack of evidence for lipocortin as the second messenger. J Biol Chem 264:6536–6544

Bishop CR, Athus JW, Boggs DR et al. (1986) Leukokinetic studies: XIII. A non steady-state evaluation of the mechainsm of cortisone-induced granulocytosis. J Clin. Invest 47:249–260

Bjork J, Goldschmidt T, Smedegard G, Arfors KE (1985) Methylprednisolone acts at the endothelial cell level reducing inflammatory responses. Acta Physiol Scand 123:221–224

Blackwell GJ (1983) Specificity and inhibition of glucocorticoid-induced macrophage secretion from rat peritoneal macrophages. Br J Pharmacol 79:587–594

Blackwell GJ, Flower RJ, Nijkamp FP, Vane JR (1978) Phospholipase A_2 activity of guinea pig isolated perfused lungs: stimulation and inhibition by antiinflammatory steroids. Br J Pharmacol 62:79–89

Blackwell GJ, Carnuccio R, Di Rosa M, Flower RJ, Parente L, Persico P (1980) Macrocortin: a polypeptide causing the anti-phospholipase effect of glucocorticoids. Nature 287:147–149

Blackwell GJ, Carnuccio R, Di Rosa M, Flower RJ, Langham CSJ, Parente L, Persico P, Russell-Smith NC, Stone D (1982) Glucocorticoids induce the formation and release of anti-inflammatory and anti-phospholipase proteins into the peritoneal cavity of the rat. Br J Pharmacol 76:185–194

Booij-Noord H, Orie NGM, de Vries K (1971) Immediate and late bronchial obstructive reactions to inhalation of house dust and protective effects od disodium cromoglycate and prednisolone. J Allergy Clin Immunol 48:344–354

Boustead CM, Walker JH, Geisow MJ (1988) Isolation and characterisation of two novel calcium-dependent phospholipid-binding proteins from bovine lung. FEBS Lett 233:233–238

Brattsand R (1989) Glucocorticosteroids for inhalation. In: Barnes PJ (ed) New drugs for asthma. IBC, London pp 117–130

Brönnegård M, Andersson O, Edwall D, Lund J, Norstedt G, Carlestedt-Duke J (1988) Human calpactin II (Lipocortin I) messenger ribonucleic acid is not induced by glucocorticoids. Mol Endocrino 2:732–739

Brown JP, Greville WH, Fincance KE (1984) Asthma and irreversible airflow obstruction. Thorax 39:131–136

Browning JL, Ward MP, Wallner BP, Pepinsky RB (1990) Studies of the structural properties of lipocortin 1 and the regulation of its synthesis by steroids. Parente L, Melli M (eds) Proceedings of the Sclavo international conference on the molecular and cell biology of IL-1, TNF and lipocortins. Wiley-Liss Prog Clin Biol Res 349:27–46
Burge PS (1982) The effects of corticosteroids on the immediate asthmatic reaction. Eur J Respir Dis 63:163–166
Carmicheal J, Paterson K, Diaz P, Crompton GK, Kay AB, Grant IWB (1981) Corticosteroid resistance in chronic asthma. Br Med J 282:1419–1422
Carnuccio R, Di Rosa M, Guerrasio B, Iuvone T, Sautebin L (1987) Vasocortin: a novel glucocorticoid-induced anti-inflammatory protein. Br J Pharmacol 90:443–445
Carnuccio R, Di Rosa M, Ialente A, Iuvone T, Sautebin L (1989) Selective inhibition by vasocortin of histamine release induced by dextran and concanavalin A from rat peritoneal cells. Br J Pharmacol 98:32–34
Chang J, Musser JH, McGregor H (1987) Phospholipase A_2: function and pharmacological regulation. Biochem Pharmacol 36:2429–2436
Cirino G, Flower RJ (1987) The inhibitory effect of lipocortin on eicosanoid synthesis is dependent upon Ca^{2+} ions. Br J Pharmacol 92:521P
Cirino G, Flower RJ, Browning JL, Sinclair LK, Pepinsky RB (1987) Recombinant human lipocortin 1 inhibits thromboxane release from guinea-pig isolated perfused lung. Nature 328:270–272
Cirino G, Peers SH, Flower RJ, Browning JL, Pepinsky RB (1989) Human recombinant lipocortin 1 has acute local anti-inflammatory properties in the rat paw oedema test. Proc Natl Acad Sci USA 86:3428–3432
Clark RAF, Gallini JI, Franci AS (1979) Effect of in vivo prednisone on in vitro eosiniphil and enutrophil adherance and chemotaxis. Blood 53:633–641
Cockcroft DW, Murdock KY (1987) Comparative effects of inhaled salbutamol, sodium cromoglycate and beclomethasone diproprionate on allergen-induced early asthmatic responses, late asthmatic responses and increased bronchial responsiveness to histamine. J Allergy Clin Immunol 79:734–740
Cohan VL, Undem BJ, Fox CC, Adkinson NF, Lichtenstein NM, Schleimer RP (1989) Dexamethasone does not inhibit the release of mediators from human mast cells residing in airway, intestine or skin. Am Rev Respir Dis 140:951–954
Collins JV, Clarke TJH, Brown D, Townsend J (1975) The use of corticosteroids in the treatment of acute asthma. QJ Med 44:259–273
Crompton MR, Moss SE, Crumpton MJ (1988) Diversity in the lipocortin/calpactin family. Cell 55:1–3
Dahl R, Johansson S-A (1982) Importance of duration of treatment with inhaled budesonide on the immediate and late bronchial reaction. Eur J Respir Dis 62:167–175
Dahl R, Pedersen B, Hagglof B (1989) Nocturnal asthma: effect of treatment with oral sustained-release terbutaline, inhaled budesonide and the two in combination. J Allergy Clin Immunol 83:811–815
Dajani BM, Sliman NA, Shubair KS, Hamzeh YS (1981) Bronchospasm caused by intravenous hydrocortisone sodium succinate (Solu-Cortef) in aspirin-sensitive asthmatics. J Allergy Clin Immunol 68:201–206
Danon A, Assouline G (1978) Inhibition of prostaglandin biosynthesis by corticosteroids requires RNA and protein synthesis. Nature 273:552–554
Davidson FF, Dennis EA, Powell M, Glenney JR (1987) Inhibition of phospholipase A_2 by lipocortins and calpactins – an effect of binding to substrate phospholipids. J Biol Chem 262:1698–1705
Davies AO, Lefkowitz RJ (1981) Agonist promoted higher affinity state of the beta adrenergic receptor in human neutrophils: modulation by corticosteroids. J Endocrinol Metab 53:703–708
De Marzo N, Fabbri LM, Crescioli S, Plabani M, Mapp CE (1988) Dose-dependent inhibitory effect of inhaled beclomethasone on late asthmatic reactions and

increased responsiveness to methacholine-induced by toluene diisocyanate in sensitised subjects. Pulm Pharmacol 1:15–20
Di Rosa M, Persico P (1979) Mechanism of inhibition of prostaglandin biosynthesis by hydrocortisone in rat leukocytes. Br J Pharmacol 66:161–163
Di Rosa M, Flower RJ, Hirate F, Parente L, Russo-Marie F (1984) Nomenclature announcement. Anti-phospholipase proteins. Prostaglandins 28:441–442
Dunsky EH, Zweiman B, Fischler E, Levy DA (1979) Early effects of corticosteroids on basophils, leukocyte histamine and tissue histamine. J Allergy Clin Immunol 63:426–432
Dutoit JI, Salome CM, Woolcock AJ (1987) Inhaled corticosteroids reduce the severity of bronchial hyperresponsiveness in asthma, but oral theophylline does not. Am Rev Respir Dis 136:1174–1178
Duval D, Lynde P, Hatzfeld A, Hatzfeld J (1986) Dexamethasone-induced stimulation of arachidonic acid release by U937 cells grow in defined medium. Biochem Biophys Res Commun 153:1267–1270
Easton JG (1981) Effect of an inhaled corticosteroid on methacholine airway reactivity. J Allergy Clin Immunol 67:388
Ellul-Micallef R (1978) Pharmacokinetics and pharmacodynamice of glucocorticosteroids. In: Jenne JW, Murphy S (eds.) Drug therapy for asthma, research and clinical practice. Marcel Dekker, New York, pp 463–516
Ellul-Micallef R (1988) Glucocorticosteroids – the pharmacological basis of their therapeutic use in bronchial asthma. In: Barnes PJ, Rodger IW, Thomson NC (eds) Asthma: basic mechanisms and clinical managment. Academic, London pp 653–692
Ellul-Micallef R, Fenech FF (1975) Intravenous prednisolone in chronic bronchial asthma. Thorax 30:312–315
Errasfa M, Russo-Marie F (1988) Rapid purification of two lipocortin-like proteins from mice lung. Biochem Biophys Res Commun 153:1271–1275
Errasfa M, Russo-Marie F (1989) A purified lipocortin shares the anti-inflammatory effect of glucocorticosteroids in vivo in mice. Br J Pharmacol 97:1051–1058
Errasfa M, Rothhut B, Fradin A, Billardon C, Junien J-L, Bure J, Russo-Marie F (1985) The presence of lipocortin in human skin fibroblasts and its regulation by anti-inflammatory steroids. Biochim Biophys Acta 874:247–254
Errasfa M, Bachelet B, Russo-Marie F (1988) Inhibition of phospholipase A_2 activity of guinea pig alveolar macrophages by lipocortin-like proteins purified from mice lung. Biochem Biophys Res Commun 153:1267–1270
Fabbri LM, Chiesura-Corona P, dal Vecchio L, Di Giagomo GR, Zocco E, de Marzo N, Maestrelli P, Mapp CE (1985) Prednisone inhibits late asthmatic reactions and the associated increase in airways responsiveness induced by toluene diisocyanate in sensitized subjects. Am Rev Respir Dis 132:1010–1014
Flower RJ (1988) Lipocortin and the mechanism of action of the glucocorticoids. Br J Pharmacol 94:987–1015
Flower RJ, Blackwell GJ (1979) Anti-inflammatory steroids induce biosynthesis of a phospholipase A_2 inhibitor which prevents prostaglandin generation. Nature 278:456–459
Fradin A, Rothhut B, Poincelot-Canton B, Errasfa M, Russo-Marie F (1988) Inhibition of eicosanoid and PAF formation by dexamethasone in rat inflammatory polymorphonuclear neutrophils may implicate lipocortin 's'. Biochim Biophys Acta 963:248–257
Fuller RW, Morris PK, Sykes RD et al. (1986) Immunoglobulin E-dependent stimulation of human alveolar macrophages: significance in Type 1 hypersensitivity. Clin Exp Immunol 65:416–426
Gambertoglio JG, Amend WJC, Benet LZ (1980) Pharmacokinetics and bioavailability of prednisone and prednisolone in healthy volunteers and patients: a reveiw. J Pharmacokinet Biopharm 8:1–52
Geisow MJ (1986) Common domain structure of Ca^{2+}-and lipid binding proteins. FEBS Lett 203:99–103

Geisow MJ, Walker JH (1986) New proteins involved in cell regulation by Ca^{2+} and phospholipids. Trends Biochem Sci 11:420–424
Goulding NJ, Smith SF, Godolphin JL, Tetley TD, Roberts CM, Guz A, Flower RJ (1989) Quantification of lipocortin 1 in bronchoalveolar lavage fluid from patients with pulmonary sarcoidosis. Br J Pharmacol 97:426P
Goulding NJ, Godolphin JL, Sharland PR, Peers SH, Samson M, Maddison PJ, Flower RJ (1990) The anti-inflammatory protein lipocortin 1 is produced by peripheral blood leukocytes in response to intravenous hydrocortisone. Lancet 335:1416–1418
Goulding NJ, Luying P, Guyre PM (1990) Characteristics of lipocortin/binding to the surface of human peripheral blood leucocytes. Biochem Soc Trans (in press)
Greaves MW, Camp RDR (1988) Prostaglandins, leukotrienes, phospholipase A_2, platelet activating factor and cytokines: an integral approach to inflammation of human skin. Arch Dermatol Res 280 [Suppl]:S33–S41
Grieninger G, Hertzberg KM, Pindy KJ (1978) Fibrinogen synthesis in serum free hepatocyte cultures: stimulation by glucocorticoids. Proc Natl Acad Sci USA 75:5506–5510
Gutteridge JMC, Stocks J (1981) Caeruloplasmin: physiological and pathological perspectives. CRC Crit Rev Clin Lab Sci 14:257–329
Guyre PM, Munck A (1988) Glucocorticoid actions on monocytes and macrophages. In: Schleimer RP, Claman HN, Oronsky AR (eds) Antiinflammatory steroids: basic and clinical aspects: Academic, New York, pp 199–225
Haigler HT, Schlaepfer DD, Burgess WH (1987) Characterisation of lipocortin 1 and an immunologically unrelated 33kDa protein as epidermal growth factor receptor/ kinase substrates and phospholipase A_2 inhibitors. J Biol Chem 262:6921–6930
Harrison BDW, Hart GJ, Ali NJ, Stokes TC, Vaughan DA, Robinson AA (1986) Need for intravenous hydrocortisone in addition to oral prednisolone in patients admitted to hospital with severe asthma without ventilatory failure. Lance ii:181–184
Hattori T, Hoffman T, Hirata F (1983) Differentiation of a histiocytic lymphoma cell line by lipomodulin, a phospholipase inhibitory protein. Biochem Biophys Res Commun 111:551–559
Haynes RC, Murad F (1985) Adrenocorticotrophic hormones: adrenocortical steroids and their synthetic analogues. In: Goodman Gilman A, Goodman LS, Rall TW, Murad F (eds) The pharmacological basis of therapeutics, 7th edn. Macmillan, New York, pp 1459–1489
Henriksen JM (1985) Effect of inhalation of corticosteroids on exercise-induced asthma: randomised double blind crossover study of budesonide in asthmatic children. Br Med J 291:248–249
Henriksen JM, Dahl R (1983) Effects of inhaled budesonide alone and in combination with low-dose terbutaline in children with exercise induced asthma. Am Rev Respir Dis 128:993–997
Hirata F (1981) The regulation of lipomodulin, a phospholipase inhibitory protein, in rabbit neutrophils by phosphorylation. J Biol Chem 256:7730-7733
Hirata F (1983) Lipomodulin: a possible mediator of the action of glucocorticoids. Adv Prostaglandin Thromboxane Leukotriene Res 11:73–78
Hirata F (1984) Modulation of beta adrenoceptors by phospholipids In: Morley J (ed) Beta adrenoceptors in asthma. Academic, New York, pp 49–53
Hirata F, Schiffmann E, Venkatasubramanian K, Salomon D, Axelrod J (1980) A phospholipase A_2 inhibitory protein in rabbit neutrophils induced by glucocorticoids. Proc Natl Acad Sci USA 77:2533–2536
Hong SC, Levine L (1976) Inhibition of arachidonic acid release from cells as the biochemical action of anti-inflammatory steroids. Proc Natl Acad Sci USA 73:1730–1734
Horn CR, Clark TJH, Cochrane GM (1984) Inhaled therapy induced morning dips in asthma. Lancet i:1143–1145

Huang K-S, Wallner BP, Mattaliano RJ, Tizard R, Burne C, Frey A, Hession C, McGray P, Sinclair LK, Chow EP, Browning JL, Ramachandran KL, Tang J, Smart JE, Pepinsky RB (1986) Two human 35KD inhibitors of phospholipase A_2 are related to substrates of $pp60^{v-src}$ and of the epidermal growth factor receptor/kinase. Cell 46:191–199
Hudson G (1966) Eosinophil granulocyte reactions. In: Yoffey JM (ed) Bone Marrow Reactions. William and Wilkins, Baltimore, pp 86–99
Hullin F, Raynal P, Ragab-Thomas JMF, Fauvel J, Chap H (1989) Effect of dexamethasone on prostaglandin synthesis and on lipocortin status in human endothelial cells. J Biol Chem 264:3506–3513
Israel RH, Poe RH, Wicks CM, Greenblatt DW, Kallay MC (1984) The protective effect of methylprednisolone on carbachol-induced bronchospasm. Am Rev Respir Dis 130:1019–1022
Iynedjian PB, Hanson RW (1977) mRNA for renal phosphoenolpyruvate carboxykinase (GTP). Its translation in a heterologous cell free system and its regulation by glucocorticoids and by changes in acid base balance. J Biol Chem 252:8398–8403
Jenkins CR, Woolcock AJ (1988) Effect of prednisolone and beclomethasone diproprionate on airway responsiveness in asthma: a comparative study. Thorax 43:378–384
Johnson MD, Gray ME, Carpenter G, Pepinsky RB, Sundell H, Stahlman MT (1989) Ontogeny of epithelial growth factor receptor/kinase and lipocortin-1, its substrate, in the ovine lung. Pediatr Res 25:535–541
Johnson MD, Gray ME, Carpenter G, Pepinsky RB, Stahlman MT (1990) Ontogeny of epidermal growth factor receptor and lipocortin 1 in foetal and neonatal human lungs. Hum Pathol 21:182–191
Juniper EF, Frith PA, Hargreave FE (1982) Long term stability of bronchial responsiveness to histamine. Thorax 37:288–291
Kaliner M, Shelhamer JH, Davis PD, Smith LJ, Venter JC (1982) Autonomic nervous system abnormalities and allergy. Ann Intern Med 96:349–357
Karasik A, Pepinsky RB, Shoelson SE, Kahn CR (1988) Lipocortins 1 and 2 as substrates for the insulin receptor kinase in rat liver. J Biol Chem 263:11862–11867
Kerrebijn KF (1990) Use of topical corticosteroids in the treatment of childhood asthma. Am Rev Respir Dis 141:S77–S81
Kerrebijn KF, von Essen-Zandvliet EEM, Neijens HJ (1987) Effect of long term treatment with inhaled corticosteroids and beta agonists on bronchial responsiveness in asthmatic children. J Allergy Clin Immunol 79:653–659
Klebanoff SJ, Clark RA (1978) The neutrophil: function and clinical disorders. Elsevier Scientific, Amsterdam.
Knudsen PJ, Dinarello CA, Strom TB (1987) Glucocorticoids inhibit transcriptional and post-transcriptional expression of Interleukin 1 in U937 cells. J Immunol 139:4129–4134
Koltai M, Kovacs Z, Nemcz G, Mecs I, Szekeres L (1987) Glucocorticoid-induced low molecular mass anti-inflammatory factors which do not inhibit phospholipase A_2. Eur J Pharmacol 134:109–112
Konig P (1988) Inhaled corticosteroids – their present and future role in the management of asthma. J Allergy Clin Immunol 82:297–306
Kraan J, van der Mark TW, Sluiter HJ, de Vries K (1985) Changes in bronchial hyperreactivity induced by 4 weeks of treatment with anti-asthma drugs in patients with allergic asthma: a comparison between budesonide and terbutaline. J Allergy Clin Immunol 76:628–636
Kraan J, Koeter GH, van der Mark T, Boorsma M, Kukler J, Sluiter HJ, de Vries K (1988) Dosage and time effects of inhaled budesonide on bronchial hyperreactivity. Am Rev Respir Dis 137:44–48
Kunos G, Kunos I, Hirata F, Ishac EJN (1985) J Allergy Clin Immunol 76:346–351
Lamanske RF, Casale TB, Kaliner M (1985) The autonomic nervous system in allergic disease. In: Kaplan AP Allergy. Churchill Livingstone, New York, pp 199–213

Lampl KL, Lichtenstein LM, Schleimer RP (1985) In vitro resistance to dexamethasone of basophils from patients receiving long-term steroid therapy. Am Rev Resp Dis 132:1015–1018

Law CM, Marchant JL, Honour JW, Preece MA, Warner JO (1986) Nocturnal adrenal suppression in asthmatic children taking inhaled beclomethasone diproprionate. Lancet i:942–944

Lenting HBM, Neys FW, van den Bosch H (1988) Regulatory aspects of mitochondrial phospholipase A_2 from rat liver: effects of proteins, phospholipids and calcium ions. Biochim Biophys Acta 961:129–138

Lundgren JD, Hirata F, Marom Z, Logun C, Steel L, Kaliner M, Shelhamer J (1988) Dexamethasone inhibits respiratory glycoconjugate secretion from feline airways in vitro by the induction of lipocortin (lipomodulin) synthesis. Am Rev Respir Dis 137:353–357

Lundgren R, Soderberg M, Horstedt P, Sterling R (1988) Morphological studies of bronchial mucosal biospies from asthmatics before and after ten years of treatment with inhaled steroids. Eur Respir J 1:883–889

Mapp C, Boschetto P, dal Vecchio L, Crescioli S, de Marzo N, Palegari N, Fabbri LM (1987) Protective effects of anti-asthma drugs on late asthmatic reactions and increased airways responsiveness induced by toluene diisocyanate in sensitised subjects. Am Rev Respir Dis 136:1403–1407

Maridonneau-Parini I, Errasfa M, Russo-Marie F (1989) Inhibition of 0_2^- generation by dexamethasone is mimicked by lipocortin 1 in alveolar macrophages. J Clin Invest 83:1936–1940

Marinkovic S, Jahreis GP, Wong GG, Baumann H (1989) Interleukin-6 modulates the synthesis of a specific set of acute phase plasma proteins in vivo. J Immunol 142:808–812

Martin GL, Atkins PC, Dunsky EH, Zweiman B (1980) Effects of theophylline, terbutaline and prednisone on antigen-induced bronchospasm and mediator release. J Allergy Clin Immunol 66;204–212

Meltzer EO, Kemp JP, Welch MJ, Orgel HA (1985) Effect of dosing schedule on efficacy of beclomethasone diproprionate aerosol in chronic asthma. Am Rev Respir Dis 131:732–736

Miele L, Cordella-Miele E, Facchiano A, Mukherjee AB (1988) Novel anti-inflammatory peptides from the region of highest similarity between uteroglobin and lipocortin 1. Nature 335:726–730

Mishler JM, Emerson PM (1977) Development of neutrophilia by serially increasing doses of dexamethasone. Br J Haematol 36:249–257

Northup JK, Valentine-Braun KA, Johnson LK, Severson DL, Hollenberg MD (1988) Evaluation of the antiinflammatory and phospholipase-inhibitory activity of calpactin II/lipocortin I. J Clin Invest 82:1347–1352

Ostergaard PA, Pedersen S (1987) The effect of inhaled disodium cromoglycate and budesonide on bronchial responsiveness to histamine and exercise in asthmatic children: a clinical comparison. In: Godfrey S (ed) Glucocorticosteroids in childhood asthma. Exerpta Medica, Amsterdam, pp 55–56

Oyanagui Y, Suzuki S (1985) Vasoregulin, a glucocorticoid-inducible vascular permeability inhibiting protein. Agents Actions 17:270–277

Parente L, Flower RJ (1985) Hydrocortisone and macrocortin inhibit the zymosan-induced release of lyso-PAF from rat peritoneal leukocytes. Life Sci 36:1225–1231

Parente L, DiRosa M, Flower RJ, Ghiara P, Meli R, Persico P, Salmon JA, Wood JN (1984) Relationship between the anti-phospolipase and anti-inflammatory effects of glucocorticoid-induced proteins. Eur J Pharmacol 99:233–239

Pash JM, Bailey JM (1988) Inhibition by corticosteroids of epidermal growth factor-induced recovery of cyclooxygenase after aspirin inactivation. FASEB J 2:2613–2646

Pepinsky RB, Sinclair LK (1986) Epidermal growth factor dependent phosphorylation of lipocortin. Nature 321:81–84

Pepinsky RB, Sinclair LK, Browning JL, Mattaliano RJ, Smart JE, Chow EP, Falbel T, Ribolini A, Garwin JL, Wallner BP (1986) Purification and partial sequence analysis of a 37kDa protein that inhibits phospholipase A_2 activity from rat peritoneal exudates. J Biol Chem 261:4239–4246

Pepinsky RB, Tizard R, Mattaliano RJ, Sinclair LK, Miller GT, Browning JL, Chow EP, Burne C, Huang K-S, Pratt D, Wachter L, Hession C, Frey AZ, Wallner BP (1988) Five distinct proteins share homology with lipocortin-1. J Biol Chem 263: 10799–10811

Pepys J, Davies RJ, Breslin ABX, Hendrick DJ, Hutchcroft BJ (1974) The effect of inhaled beclomethasone diproprionate (Becotide) and sodium cromoglycate on asthmatic reactions to provocation tests. Clin Allergy 4:13–24

Persson CGA (1989) Historical perspectives – glucocorticoids for asthma – early contributions. Pulmon Pharmacol 2:163–166

Petroni KC, Shen L, Guyre PM (1988) Modulation of human polymorphonuclear leukocyte IgE Fc receptors and Fc receptors mediated functions by IFN-gamma and glucocorticoids. J Immunol 140:3467–3472

Powell MA, Glenney JR (1987) Regulation of calpactin I phospholipid binding by calpactin I light chain binding and phosphorylation by p60v-src. Biochem J 247:321–328

Prahl P, Jensen T (1987) Decreased adreno-cortical suppression utilising the Nebuhaler for inhalation of steroid aerosols. Clin Allergy 17:393–398

Raghavachar A, Fleischer S, Frickhofen N, Heimpl H, Fleische B (1987) T-lymphocyte control of human eosinophilic granuloproteins. J Immunol 139:3753–3758

Reed CE (1990) Aerosol glucocorticoid treatment of asthma: adults. Am Rev Reapir Dis 140:S82–S88

Reed JC, Abidi AH, Alpers JD, Hoover RS,Robb RJ, Nowell PC (1986) Effect of cyclosporin A and dexamethasone on interleukin 2 receptor gene expression. J Immunol 137:150–154

Reinberg A, Gervais P, Chaussade M, Fraboulet G, Duburque B (1983) Circadian changes in effectiveness of corticosteroids in eight patients with allergic asthma. J Allergy Clin Immunol 71:425–433

Ringold GM (1985) Steroid hormone regulation of gene suppression. Annual Rev Pharmacol Toxicol 25:529–566

Russo-Marie F, Paing M, Duval D (1979) Involvement of glucocorticoid receptors in steroid-induced inhibition of prostaglandin secretion. J Biol Chem 254:8498–8504

Ryan G, Latimer KM, Roberts RS, Hargreave FE (1985) Effect of beclomethasone diproprionate on bronchial responsiveness to histamine in controlled non-steroid dependent asthma. J Allergy Clin Immunol 75:25–30

Saavedres-Delgado AMP, Mathews KP, Pan PM, Kay DR, Muilenberg ML (1980) Dose response studies of the suppression of whole blood histamine and basophil counts by prednisone. J Allergy Clin Immunol 66:464–471

Samuelson WM, Davies AO (1984) Hydrocortisone induced reversal of beta adrenergic receptor uncoupling. Am Rev Respir Dis 130:1023–1026

Sanjar S, Aoki S, Boubekeur K, Chapman ID, Smith D, Kings MA, Morley J (1990) Eosinophil accunulation in pulmonary airways of guinea pigs induced by exposure to an aerosol of platelet activating factor: effect of anti-asthma drugs. Br J Pharmacol 99:267–272

Sapolsky R, Rivier C, Yamamoto G, Plotsky P, Vale W (1987) Interleukin-1 stimulates the secretion of hypothalamic corticotrophin releasing factor. Science 238:522–524

Saris CJM, Tach BF, Kristensen T, Glenney JR, Hunter T (1986) The cDNA sequence for the protein-tyrosine kinase substrate p36 (calpactin 1 heavy chain) reveals a multidomain protein with internal repeats. Cell 46:201–212

Schlaepfer DD, Haigler HT (1987) Characterisation of Ca^{2+}-dependent phospholipid binding and phosphorylation of lipocortin 1. J Biol Chem 262:6931–6937

Schlaepfer DD, Haigler HT (1988) In vitro protein kinase c phosphorylation sites of placental lipocortin. Biochemistry 27:4253–4258
Schleimer RP (1990) Effects of glucocorticoids on inflammatory cells. Am Rev Respir Dis [Suppl] 141:559–569
Schleimer RP, MacGlashan DW, Gillespie E, Lichtenstein LM (1982) Inhibition of basophil histamine release by anti-inflammatory steroids II. Studies on the mechanism of action. J Immunol 129:1632–1636
Schleimer RP, Freeland HS, Peters SP, Brown KE, Derse CP (1989) An assessment of the effects of glucocorticoids on degranulation, chemotaxis, binding, to vascular endothelium and formation of leukotriene B_4 by purified human neutrophils. J Pharmacol Exp Ther 250:598–605
Schutz G, Killewich L, Chen G, Feigelson P (1975) Control of the mRNA for hepatic tryptophan oxygenase during hormonal and substrate induction. Proc Natl Acad Sci USA 72:1017–1020
Smillie F, Peers SH, Elderfield AJ, Bolton C, Flower RJ (1989) Differential regulation by glucocorticoids of intracellular lipocortin 1, 2 and 5 in rat mixed peritoneal leukocytes. Br J Pharmacol 97:425P
Smith SF, Goulding NJ, Tetley TD, Godolphin JL, Guz A, Flower RJ (1989a) A possible anti-inflammatory mechanism for glucocorticoids in the human lung. Clin Sci 77 [Suppl 21]:20P
Smith SF, Tetley TD, Guz A, Flower RJ (1989b) Inflammation in the lung: a novel role for the neutrophil. Am Rev Respir Dis 139:A303
Smith SF, Tetley TD, Guz A, Flower RJ (1990) Detection of lipocortin 1 in human lung lavage fluid: lipocortin degradation as a possible proteolytic mechanism in the control of inflammatory mediators and inflammation. Environ Health Perspect 85:135–144
Sorenson DK, Kelly TM, Murray DK, Nelson DH (1988) Corticosteroids stimulate an increase in phospholipase A_2 inhibitor in human serum. J Steroid Biochem 29:271–273
Sotomayor H, Badier M, Vervloet D, Orehek J (1984) Seasonal increase of carbachol airway responsiveness in patients allergic to grass pollen. Am Rev Respir Dis 56–58
Svendsen UG, Frolund L, Madsen F, Nielsen NH, Holstein-Rathlou N-H, Weeke B (1987) A comparison of the effects of sodium cromoglycate and beclomethasone diproprionate on pulmonary function and bronchial hyperreactivity in subjects with asthma. J Allergy Clin Immunol 80:68–74
Svensjo E, Roempke K (1985) Time dependent inhibition of bradykinin and histamine-induced microvascular permeability increase by local glucocorticoid treatment. Prog Respir Res 19:773–780
Thorne KJI, Richardson BA, Butterworth AE, Higginbottom TW (1988) Effect of drugs used in the treatment of asthma on the production of eosinophil-activating factor by monocytes. Int Arch Allergy Appl Immunol 85:257–259
Tiffeneau R, Dunoyer P (1956) Action de la cortisone sur l'hypersensibilite cholinegique pulmonaire do l'asthmatique. Presse Med 54:719–722
Toogood JH, Baskerville J, Jennings B, Lefcoe NM, Johansson S-A (1984) Use of spacers to facilitate inhaled corticosteroid treatment of asthma. Am Rev Respir Dis 129:723–729
Toogood JH, Jeenings B, Baskerville JC (1985) Aerosol corticosteroids. In: Weisse EB, Segal MS, Stein M (eds) Bronchial asthma: mechanisms and therapeutics. Little Brown, Boston, pp 698–713
Toogood JH, Crilly RG, Jones G, Nadeau J, Wells GA (1988) Effect of high dose inhaled budesonide on calcium and phosphate metabolism and the risk of osteoporosis. Am Rev Respir Dis 138:57–61
Touqui L, Rothhut B, Shaw AM, Fradin A, Vargaftig BB, Russo-Marie F (1986) Platelet activation – a role for a 40K anti-phospholipase A_2 protein indistinguishable from lipocortin. Nature 321:177–180

Travis J, Salvesen GS (1983) Human plasma proteinase inhibitors. Annual Rev Biochem 52:655–709

Tsurufuji S, Sugio K, Takemasa F (1979) The role of glucocorticoid receptor and gene expression in the antiinflammatory action of dexamethasone. Nature 280:408–410

Tsurufuji S, Sugio K, Takemasa F, Yoshizawa S (1980) Blockade by anti-glucocorticoids, actinomycin-D and cycloheximide of anti-inflammatory action of dexamethasone against bradykinin. J Pharmacol Exp Ther 212:225–231

Vannice JL, Taylor JM, Ringold GM (1984) Glucocorticoid-mediated induction of 1-acid glycoprotein: evidence for hormone regulated RNA processing. Proc Natl Acad Sci USA 81:4241–4245

Varticovski L, Chahwala SB, Whitman M, Cantley L, Schindler D, Chow EP, Sinclair LK, Pepinsky RB (1988) Location of sites in human lipocortin 1 that are phosphorylated by protein tyrosine kinases and protein kinases A and C. Biochemistry 27:3682–3690

Venezio FR, Westerfelder GO, Phair JP (1982) The adherance of polymorphonuclear leukocytes in patients with sepsis. J Infect Dis 145:351–357

Wallner BP, Mattaliano RJ, Hession C, Cate RL, Tizard R, Sinclair LK, Foeller C, Chow EP, Browning JL, Ramachandran KL, Pepinsky RB (1986) Cloning and expression of human lipocortin, a phospholipase A_2 inhibitor with potential anti-inflammatory activity. Nature 320:77–81

Werb Z (1979) Hormone receptors and hormonal regulation of macrophage physiological functions. In: van Furth R (ed) Mononuclear phagocytes, part I: functional aspects. Nijhoff, Dordrecht, pp 809–829

William F, Mroczkowski B, Cohen S, Kraft AS (1988) Differentiation of HL-60 cells is associated with an increase in the 35kDa protein lipocortin-1. J Cell Physiol 137:402–410

Williams TJ, Yarwood H (1990) Effect of glucocortiociods on microvascular permeability. Am Rev Respir Dis [Suppl] 141:539–543

Wolf JD, Rosenthal RR, Bleeker E, Laube B, Norman PS, Permutt S (1979) The effect of corticosteroids on cholinergic hyperreactivity. J Allergy Clin Immunol 63:1634

Yarwood H, Nourshargh S, Brain SD, Williams TJ (1988) Suppression of neutrophil accumulation and neutrophil-dependent oedema by dexamethasone in rabbit skin. Br J Pharmacol 95:535P

Zweiman B, Slott RI, Atkins PC (1976) Histologic studies of human skin test responses to ragweed and compound 48/80. III. Effects of alternate day steroid therapy. J Allergy Clin Immunol 58:657–663

CHAPTER 9

Pharmacology of Prophylactic Anti-Asthma Drugs

L.G. Garland

Among the different classes of drugs used in the treatment of asthma, the bronchodilators (β_2-adrenoceptor agonists, muscarinic receptor antagonists, methylxanthines) bring symptomatic relief but contribute little to ameliorating the underlying bronchial inflammation. Glucocorticoids (see Chap. 8) exert a potent antiinflammatory effect and, particularly when inhaled, have both therapeutic and prophylactic value. The drug that is now accepted clinically as having a prophylactic effect in asthma, whilst only rarely bringing symptomatic relief, is cromoglycate; the recently introduced nedocromil is another example of the same class of compound. Ketotifen has a different pharmacological profile and has gained some acceptance as a prophylactic agent in asthma. This chapter will only review the pharmacology of drugs in clinical use (cromoglycate, nedocromil, ketotifen), although there is much current research directed towards finding other new types of antiinflammatory agents to be used as first-line durgs in mild asthma, leaving glucocorticoids for treatment of severe asthma.

A. Cromoglycate and Nedocromil

The unusual way in which the activity of cromoglycate was discovered has been written about before and is probably relayed most accurately by those involved in its discovery (Altounyan 1980; Suschitzky and Sheard 1984; Richards et al. 1986). Suffice to say, antiasthmatic activity was identified first in humans; subsequent animal experiments revealed no bronchodilator activity (expect in vitro at very high concentrations) but an antiallergic action which was thought to explain the clinical effect of the compound. As more data has been accrued, particularly from the clinic, interpretation of the action of cromoglycate has changed, but the pharmacological basis for its action is still not well-defined. In this review cromoglycate and nedocromil will be discussed as two examples from the same class of compounds in order to attempt some form of pharmacological classification.

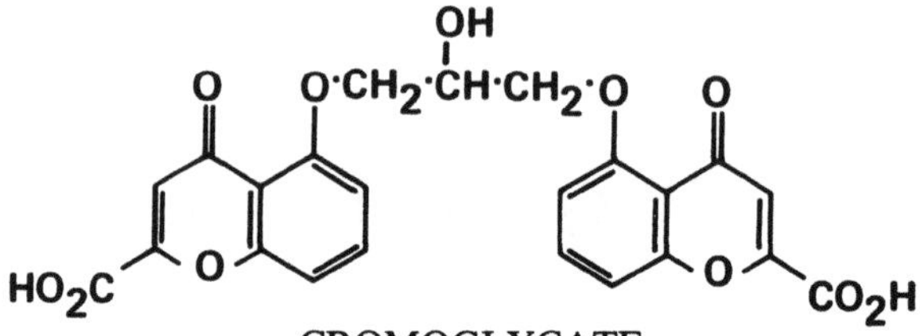

CROMOGLYCATE

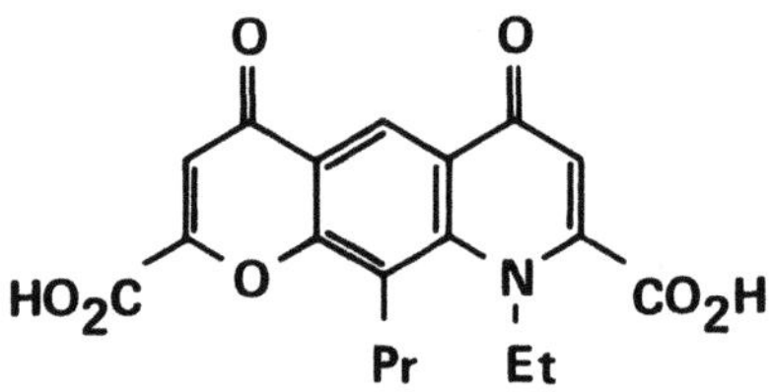

NEDOCROMIL

I. Cellular Pharmacology

1. Mast Cell and Basophils

a) Mast Cell Heterogeneity

Inhibition of mast cell secretion by cromoglycate was reviewed in an earlier volume of the *Handbook of Experimental Pharmacology* (Vol. 50, GARLAND et al. 1979). Results available at that time indicated that the effect of cromoglycate varied among mast cells from different tissues and between tissue mast cells and circulating basophils. Hence, antigen-induced release of histamine from IgE-sensitised rat skin or rat peritoneal mast cells was inhibited fully whilst release from IgE-sensitised human skin was not inhibited at all. Similarly, IgE-dependent release from human peripheral blood basophils was resistant to inhibition by cromoglycate. An intermediate effect was found using IgE-sensitised human lung fragments where the maximum inhibition of anaphyactic histamine release was ≤50%, even with high concentrations (100 μM) of cromoglycate.

Continued investigation has increased evidence for the variable efficacy of cromoglycate among different populations of mast cells. The compound is a potent inhibitor of anaphylactic histamine release from serosal mast cells of the rat, moderately effective against these cells from the hamster but totally inactive in mouse cells (PEARCE et al. 1985). It is also totally ineffective in rat intestinal mucosal mast cells and connective tissue mast cells of the guinea-pig. Furthermore, cromoglycate blocked degranulation of guinea-pig basophils, but not mast cells, stimulated by complement fragment C5a (GOLDEN et al. 1987). Further examples of mast cell specificity are described by PEARCE (1983).

Weak inhibition of histamine release from human lung parenchymal mast cells has been confirmed (BUTCHERS et al. 1979; CHURCH et al. 1983b; CHURCH and YOUNG 1983) but data from a large number of experiments emphasises the great variability among lung fragments from different individuals (ranging from 0% to 90% inhibition with 20 μM cromoglycate; CHURCH 1986). Thus, while it is interesting that anaphylactic histamine release from human lung mast cells in bronchoalveolar lavage fluid appears to be inhibited more than release from parenchymal cells (PEARCE et al. 1985), these data may be too preliminary for far-reaching conclusions to be drawn about differences between these different cell types.

Nedocromil is equipotent with cromoglycate as an inhibitor of IgE-dependent mediator release from rat peritoneal mast cells (EADY et al. 1985; WELLS et al. 1986) but is about ten times more potent as an inhibitor of release from human lung mast cells. In each case dose-response curves for the two compounds achieved similar maxima. Antigen-induced histamine release from sensitised monkey bronchoalveolar lavage mast cells was inhibited much more effectively by nedocromil than cromoglycate, the IC_{30} value (5×10^{-6} *M*) being 190 times lower than that for cromoglygate and the maximum inhibition (60%) being twice that achieved by cromglygate (WELLS et al. 1986). Furthermore, nedocromil, like cromoglycate, appears to be strikingly more effective as an inhibitor of IgE-dependent histamine release from human lung mast cells recovered from bronchoalveolar lavage than from cells recovered from dispersed parenchyma (LEUNG et al. 1986). The mast cells present in bronchoalveolar lavage fluid have staining characteristics typical of mucosal mast cells rather than connective tissue mast cells (FLINT et al. 1985). Thus, nedocromil appears to have greater efficacy than cromoglycate as an inhibitor of lung mucosal mast cell secretion. Like cromoglycate, nedocromil blocks the antigen-induced synthesis and release of eicosanoids from IgE-sensitised lung bronchoalveolar lavage cells. However, nedocromil was significantly more active (EADY, 1986).

b) Secretogogues

The apparent selectivity of action of cromoglygate on inhibition of secretion provoked by some, but not all, stimuli has received much attention experimentally (reviewed by GARLAND et al. 1979). In such experiments, as when studying apparent selectivity among cell type, it is essential to recognise that inhibition by cromoglycate varies inversely with the strength of the secretory stimulus. This was first shown in vitro by varying the concentration of antigen used to stimulate IgE-sensitised lung fragments (ASSEM and MONGAR 1970) and confirmed using either dextran or antigen to stimulate histamine release from rat peritoneal mast cell (GARLAND and MONGAR 1974;

PEARCE and RAFII-TABOR 1983). Furthermore, phosphatidylserine increased the response of rat peritoneal mast cells stimulated by either dextran or antigen, and this overcame the inhibition by cromoglycate (GARLAND and MONGAR 1974; READ et al. 1977; PEARCE and RAFII-TABOR 1983). Thus, it is important to maintain a constant stimulus strength when assessing the differential activity of cromoglycate or nedocromil, either among mast cell types or between different secretogogues.

When secretory responses, provoked by antigen, dextran or the calcium ionophore A23187 were held constant, cromoglycate ($\leq 30\ \mu M$) did not block the response to A23187 but abolished responses to the other two stimuli (GARLAND and MONGAR 1976). Higher concentrations of cromoglycate (30–1000 μM) inhibited release provoked by ionophores (A23187, ionomycin, chlortetracycline), and inhibition varied inversely with the strength of the stimulus (PEARCE and TRUNEH 1981; WHITE and PEARCE 1983; TRUNEH and PEARCE 1984). In these experiments cromoglycate slowed the initial rate of release (at 20 s) but did not reduce the maximum release (at 2 min). However, there is now convincing evidence that cromoglycate is particularly effective as an inhibitor of release stimulated by calcium ionophores at low extracellular calcium concentrations, when secretion is presumably stimulated by release of calcium from intracellular stores (GARLAND and HODSON 1984; TRUNEH and PEARCE 1984). Under these conditions, cromoglycate was not only effective at low concentrations (10–30 μM) but also reduced the maximum secretory response. These observations were extended by WHITE et al. (1984) who used the fluorescent indicator Quin 2 to show that when cromglycate inhibited IgE-dependent histamine release the concomitant rise in intracellular free calcium was also reduced; this reduction occurred both in the absence and presence of EGTA, suggesting an effect on calcium mobilisation from intracellular stores. In contrast, A23187-induced histamine release was inhibited without an effect on the concomitant rise in cytoplasmic calcium. This is umambiguous evidence that inhibition of secretion by cromoglycate can also occur by a mechanism separate from inhibition of calcium mobilisation.

2. Neutrophils, Eosinophils, Monocytes, and Platelets

The emphasis on a specific mast cell action for cromoglycate prevalent 10 years ago has been replaced by evidence of a broader antiinflammatory action. Increase in airway resistance, provoked in asthmatics by either allergen inhalation or exercise, is accompanied by activation of circulating blood neutrophils and monocytes, and this activation is prevented by inhalation of cromoglycate prior to challenge (PAPAGEORGIOU et al. 1983; DURHAM et al. 1984). Furthermore, clinical improvement of asthmatics treated with cromoglycate is associated with a reduction in eosinophils found in bronchoalveolar lavage fluid (DIAZ et al. 1984). Observations such as these, suggesting that cromoglycate may act on leukocytes apart from mast cells, have been extended by experiments in vitro. The observation that IgE-dependent responses of alveolar macrophages from asthmatics were inhibited in vitro by low concentration of cromoglycate (JOSEPH et al. 1981) were not reproducible (FULLER and MACDERMOT 1986). However, other reports showed that responses of neutrophils, eosinophils and monocytes were inhibited by low concentrations of cromoglycate. Thus, KAY et al. (1987) showed tha f-met-leu-phe (fMLP) stimulation of leukocytes increased expression of membrane receptors for complement (C3b) and IgG (F_c) and increased cytotoxicity for complement-coated schistosomula: each of these fMLP-induced responses was fully inhibited by cromoglycate (IC_{50}~0.01 μM), inhibition developing to maximum during 30 min preincubation prior to stimulation with fMLP. Similar observations were reported by MOQBEL et al. (1986a). Responses of neutrophils from non-asthmatics stimulated in vitro were also inhibited by cromglycate (SKEDINGER et al. 1987). Thus, the drug (2–20 μM) inhibited chemotaxis stimulated by either fMLP or opsonised zymosan but not by leukotriene (LT) B_4; this differential effect was observed despite the responses to each stimulus being of comparable magnitude. The maximum effect (~ 50% inhibition) was observed after 30 min preincubation and was reversed by washing away the inhibitor. Cromoglycate had no effect on fMLP receptor density but inhibited the increase in intracellular free calcium induced by opsonised zymosan in neutrophils suspended in calcium-free medium, suggesting an effect on calcium release from intracellular stores; the LTB_4-induced rise in intracellular calcium was not blocked by cromoglycate.

Nedocromil also inhibits responses of human neutrophils and eosinophils in vitro, as assessed using fMLP-induced enhancement of C3b and IgG (Fc) receptor expression and cytotoxicity (MOQBEL et al. 1986b), and inhibits the release of granule protein from human eosinophils stimulated with sepharose C3b (SPRY et al. 1986). Furthermore, nedocromil and cromoglycate both inhibit several IgE-dependent responses of rat peritoneal macrophages and rat platelets, including superoxide production (JOSEPH et al. 1986; TSICOPOULOS et al. 1988). Where nedocromil and cromoglycate have both been evaluated the two compounds generally appear to have comparable activity. However, one apparent difference is that degranulation of rabbit neutrophils stimulated by phorbol ester was inhibited by nedocromil but not by cromoglycate (BRADFORD and RUBIN 1986). Another difference may have been revealed by the observation that nedocromil but not cromoglycate inhibited in vitro responses of platelets from aspirin-sensitive asthmatics (JOSEPH et al. 1988). In these experiments aspirin stimulated the generation of cytotoxic substances and oxygen-derived free radicals from platelets of aspirin-sensitive asthmatics. This response was inhibited by preincubating platelets with either nedocromil or cromoglycate, although nedocromil (IC_{50} = 2 nM) was 500 times more potent than cromolycate in this experimental system. Furthermore, inhalation of nedocromil (4 mg) blocked the cytotoxic activity of platelets stimulated by incubation with aspirin ex vivo. Consistent with these observations, inhalation of cromoglycate (20 mg) before and after ingestion of aspirin prevented bronchospasm, while inhalation of cromoglycate after aspirin ingestion delayed progress of the ongoing bronchospasm (MARTELLI and USANDIVARAS 1977).

II. Biochemical Pharmacology

Cromoglycate and nedocromil are both hydrophilic compounds that are unlikely to cross cell membranes other than by a specific transport process. Indeed, cromoglycate inhibited mast cell secretion when linked to polymer beads that prevented cell penetration (MAZUREK et al. 1980a). Thus, the drug acts at the cell membrane and not by directly inhibiting intracellular enzymes, such as cyclic nucleotide phosphodiesterase(s), as was once suggested (FOREMAN and GARLAND 1976). A binding protein for cromoglycate has been isolated from basophil membranes (MAZUREK et al. 1982) and evidence was presented to suggest that this protein is part of a calcium channel associated with IgE (Fc) receptors (MAZUREK et al. 1983a,b; 1984; CORCIA et al. 1986). These observations may indicate a mechanism by which cromoglycate prevents transmembrane flux of calcium ions. However, evidence discussed above (Sect. A.I.1.b) showed that the drug also acts in leukocytes by a mechanism separate from inhibition of calcium influx. A key observation concerning the mechanism of action of cromoglycate is that the drug stimulates phosphorylation of an intracellular 78 kDa protein (THEOHARIDES et al. 1980). This cellular response reflected several properties of cromoglycate, assessed by measuring inhibition of mast cell secretion, i.e. both had the same IC_{50} value, were relatively transient and exhibited tachyphylaxis. Stimulation of protein phosphorylation in mast cells has been confirmed (WELLS and MANN 1983) and has been observed also with nedocromil (WELLS et al. 1986). The identity of the 78 Kda protein phosphorylated in these experiments is unknown and comparison with the substrate for cGMP-dependent kinase on the basis of molecular weight alone (WELLS and MANN 1983) could be misleading.

Suggestions in the literature that cromoglycate acts by inhibiting protein kinase (PK)C (SAGI-EISENBERG 1985; LUCAS and SHUSTER 1987) deserve some comment. If present in the ionised form cromoglycate will not penetrate the cell membrane to interact with PKC. Nevertheless, it has been suggested that, when complexed with calcium (MAZUREK et al. 1980b), cromoglycate may associate with membrane lipid. Thus, we investigated PKC inhibition by cromoglycate in a lipid/Triton X-100 system used previously to assess molecular requirements for PKC activation by diaclyglycerol analogues (BONSER et al. 1988). In this system, which contains $CaCl_2$ at 0.1 mM, cromoglycate had no effect on the activity of PKC (BONSER and GARLAND, unpublished). Taken together with the observation that cromoglycate affects mast cell secretion in the absence of extracellular calcium (and hence absence of penetration into membrane lipids) there is no reason to believe the drug interacts directly with PKC. However, the observations that cromoglycate and nedocromil inhibit a phorbol ester-stimulated response of lizard melanocytes (LUCAS and SHUSTER 1987) should be considered along with the results of BRADFORD and RUBIN (1986) mentioned above (Sect. A.2). In this context it is also interesting to note that cromoglycate inhibits clustering of glycosphingolipids in the plasma membrane (CURTAIN et al. 1981) which may influence cellular PKC activity.

III. Animal Pharmacology

1. Anaphylactic Bronchospasm

The history of this area of pharmacology records that IgE-dependent anaphylactic reactions in skin or lungs of various species (especially the rat) failed to identify compounds with cromoglycate-like antiasthmatic activity in humans (SUSCHITZKY and SHEARD 1984; GARLAND and HODSON 1984). However, anaphylactic bronchospasm (in guinea-pigs, sheep, dogs and monkeys) has remained a laboratory model for investigation of asthma, with the efficacy of therapeutically useful drugs like cromoglycate being used to indicate validity of the model. In monkeys sensitised by infestation with *Ascaris suum*, antigen-induced bronchospasm was not reliably inhibited by cromoglycate given either intravenously or orally. However, this model did have some predictive value since analogues of cromoglycate that were known to be more effective in humans were found to protect monkeys from immediate bronchospasm. Thus, nedocromil was active in this model, whereas cromoglycate was not (EADY et al. 1985; JACKSON and EADY 1986). Similarly, in dogs sensitised to *Ascaris suum*, bronchospasm provoked by inhalation of Ascaris antigen was not prevented by cromoglycate but certain analogues that were more effective in humans did block the immediate response in this model (see review by RICHARDS et al. 1986). Thus, while many studies have indicated that mast cell stabilisation alone is not an important feature for prophylactic antiasthma drugs, the different efficacies of cromoglycate and nedocromil as inhibitors of IgE-dependent, immediate bronchospasm in monkey and dog did appear to correspond with their efficacies in humans. This may reflect the different efficacies of the two drugs as inhibitors of bronchoalveolar lavage mast cell secretion described above (Sect. A.I.1.a).

Inhibition of the late phase of bronchospasm that occurs between 4 – 12 h after allergen inhalation may be relevant to the broader antiinflammatory action of cromoglycate-like drugs. Inhibition of late phase bronchospasm in humans was reported by PEPYS et al. (1968) soon after cromoglycate was introduced for clinical use. It should be noted in passing that interpretation of drug effects in animal (or human) models that require many hours to develop requires prior understanding of the pharmacokinetic properties of the drug in the species of choice. Variations in rate of clearance after inhalation, for example, will affect drug actions on the late response more than on the immediate response. In sheep sensitised to *Ascaris*, pretreatment with aerosols of cromoglycate or nedocromil (20 mg inhaled) reduced both early and late phase (7 h) bronchospasm following antigen inhalation. Nedocromil appeared to be less effective than cromoglycate but the difference was not significant (ABRAHAM et al. 1987). HUTSON et al. (1988) have described a model using conscious guinea-pigs actively sensitised by inhalation of ovalbumin aerosol on two occasions 7 days apart, in which the third exposure to ovalbumin aerosol provoked three phases of bronchospasm (at 2, 17 and 72 h). The response at 17 h was accompanied by infiltration of neutrophils and eosinophils into the bronchial lumen; by 72 h eosinophil accumulation had increased but neutrophil levels had returned towards normal. When administered both 15 min before and 6 h after antigen challenge, cromoglycate inhibited all phases of bronchospasm and reduced the accumulation of neutrophils at 17 h and of eosinophils at 72 h. The authors proposed that the similarities between the actions of drugs in this model and their clinical effects support its validity as a model of allergen-induced asthma.

2. Sensory Reflexes

Apart from inhibiting responses of leukocytes, including mast cells, another significant action of cromoglycate is on certain sensory nerve endings. Thus, in dogs low intravenous doses of cromoglycate (20 μg/kg) provoked reflex bradycardia and hypotension through stimulation of sensory receptors in the pulmonary and coronary circulation (COX et al. 1970). Given by aerosol the drug significantly inhibited the vagal reflex component of histamine-induced bronchospasm, and, given intravenously, it reversed the sustained reflex bronchospasm provoked by histamine aerosol. Since it did not inhibit bronchospasm due to direct vagal stimulation, it was proposed that cromoglycate attenuated the activity of lung irritant receptors (JACKSON and RICHARDS 1977). However, a more detailed investigation of lung irritant receptors showed that cromoglycate had no effect on either resting discharge rate or their response to histamine. Application of local anaesthetic to the pericardium blocked reflex hypotension and the reversal of reflex bronchospasm following intravenous cromoglycate. Thus, it appears that in the dog cromoglycate reverses reflex bronchospasm by stimulating sensory receptors in the left ventricle of the heart, not by inhibiting irritant receptor discharge (DIXON et al. 1979). This is consistent with the report (BERGREN et al. 1985) that cromoglycate had no effect on histamine-induced responses of rapidly adapting ("irritant") receptors in dog lung, and the same mechanism may also explain blockade by cromoglycate of reflex cardiovascular responses provoked by phenylbiguanide in guinea-pigs (BIGGS and GOEL 1985). A related observation in dogs is that cromoglycate prevented hypoxia-induced pulmonary hypertension by increasing vagal tone due to pulmonary chemoreceptor discharge (RENGO et al. 1979). A blocking action on C-fibre nerve endings was indicated when low doses (100 μg kg, i.v.) of cromoglycate limited the increase in discharge rate provoked by capsaicin (DIXON et al. 1980). This observation was used by CHIAVARELLI et al. (1981) to interpret the blockade by cromoglycate of prostacyclin-induced bradycardia as being through an action on C-fibre nerve endings. However, this has not been a consistent observation since others found that cromoglycate did not desensitise bronchial or pulmonary C-fibres to the effects of capsaicin (COLERIDGE et al. 1982; DAVIS et al. 1982).

Inhalation of ozone causes epithelial damage, neutrophil infiltration and increased sensitivity of vagal afferent sensory nerve endings that contribute to the subsequent increase in airway reactivity to stimuli such as histamine (FABBRI et al. 1984; HOLTZMAN et al. 1983). Thus, attenuation by cromoglycate of ozone-induced changes in guinea-pig lung function (MILLER et al. 1988) may have been through an effect on sensory nerve function. Similarly, the increase in airway reactivity in dogs following inhalation of SO_2 was also attenuated by cromoglycate (JACKSON unpublished, ref, EADY 1986).

A detailed comparison between the effects of cromoglycate and nedrocromil has not been reported. However, like cromoglycate, nedocromil reduced SO_2-induced airway hyperreactivity in dogs (JACKSON et al. 1986a; JACKSON and EADY 1988). The effect of nedocromil was attributed to an antiinflammatory effect of the drug. In contrast to cromoglycate, nedocromil attenuated citric acid-induced cough in dogs (JACKSON 1988). This effect was interpreted as being through inhibition of sensory nerve activity in the lung, but it cannot be ruled out that, as with cardiovascular reflexes, inhibition of the cough reflex is an indirect effect following stimulation by cromoglycate of sensory nerve endings.

IV. Human Pharmacology

Observations in humans have contributed greatly to the pharmacological assessment of cromoglycate and nedocromil. Inhibition of allergen-induced bronchospasm, both immediate and delayed responses, by either cromoglycate or nedocromil inhaled prior to allergen is consistent with an effect on release of inflammatory mediators and reduction of leukocyte infiltration into the airways, rather than antagonism of specific mediators. Certainly, cromoglycate has no inhibitory effect in humans on bronchospasm provoked by inhalation of LTD_4 (HOLROYDE et al. 1981; ROBERTS et al. 1986).

The late-onset skin response (erythema and hyperalegesia) at 3–6 h following intradermal injection of platelet activating factor (PAF) was blocked in some, but not all, subjects by the co-injection of cromoglycate (BASRAN et al. 1983; ARCHER et al. 1985). Also, the increase in nasal airway resistance provoked by PAF was significantly reduced by cromoglycate insufflation for 3 days prior to challenge (KARLSSON and PIPKORN 1984). As there is no evidence from animal studies that cromoglycate is a PAF-receptor antagonist, these observations indicate that responses to PAF occur indirectly through either inflammatory cells or neural reflexes that are sensitive to inhibition by cromoglycate.

A similarly imprecise conclusion usually has to be drawn from experiments where bronchospasm induced by an inhaled spasmogen is blocked by cromoglycate. For example, inhalation of adenosine causes bronchospasm in asthmatic but not usually in healthy subjects (CUSHLEY et al. 1983); it has been shown that this response is significantly reduced by prior inhalation of either cromoglycate or nedocromil, with nedocromil being significantly more effective than cromoglycate (CUSHLEY and NOLGATE 1985; ALTOUNYAN et al. 1986b; CRIMI et al. 1986; 1988a, b; PHILLIPS et al. 1988). Similar observations have been made in animals (PAUWELS and VAN DER STRAETEN 1986). However, the pharmacological interpretation is imprecise because adenosine not only augments the ongoing release of mediators from human basophils and mast cells (CHURCH et al. 1983a,c) and contracts airway smooth muscle from asthmatics in vitro (DAHLEN et al. 1983) but also stimulates sensory nerve endings (BLEEHEN and KEELE 1977). Thus, any or all of the major pharmacological properties of cromoglycate discussed above (see Sect. A.I and II) might account for these observations in adenosine-challenged asthmatics. There is, however, no published evidence to support the suggestion (CRIMI et al. 1988a) that cromoglycate is an antagonist of adenosine receptors.

The increased sensitivity of the airways to a range of stimuli is a characteristic feature of bronchial asthma, although the causes of such bronchial hyperreactivity are poorly understood. Like the inhaled steroids, cromoglycate appears to have long-term effects on airway reactivity (ALTOUNYAN 1970; DICKSON 1970). Studies of airway reactivity often involve monitoring responses of asthmatic's airways to inhaled spasmogens such as histamine and methacholine. Cromoglycate does not antagonise the receptors for such spasmogens but changes the reactivity of the airways by another mechanism. Early studies showing an improvement in airway hyperreactivity were in atopic asthmatics. Hence the effect of cromoglycate therapy in such subjects is consistent with a reduction in release of inflamatory mediators and a subsequent fall in mucosal oedema, smooth muscle tone and possibly also sensory afferent nerve activity in the lungs. Several careful studies support this interpretation, as cromoglycate or nedocromil protected allergic subjects from the increased responsivness to inhaled histamine that occurred during the pollen season (STAFFORD et al. 1984; LOWHAGEN and RAK 1985a; ALTOUNYAN et al. 1986a; DORWARD et al. 1986). However, when atopic patients with a history of perennial asthma were studied in the pollen-free season, cromoglycate therapy had no significant effect on bronchial reactivity to inhaled histamine (RYO et al. 1976; LOWHAGEN and RAK 1985b; LAITINEN et al. 1986). These studies indicate that the effect of either cromoglycate or nedocromil on histamine-induced bronchial hyperreactivity can be interpreted properly only when environmental allergen exposure is controlled and suggests that the drugs have no effect on basal reactivity to histamine. The same also appears to be true when studying airway hyperreactivity to cholinergic agonists. Allergen-induced increase in bronchial reactivity to inhaled methacholine was clearly blocked by cromoglycate (MATTOLI et al. 1986), but the drug was not effective out of season when the allergic asthmatic was not subjected to allergen challenge (RYO et al. 1976). This probably explains why bronchial hyperreactivity to methacholine or carbachol has not been prevented consistently by cromglycate when exposure to environmental allergens has not been controlled (WOENNE et al. 1979; FABBRI et al. 1983; BONER et al. 1987; BARBATO et al. 1987; KRAEMER et al. 1987). It is interesting to note that patients with cystic fibrosis exhibit some of the clinical features of asthma, including bronchial hyperreactivity to inhaled methacholine, and this was substantially reduced by cromoglycate (MITCHELL 1985).

DAVIES (1968) was the first to report that cromoglycate protects against exercise-induced asthma. Subsequently, this has been a fequent finding (GODFREY et al. 1973; MORTON and FITCH 1974; KENNEDY et al. 1980; BEN-DOV et al. 1983; BONER et al. 1985; PATEL and WALL 1986; TAN and LIM 1987) and has been supported by recent observations with nedocromil (KONIG et al. 1987; BAUER 1986; DEBELIC 1986; SHAW and KAY 1986; THOMSON and ROBERTS 1986). Interestingly, this is one of the few responses in humans in which tachyphylaxis has been noted after multiple dosing with cromoglycate (BROOKS et al. 1986). Since exercise-induced bronchospasm is not accompanied by a rise in plasma levels of known inflammatory mediators, the protective effect of the drugs may not be through their action on inflammatory cells. Thus, the alternative interpretation is that exercise-induced bronchospasm is prevented by an action on sensory reflexes in the lung. Protection by muscarinic receptor antagonists indicates that cholinergic reflexes are involved in exercise-induced bronchospasm. Like atropine, inhalation of cromoglycate also blocked bronchospasm induced by inhalation of hypotonic mist (FULLER and COLLIER 1984; DEL BUFALO et al. 1988). This effect, which was immediate in onset, is consistent with attenuation of sensory nerve function but is not uniform among airway reflexes, as the cough response stimulated by either hypotonic mist or capsaicin (COLLIER and FULLER 1984) was not blocked by cromoglycate, although the drug appeared to improve cough stimulated by cigarette smoke (LEITCH et al. 1984). Furthermore, the action of cromoglycate on sensory reflexes was distinguished from that of atropine by studying bronchospasm induced by cold air. Hyperventilation with freezing air provokes bronchospasm in both healthy and asthmatic subjects, and this is not accompanied by release into the plasma of mast cell-derived mediators (DEAL et al. 1980). Generally, asthmatics are more sensitive than healthy subjects to this stimulus. The response of asthmatics to cold air challenge was reduced by pretreatment with inhaled cromoglycate (BRESLIN et al. 1980; GRIFFIN et al. 1983; LATIMER et al. 1983; PICHURKO et al. 1984; RAFFERTY et al. 1985; JUNIPER et al. 1986, 1987). In contrast to cromoglycate, atropine did not reduce airway responses stimulated by freezing air. Cromoglycate was ineffective when sprayed into the oropharynx, indicating a site of action in the lungs. It is unlikely to act through preventing release of inflammatory mediators which appear not to be involved in the response. Furthermore, BRESLIN et al. (1980) concluded that cromoglycate did not affect heat loss from the airways but PICHURKO et al. (1984) found that it increased airway temperature by an action (direct or indirect) on the bronchial vasculature. However, the most probable explanation for the action of cromoglycate is the attenuation of non-cholinergic bronchial reflexes. Inhibition of cold air challenge has also been seen after inhalation of nedocromil (del BONO et al. 1986; JUNIPER et al. 1987) and this drug was found to be more effective than cromoglycate.

Inhalation of SO_2 gas during moderate exercise stimulates bronchospasm in asthmatics by a mechanism involving parasympathetic pathways (NADEL et al. 1965; SHEPPARD et al. 1980) and possibly sensory reflexes. Inhalation of cromoglycate prior to SO_2 substantially inhibited the ensuing bronchospasm (SHEPPARD et al. 1981; TAN et al. 1982). The inhibitory effect of cromoglycate (20 and 200 mg) appeared to be dose-dependent (MYERS et al. 1986a), and this was confirmed by KOENIG et al. (1988) who studied 20, 40 and 60 mg doses and found that 60 mg completely abolished pulmonary function changes induced by SO_2 in atopic subjects. The combination of cromoglycate (200 mg) and the muscarinic receptor antagonist ipratropium (200 μg) inhibited SO_2-induced bronchospasm more than either agent alone (MYERS et al. 1986b).

There is no direct evidence in humans that cromoglycate attenuates sensory nerve function in the lungs or any other tissue. However, COLLIER and FULLER (1983) found that intravenous infusion of the drug increased heart rate and blood pressure and caused a sensation of warmth without changes in skin blood flow or temperature. These observations were interpreted to suggest that cromoglycate may stimulate certain sensory nerve endings in humans. This effect was subject to tachyphylaxis. Thus, as in animals, it is possible that inhibition of sensory nerve function in the lung is secondary to stimulation of sensory nerve endings elsewhere in the body.

B. Ketotifen

As in the case of cromoglycate, ketotifen is not well-characterised pharmacologically. It has been claimed to have prophylactic activity in asthma, but this has not been a consistent finding and the mechanism by which this might occur is still a subject of debate.

I. Pharmacology In Vitro

Ketotifen is a potent inhibitor of histamine-induced contractions of smooth muscle, including guinea-pig trachea where it is active at concentrations of 10^{-9}–10^{-7} *M*. Depression of the maximum response to histamine indicates

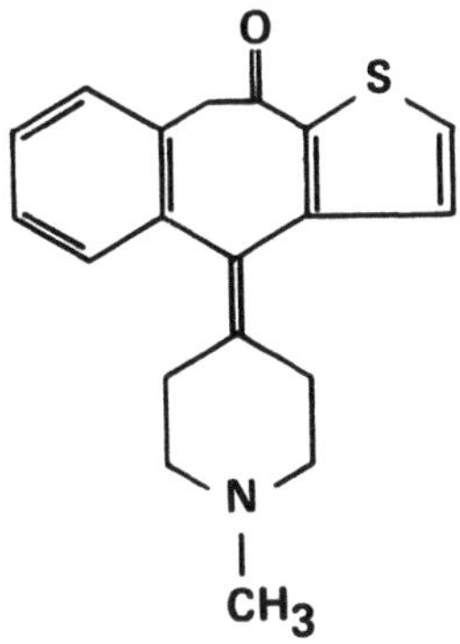

KETOTIFEN

that the compound is not a competitive antagonist of H_1 receptors (MORLEY and CHAPMAN 1989). Nevertheless, inhibition of histamine responses occurs fairly selectively, concentrations of 10^{-6}–10^{-4} M being required before other non-competitive actions of ketotifen become apparent. A range of non-specific actions have been reported to occur at these higher concentrations, including inhibition of smooth muscle contraction induced by 5-hydroxytrypatamine (serotonin, 5HT), bradykinin, acetylcholine, prostaglandin $F_{2\alpha}$, peptidoleukotrienes (SRS-A), barium chloride, potassium chloride and electrical stimulation (POLSON et al. 1982; ARRUZAZABALA and GONZALEZ 1984; 1985; CRAPS and NEY 1984; LOFTUS et al. 1985). Surprisingly, the effect of ketotifen on PAF-induced smooth muscle responses in vitro have not been well-documented.

Ketotifen at these relatively high concentrations exerts a range of inhibitory effects on leukocyte function. The most sensitive is inhibition of LTC_4-induced glass adherence of leukocytes (from asthmatics) that was abolished by incubation with ketotifen 2 μM (FINK et al. 1986). Higher concentrations of ketotifen (20–240 μM) inhibit oxygen radical production from human alveolar macrophages stimulated with either phorbol ester or opsonised zymosan (KAKUTA et al. 1988) and from human neutrophils stimulated by chemotactic peptide or calcium ionophore (KATO et al. 1985). Very high concentrations ($>$100 μM) have been reported to inhibit degranulation and mediator release from rat mast cells (TRUNEH et al. 1982; GUSHCHIN and ZEBREV 1986, NEMETH et al. 1987; HACHISUKA et al. 1988) and anaphylactic mediator release from human lung parenchyma (GREENWOOD 1982) and human eosinophils (PODLESKI et al. 1984).

Bearing in mind that the dose of ketotifen administered clinically is rather low (2–6 mg/day), the activities reported to occur at these high concentrations in vitro must be considered irrelevant to its action in humans, unless the drug is accumulated in cell membranes in vivo.

II. Animal Pharmacology

1. Antihistamine, Antiallergic Properties

Preliminary studies in vivo provided results that were consistent with properties of ketotifen observed in vitro. That is, the drug blocked responses (cutaneous vascular permeability, brochospasm) provoked by histamine at very much lower doses than were required when the same responses were provoked by other agents. (GREENWOOD 1982; CRAPS and NEY 1984). A separate antiallergic property seemed to account for inhibition of rat passive cutaneous anaphylaxis (GREENWOOD 1982), whereas inhibition of lung anaphylaxis was probably due to antagonism of the action of released spasmogens (ARMOUR and TEMPLE 1982; OTTENHOF et al. 1985).

2. Inhibition of PAF-Induced Airway Hyperreactivity

Assessment of the action of ketotifen took a new direction following investigations of the possible role of PAF as an inflammatory mediator in asthma. In summary, PAF provoked inflammatory changes in the lungs of experimental animals including bronchospasm, mucosal oedema, cellular infiltration (especially of eosinophils), and non-specific airway hyperreactivity to other spasmogens (VARGAFTIG et al. 1980; ARNOUX et al. 1988; DENJEAN et al. 1984; PAGE et al. 1985a,b,c; MAZZONI et al. 1985). The intravenous injection of PAF into spontaneously breathing guinea-pigs provoked bronchospasm that was inhibited by low doses of ketotifen. Given intravenously, the doses required were comparable to those that blocked histamine-induced bronchospasm (PAGE et al. 1985b). Other histamine H_1 antagonists did not inhibit PAF-induced bronchospasm (ARNOUX et al. 1985; VARGAFTIG et al. 1982). Hence, this action of ketotifen may be regarded as being distinct from histamine receptor blokcade (MORLEY et al. 1986). Furthermore, the non-specific and sustained airway hyperreactivity that followed intravenous infusion of PAF was prevented by the simultaneous infusion of ketotifen (PAGE et al. 1985). However, subsequent treatment with ketotifen did not reverse established hyperreactivity. Ketotifen also blocked PAF-induced eosinophil infiltration into the airway lumen (ARNOUX et al. 1988; MORLEY et al. 1988b), another effect of ketotifen that is not shared by other histamine H_1 receptor antagonists. Development of non-specific airway hyperreactivity due to intravenous PAF was associated with the accumulation of platelets in the microvasculature of the pulmonary circulation (DEWAR et al. 1984; ARNOUX et al. 1988; ROBERTSON and PAGE 1987). Ketotifen inhibited this accumulation of platelets when PAF was given by inhalation (ARNOUX et al. 1988). Ketotifen also inhibited airway hyperreactivity that was not associated with platelet activation, such as that due to intravenous (±) isoprenaline and allergen (MORLEY and SANJAR 1987).

With the identification of platelet-independent airway hyperreactivity that is sensitive to ketotifen, it is now difficult to assess the importance of the platelet-derived factor which induces airway hyperreactivity and whose production is blocked by ketotifen (SMITH et al. 1989; SANJAR et al. 1989; MORLEY et al. 1989). Current emphasis seems to be directed towards the effects of ketotifen on eosinophil accumulation and airway hyperreactivity as being properties of the drug that account for its prophylactic activity in asthma.

III. Human Pharmacology

Apart from double-blind, placebo-controlled clinical trials in asthma therapy or prophylaxis, a number of studies in humans have sought to throw light on the mechanism of action of ketotifen. These have been of variable quality and only placebo-controlled studies are considered here.

1. Blockade of Responses to Spasmogens and Allergens

The most clear-cut action demonstrated by ketotifen in humans is histamine H_1 receptor antagonism. Thus, skin wheals and bronchospasm provoked by histamine were potently blocked by the drug (PHILLIPS et al. 1983; ADACHI et al. 1984; ESAU et al. 1984; LISBOA et al. 1985; TAMURA et al. 1986; TJWA et al. 1986). In contrast, bronchospasm provoked by methacholine was not blocked by ketotifen, even after administration (2 mg/day po) for 12 weeks (LISBOA et al. 1985; TJWA et al. 1986). A double-blind, randomised cross-over study in normal subjects showed that acute dosing with ketotifen (3 × 2 mg during 14 h prior to test) did not block PAF-induced bronchospasm or the accompanying flushing, coughing and neutropaenia. Furthermore, ketotifen had no effect on the PAF-induced increase in airway responsiveness to inhaled methacholine. In the skin, ketotifen inhibited the PAF-induced wheal and flare response which is probably histamine-dependent (CHUNG et al. 1988).

The effect of ketotifen on allergen-induced responses has not been consistent. Three separate studies of allergen-induced bronchospasm in atopic subjects indicated that the drug had either no protective effect or a reduction of immediate bronchospasm, consistent with histamine blockade (ESAU et al. 1984; ADACHI et al. 1984; TAMURA et al. 1986). Similarly, blockade of allergen-induced skin responses could not be distinguished from the antihistamine property of the drug (PHILLIPS et al. 1983). Furthermore, inhibition by ketotifen of aspirin-induced broncospasm in sensitive subjects (DELANEY 1983) can probably also be attributed to potent histamine H_1 receptor blockade as clemastine was also found to be effective in this condition (SZCZEKLIK and SERWONSKA 1979). However, inhibition of urticarias seems to be associated with inhibition of histamine release and blockade of the action of histamine (HUSTON et al. 1986).

2. Reversal of β-Adrenergic Receptor Tachyphylaxis

A property of ketotifen that, at present, seems quite separate from the others described above is its ability to abrogate the tachyphylaxis of β-adrenergic receptors that occurs after exposure to high doses of β-adrenergic agonists. Initial experiments in rats showed that pretreatment with isoprenaline for 7 days suppressed the antiallergic effect of a further dose given prior to antigen on day 7. Ketotifen (1 mg/kg po) given on day 7 restored the antiallergic activity of isoprenaline. Furthermore, continuous delivery of ketotifen along with isoprenaline prevented the development of tachyphylaxis. Of the other drugs tested, only dexamethasone had a similar effect (Bretz et al. 1983). In healthy volunteers repeated dosing with terbutaline (5 mg po, t.i.d., 9 days) caused a decrease in lymphoeyte β_2-adrenergic receptor density, together with decreased cyclic adenosine monophosphate (cAMP) responses to isoprenaline. Reversal to normal of both parameters after cessation of terbutaline was accelerated from 4 days to 1 day by ketotifen (1 mg po, b.i.d). A similar effect was observed with prednisone (100 mg i.v.). Furthermore, ketotifen (1 mg po, b.i.d) completely prevented terbutaline-induced decreases in lymphocyte β_2-adrenergic receptor density and responsiveness (Brodde et al. 1985). Ketotifen is a weak inhibitor of cyclic nucleotide phosphodiesterase (Martin and Romer 1978), but this property does not explain reversal of tachyphylaxis.

It has been recognised for some years that both β_2-adrenergic receptor density and responsiveness are diminished in lymphocytes of asthmatics compared with healthy subjects. This is probably the result of prolonged use of bronchodilator therapy. In asthmatics, ketotifen (1 mg po, b.i.d.) for 6 days increased lymphocyte β_2-adrenergic receptor density and responsiveness to values within the normal range. This improvement was accompanied by a significant increase in peak expiratory flow rate before and after inhalation of salbutamol (Brodde et al. 1988). Pauwels et al. (1988) were unable to determine whether ketotifen restored the sensitivity of desensitised β-adrenergic receptors in healthy volunteers, but confirmed that treatment with ketotifen (1 mg po, b.i.d., 3 weeks) prevented the development of β-adrenergic receptor tachyphylaxis in the airways. A recent double-blind, placebo-controlled study in asthmatic children showed that ketotifen (1 mg po, b.i.d. 1 week) restored the up-regulation of β-adrenergic receptors in response to exercise that before treatment was impaired (Reinhardt et al. 1988).

3. Double-Blind Placebo-Controlled Clinical Trials

Of the clinical trials with ketotifen that have been reported in the literature, nine were double-blind and placebo-controlled. These are summarised in Table 1. Two trials were excluded because ketotifen was compared with cromoglycate rather than placebo. In these, ketotifen was found to be not significantly different from cromoglycate (Clarke and May 1980; Paterson et al. 1983). However, this was considered an inappropriate comparison since the condition can improve spontaneously and both drugs may have been either ineffective or effecitve.

Table 1. Summary of double-blind, placebo-controlled clinical trials with ketotifen

Reference	Ketotifen		Patient			Stimulus	Signifance of treatment	Comment
	Dose	Duration	*n*	Age (yrs)	Atopy			
Wells and Taylor 1979	1 mg t.i.d.	1 day	10	7–13	+	Inhaled allergen	—	DSCG-significant protection
Graff-Lonnevig and Hedlin 1985	1 mg b.i.d.	12 weeks	18	7–13	+	Inhaled methacholine	—	No effect on hyperreactivity to methacholine
Loftus and Price 1987	1 mg b.i.d.	24 weeks	47	2–6	+		—	
Rafferty et al. 1987	1 mg b.i.d.	12 weeks	8	Adult	?	Isocapnic hyperventilation	—	
Tan and Lim 1987	1 mg b.i.d.	1 week	10	Adult	?	Exercise	—	DSCG-significant protection
Tinkelman et al. 1985	1 mg b.i.d.	8 weeks	229	≥12	+		+	Decrease in symptoms and concomitant medication
Broberger et al. 1986	1 mg b.i.d.	2 weeks	32	10.5	+	Inhaled allergen	+	Rhinoconjunctivitis also improved
Boner et al. 1986	1 mg b.i.d.	4 weeks	26	1–63	+	Food allergen	+	
St Pierre et al. 1985	1 mg b.i.d.	1 week	11	Adult	?	Cold	+	Cold urticaria improved
Medici et al. 1989	2 mg/day	12 weeks	229	6–51	+		+	

DSCG, disodium cromoglycate.

The conclusion that can be drawn from the data in Table 1 is that ketotifen does not have an undisputed effect in the treatment of asthma. An argument that is frequently heard is that ketotifen needs to be given for prolonged periods before its effect is evident. This allows the study of WELLS and TAYLOR (1979) to be discounted, but does not account for the non-significant effects seen in the other studies. However, it is noteworthy that the multicentre studies of TINKELMAN et al. (1985) and MEDICI et al. (1989), which had the largest number of patients, indicated that ketotifen had a significant effect. Also, prolonged ingestion of ketotifen prior to allergen challenge had significant effects. Recent clinical trials (RACKMAN et al. 1989) indicate that ketotifen has striking benefit in therapy of children with asthma. Thus, in summary, the clinical value of ketotifen in the treatment of asthma is at present not widely-recognised in the West. If the drug has true value in the prophylaxis of asthma, further objective assessment is required for this to be illustrated unambiguously.

C. Pharmacological Analysis

I. Cromoglycate and Nedocromil

In the 22 years since the antiasthmatic action of cromoglycate was first described, pharmacological reports about the drug have been predominantly phenomenological. Some attempts have been made to evaluate its biochemical and molecular properties, but these have not lead to a clear understanding about its mechanism of action. This probably accounts for the failure to discover other molecules with the same, or improved, properties.

The survey of the human phamacology of cromoglycate and nedocromil illustrates the problem of interpretation that is posed when drugs are not well-classified pharmacologically. There is no doubt that the drugs have a clinically useful effect. From experiments in vitro and in animals they are known to prevent mediator release and suppress other activities of leukocytes and platelets and, in addition, attenuate the function of certain sensory nerves possibly by stimulating others. Thus, if a role for inflammatory cells is not evident as, for example, in bronchospasm provoked by exercise, hypotonic mist, cold air or SO_2, then one concludes that cromoglycate and nedocromil are effective by acting on sensory nerves. Suppression of airway hyperreactivity in atopic subjects during the pollen season may involve actions both on inflammatory cells and sensory nerve reflexes. Unfortunately, a more precise interpretation of the human pharmacology of cromoglycate and nedocromil cannot be made at present.

Possibly the key observation of the studies described above (see Sect. A.I, II and III) is that cromoglycate and nedocromil stimulate the phosphorylation of an intracellular protein. This indicates that these drugs should not be considered as essentially passive blocking agents (e.g. for certain calcium channels). Rather, they are active molecules that stimulate biochemical changes in responsive cells. That is, cromoglycate and nedocromil should be considered as agonists. Consistent with this view are the observations that cromoglycate acts on the outside of the cell membrane without penetrating the cell. Many of its action, including stimulation of protein phosphorylation, exhibit tachyphylaxis which is frequently seen with agonists. Furthermore, cross-tachyphylaxis has been observed with related molecules (discussed by GARLAND et al. 1979). There are several reports that cromoglycate promotes rather than inhibits histamine release from mast cells. This was seen with low concentrations of cromogylcate for both IgE-dependent release and A23187-induced release from rat peritoneal mast cells (GARLAND 1973; GARLAND et al. 1979) and with a range of cromoglycate concentrations for IgE-dependent release from human lung parenchyma (CHURCH and GRADIDGE 1978). Also, cromoglycate stimulates sensory afferent endings of certain neuronal reflexes and this may also be subject to tachyphylaxis. Thus, it is reasonable to argue that inhibition of cellular responses by cromoglycate (and nedocromil) is a functional antagonism. That is, cromglycate stimulates biochemical changes that oppose effects stimulated by, for example, cross-linking of IgE (Fc) receptors. This would explain why the apparent efficacy of cromoglycate changes with the strength of the opposing stimulus, such as when antigen concentration is varied. This type of interaction between functional antagonists is well-known and has been analysed by VAN DEN BRINK (1973a,b).

Other observations that should be considered in this analysis of cromoglycate and nedocromil as agonists is the heterogeneity of effects among different types of mast cells and other leukocytes. This is reminiscent of the tissue variation, dependent on receptor number, that is observed when the intrinsic efficacy of the agonist is low (see KENAKIN 1987). Indeed, where dose-response curves for cromoglycate and nedocromil are compared there are several examples where the maximum response for nedocromil is greater than that for cromglycate. That is, both drugs would appear to be partial agonists, but the intrinsic efficacy of nedocromil is greater than that of cromoglycate.

The biochemical properties attributed to cromoglycate might also be explained if the drug is an agonist that stimulates protein phosphorylation. It is known that inhibition of intracellular cAMP phosphodiesterase and PKC can occur indirectly as a result of phosphorylation of the respective proteins, such as occurs subsequent to receptor stimulation. Changes in calcium homoeostasis (e.g. increased calcium efflux) or blockade of receptor-operated calcium channels might also occur as a consequence of kinase activation following receptor stimulation by cromoglycate.

The evaluation of an hypothesis such as this is likely to be more rewarding for the pharmacologist interested in this subject than the continued accumulation of phenomenological observations of the actions of cromoglycate and nedocromil.

II. Ketotifen

The pharmacological properties of ketotifen that occur at low concentrations in vitro or at low doses in vivo are: (1) selective but non-competitive blockade of histamine H_1 receptors; (2) reversal of β-adrenergic receptor tachyphlaxis; (3) inhibition of airway hyperreactivity in guinea-pigs provoked by a range of intravenous stimuli; (4) inhibition of eosinophil accumulation in lungs of animals exposed to PAF. It is not possible to provide a single pharmacological analysis to explain these properties. However, they are likely to result from effects on receptor transduction or other cellular control pathways. In particular, reversal of β-receptor tachyphylaxis by ketotifen might be due to an action of the drug on the so-called "β-adrenergic receptor kinase" (BARK), described by BENOVIC et al. 1987.

In conclusion, two properties of ketotifen (H_1 receptor antagonism and reversal of β-adrenergic receptor tachyphylaxis) have been shown to occur in humans. Results from the well-controlled study of PAF-induced airway hyperreactivity in humans did not support results from animal studies, but in humans PAF was inhaled rather than given intravenously. Therefore, the value of the animal data is uncertain. It remains to be seen whether ketotifen reduces eosinophil migration into the lungs of asthmatics.

References

Abraham WM Stevenson JS, Chapman GA, Tallent MW, Jackowski J (1987) The effect of nedocromil sodium and cromolyn sodium on antigen-induced responses in allergic sheep in vivo and in vitro. Chest 92:913–917

Adachi M, Kobayashi H, Aoki N, Iijima M, Kokubu F, Furuya A, Takahashi T (1984) A comparison of the inhibitory effects of ketotifen and disodium cromoglycate on bronchial responses to house dust, with special reference to the late asthmatic response. Pharmatherapeutica 4:36–42

Altounyan REC (1970) Changes in histamine and atropine responsiveness as a guide to diagnosis and evaluation of therapy in obstructive airways disease. In: Pepys J, Frankland AW (eds) Disodium cromoglycate in allergic airways disease. Butterworths, London

Altounyan REC (1980) Review of clinical activity and mode of action of sodium cromoglycate. Clin Allergy 10:481–489

Altounyan REC, Cole M, Lee TB (1986a) Effect of nedocromil sodium on changes in bronchial hyperreactivity in non-asthmatic, atopic rhinitic subjects during the grass pollen season. Eur J Respir Dis 69 [Suppl 147]:271–273

Altounyan RCE, Lee TB, Rocchiccioli KMS, Shaw CL (1986b) A comparison of the inhibitory effects of nedocromil sodium and sodium cromoglycate on adenosine monophosphate-induced bronchoconstriction in atopic subjects. Eur J Respir Dis 69:277–279

Archer CB, Page CP, Paul W, Morley J, MacDonald DM (1985) Actions of disodium cromoglycate (DSCG) on human skin responses to histamine, codeine and PAF-acether. Agents Actions 16:6–8

Armour C, Temple DM (1982) The modification by ketotifen of respiratory responses to histamine and antigen in guinea-pigs. Agents Actions 12:285–288

Arnoux B, Denjean A, Page CP, Morley J, Benveniste J (1985) Pulmonary effects of platelet activating factor in a primate are inhibited by ketotifen. Am Rev Respir Dis 131 [Suppl]:A2

Arnoux B, Denjean A, Page CP, Nolibe D, Morley J, Benveniste J (1988) Accumulation of platelets and eosinophils in baboon lung after paf-acether challenge. Am Rev Respir Dis 137:855–860

Arruzazabala ML, Gonzalez R (1984) Protective effects of ketotifen on guinea pig trachea. Respiration 45:50–55

Arruzazabala ML, Gonzalez R (1985) Pharmalogical characterization of ketotifen effects on guinea pig ileum. Arch Immunol Ther Exp (Warsz) 33:353–359

Assem ESK, Monger JL (1970) Inhibition of allergic reactions in man and other species by cromoglycate. Int Arch Allergy Appl Immunol 38:68–77

Barbato A, Pisetta F, Mesirca P, Ragusa A, Zacchello F (1987) Sodium cromoglycate and carbachol-induced bronchoprovocation in asthmatic children. Pediatr Pulmonol 3:161–165

Basran GS Page CP, Paul W, Morley J (1983) Cromoglycate (DSCG) inhibits responses to platelet-activating factor (PAF-acether) in man: an alternative mode of action for DSCG in asthma. Eur J Pharmacol 86:143–144

Bauer CP (1986) The protective effect of nedocromil sodium in exercise-induced asthma. Eur J Respir Dis 69:[Suppl 147] 252–254

Ben-Dov I, Bar-Yishay E, Godfrey S (1983) Heterogeneity in the response of asthmatic patients to pre-exercise treatment with cromolyn sodium. Am Rev Respir Dis 127:113–116

Benovic JL, Mayor F, Staniszewski C, Lefkowitz RJ Caron MG (1982) Purification and characterization of the β-adrenergic receptor kinase. J Biol Chem 262:9026–9032

Bergren DR, Myers DL, Mohrman M (1985) Activity of rapidly-adapting receptors to histamine and antigen challenge before and after sodium cromoglycate. Arch Int Pharmacodyn 273:88–89

Biggs DF, Goel V (1985) Mechanisms of action of sodium cromoglycate. Can J Physiol Pharmacol 63:760–765

Bleehen T, Keele CA (1977) Observations on the algogenic actions of adenosine compounds on the human blister base preparation. Pain 3:367–377

Boner AL, Vallone G, Andreiol A, Biancotto R, Warner JO (1987) Nebulised sodium cromoglycate and verapamil in methacholine induced asthma. Arch Dis Child 62:264–268

Boner AL, Niero E, Grigolini C, Valletta EA, Biancotto R, Gaburro D (1985) Inhibition of exercise-induced asthma by three forms of sodium cromoglycate. Eur J Respir Dis 66:21–24

Boner AL, Richelli C, Antolini I, Vibelli C, Andri L (1986) The efficacy of ketotifen in a controlled double-blind food challenge study in patients with food allergy. Ann Allergy 57:61–64

Bonser RW, Thompson NT, Hodson HF, Beams RM, Garland LG (1988) Evidence that a second stereochemical centre in diacylglycerols defines interaction at the recognition site on protein kinase C. FEBS Lett 234:341–344

Bradford PG, Rubin RP (1986) The differential effects of nedrocromil sodium and sodium cromoglycate on the secretory response of rabbit peritoneal neutrophils. Eur J Respir Dis 69 [Suppl 147]:238–240

Breslin FJ, McFadden ER, Ingram RH (1980) The effects of cromolyn sodium on the airway response to hyperpnea and cold air in asthma. Am Rev Respir Dis 122:11–16

Bretz U, Martin U, Mazzoni L, Ney UM (1983) β-adrenergic tachyphylaxis in the rat and its reversal and prevention by ketotifen. Eur J Pharmacol 86:321–328

Broberger U, Graff-Lonneving V, Lilja G, Rylander E (1986) Ketotifen in pollen-induced asthma: a double-blind placebo-controlled study. Clin Allergy 16:119–127
Brodde O-E, Brinkmann M, Schemuth R, O'Hara N, Daul A (1985) Terbutaline-induced desensitization of human lymphocyte β2-adrenoceptors. J Clin Invest 76:1096–1101
Brodde O-E, Howe U, Egerszegi S, Konietzko N, Michel MC (1988) Effect of prenisolone and ketotifen on β2-adrenoceptors in ashmatic patients receiving β2-bronchodilators. Eur J Clin Pharmacol 34:145–150
Brooks CD, Nelson AL, Metzler C (1986) Attentuation of protection with repeated cromolyn dosing before exercise challenge in subjects with asthma. J Clin Pharmacol 26:91–96
Butchers PR, Fullarton JR, Skidmore IF, Thompson LE, Vardey CJ, Wheeldon A (1979) A comparison of the anti-anaphylactic activities of salbutamol and disodium cromoglycate in the rat, the rat mast cell and in human lung tissue. Br J Pharmacol 67:23–32
Chiavarelli M, Moncada S, Mullane K (1981) Prostacyclin-induced bradycardia is dependent on the basal heart rate and is antagonised by sodium cromoglycate. Br J Pharmacol 74:252
Chung KF, Minette P, McCusker M, Barnes PJ (1988) Ketotifen inhibits the cutaneous but not the airway responses to platelet-activating factor in man. J Allergy Clin Immuol 81:1192–1198
Church MK (1986) Is inhibition of mast cell mediator release relevant to the clinical activity of anti-allergic drugs? Agents Actions 18:288–293
Church MK, Gradidge CF (1978) Potentiation of histamine release by sodium cromoglycate. Life Sci 23:1899–1904
Church MK, Young KD (1983) The characteristics of inhibition of histamine release from human lung fragments by sodium cromoglycate, salbutamol and chlorpromazine. Br J Pharmacol 78:671–679
Church MK, Holgate ST, Hughes PJ (1983a) Adenosine inhibits and potentiates IgE-dependent histamine release from human basophils by an A_2-receptor mediated mechanism. Br J Pharmacol 80:719–726
Church MK, Holgate ST, Pao G J-K (1983b) Histamine release from mechanically and enzymatically dispersed human lung mast cell: Inhibition by salbutamol and cromoglycate. Br J Pharmacol 79:374p
Church MK, Hughes PJ, Holgate ST (1983c) Adenosine modulation of histamine release from human basophils and mast cells. Fed Proc 42:1342
Clarke CW, May CS (1980) A comparison of the efficacy of ketotifen (HC 20–511) with sodium cromoglycate (SCG) in skin test positive asthma. Br J Clin Pharmacol 10:473–476
Coleridge JC, Coleridge HM, Roberts AM, Kaufman MP, Baker DG (1982) Tracheal contraction and relaxation initiated by lung and somatic afferents in dogs. J Appl Physiol 52:984–990
Collier JG, Fuller RW (1983) Evidence for an effect of sodium cromoglycate on sensory nerves in man. Br J Clin Pharmacol 16:639–643
Collier JG, Fuller RW (1984) Capsaicin inhalation in man and the effects of sodium cromoglycate. Br J Pharmacol 81:113–117
Corcia A, Schweitzer-Stenner R, Pecht I, Rivany B (1986) Characterisation of the ion channel activity in planar bilayers containing IgE-Fc receptor and the cromolyn-binding protein. EMBO J 5:849–854
Cox JSG, Beach JE, Blair AMJN, Clarke AJ, King J, Lee TB, Loveday DEE, Moss GF, Orr TSC, Ritchie JT, Sheard P (1970) Diodium cromoglycate (Intal). Adv Drug Res 5:115–196
Craps LP, Ney UM (1984) Keotofifen: current views on its mechanism of action and their therapeutic implications. Respiration 45:411–421
Crimi N, Palermo F, Oliveri R, Cacopardo B, Vanchieri C, Mistretta A (1986)

Adenosine-induced bronchoconstriction: Comparison between nedocromil sodium and sodium cromoglycate. Eur J Respir Dis 69 [Suppl 147]:258–262

Crimi N, Palermo F, Vancheri C, Oliveri R, Distefano SM, Polosa R, Mistretta A (1988a) Effect of sodium cromoglycate and nifedipine on adenosine-induced bronchoconstriction. Respiration 53:74–78

Crimi N, Palermo F, Oliveri R, Vancheri C, Polosa R, Palermo B, Maccarrone C, Mistretta A (1988b) Comparative study of the effects of nedocromil sodium (4 mg) and sodium cromoglycate (10 mg) on adenosine-induced bronchoconstriction in asthmatic subjects. Clin Allergy 18:367–374

Curtain C, Looney FD, Smelstorius JA (1981) Glycosphingolipid clustering and mast cell degranulation. Int Arch Allergy Appl Immunol 65:34–41

Cushley MJ, Holgate ST (1985) Adenosine-induced bronchoconstriction in asthma: role of mast cell-mediator release. J Allergy Clin Immunol 75(2):272–278

Cushley MF, Tattersfield AE, Holgate ST (1983) Inhaled adenosine and guanosine on airway resistance in normal and asthmatic subjects. Br J Clin Pharmacol 15, 161–165

Dahlen-S-E, Hansson G, Hedqvist P, Bjorck T, Granstrom E, Dahlen B (1983) Allergen challenge of lung tissue from asthmatics elicits bronchial contraction that correlates with the release of leukotrienes C_4, D_4, and E_4. Proc Natl Acad Sci USA 80:1712–1716

Davies SE (1968) Effect of disodium cromoglycate on exercise-induced asthma. Br Med J 3:593–594

Davis B, Roberts AM, Coleridge HM, Coleridge JC (1982) Reflex tracheal gland secretion evoked by stimulation of bronchial C-fibers in dogs. J Appl Physiol 53:985–991

Deal EC, Wasserman Sl, Soter NA, Ingram RH, McFadden ER (1980) Evaluation of the role played by the mediators of immediate hypersensitivity in exercise-induced asthma. J Clin Invest 65:659–665

Debelic M (1986) Nedocromil sodium and exercise-induced asthma in adolescents. Eur J Respir Dis 69 [Suppl 147]:266–267

Del Bono L, Dente FL, Patalano F, Del Bono N (1986) Protective effect of nedocromil sodium and sodium cromoglycate on bronchospasm induced by cold air. Eur J Respir Dis 69 [Suppl 147]:268–270

Del Bufalo D, Fasano L, Patalano F, Ruggieri F, Gunella G (1988) Prevention of non-specific bronchial hyperreactivity. Dose-dependent effect of sodium cromoglycate metered dose aerosol. Allergol Immunopathol (Madr) 16:77–80

Delaney JC (1983) The effect of ketotifen on aspirin-induced asthmatic reactions. Clin Allergy 13:247–251

Denjean A, Arnoux B, Lockhart A (1984) Modification of alveolar cell population after PAF-acether induced bronchoconstriction in baboons. Am Rev Respir Dis 129 [Suppl]:A3

Dewar A, Archer CB, Paul W, Page CP, MacDonald DM, Morley J (1984) Cutaneous and pulmonary histopathological responses to platelet activating factor (Paf acether) in the guinea pig. J Pathol 144:25–34

Diaz P, Galleguillos FR, Gonzalez MC, Pantin CF, Kay AB (1984) Bronchoalveloar lavage in asthma: the effect of disodium cromoglycate (cromolyn) on leukocyte counts, immunoglobulins, and complement. J Allergy Clin Immuol 74: 41–48

Dickson W (1970) A one year's trial of Intal compound in 24 children with severe asthma. In: Pepys J, Frankland AW (eds) Disodium cromoglycate in allergic airways disease. Butterworths, London, p 105

Dixon M, Jackson DM, Richards IM (1979) The effects of sodium cromoglycate on lung irritant receptors and left ventricular cardiac receptors in the anaesthetized dog. Br J Pharmacol 67:569–574

Dixon M, Jackson DM, Richards IM (1980) The action of sodium cromoglycate on 'C' fibre endings in the dog lung. Br J Pharmacol 70:11–13

Dorward AJ, Roberts JA, Thomposn NC (1986) A preliminary report on the effect of nedocromil sodium on histamine airway responsiveness in patients allergic to grass pollen. Eur J Respir Dis 69 [Suppl 147]:299–301
Durham SR, Carroll M, Walsh GM, Kay AB (1984) Leukocyte activation in allergen-induced late-phase asthmatic reactions. N Engl J Med 311:1398–1402
Eady RP (1986) The pharmacology of nedocromil sodium. Eur J Respir Dis 61 [Suppl 147]:112–119
Eady RP, Greenwood B, Jackson DM, Orr TSC, Wells E (1985) The effect of nedrocromil sodium and sodium cromoglycate on antigen-induced bronchoconstriction in the Ascaris-sensitive monkey. Br J Pharmacol 85:323–325
Easu S, del Carpio J, Martin JG (1984) A comparison of the effects of ketotifen and clemastine on cutaneous and airway reactivity to histamine and allergen in atopic asthmatic subjects. J Allergy Clin Immunol 74:270–274
Fabbri LM, Mapp CE, Hendrick DJ (1983) Effect of cromolyn on carbachol-induced bronchoconstriction. Ann Allergy 50:195–198
Fabbri LM, Aizawa H, Alpert SE, Walters EH, O'Bryne PM, Gold BD, Nadel JA, Holtzman MJ (1984) Airway hyperresponsiveness and changes in cell counts in bronchoalveolar lavage after ozone exposure in dogs. Am Rev Respir Dis 129:288–291
Fink A, Bibi H, Eliraz A, Schlesinger M, Bentwich Z (1986) Ketotifen, disodium cromoglycate, and verapamil inhibit leukotriene activity: determination by tube leukocyte adherence inhibition assay. Ann Allergy 57:103–106
Flint KC, Leung KBP, Pearce FL, Hudspith BN, Brostoff J, Johnson N Mcl (1985) Human mast cells recovered by bronchoalveolar lavage: their morphology, histamine release and the effects of sodium cromoglycate. Clin Sci 68:427–432
Foreman JC, Garland LG (1976) Cromoglycate and other antiallergic drugs: a possilbe mechanism of action. Br Med J 1:820–821
Fuller RW, Collier JG (1984) Sodium cromoglycate and atropine block the fall in FEV_1 but not the cough induced by hypotonic mist. Thorax 39:766–770
Fuller RW, MacDermot J (1986) Stimulation of IgE-sensitized human alveolar macrophages by anti-IgE is unaffected by sodium cromoglycate. Clin Allergy 16:523–526
Garland LG (1973) Effect of cromoglycate on anaphylactic histamine release from rat peritoneal mast cells. Br J Pharmacol 49:128–130
Garland LG, Hodson HF (1984) Antiallergic drugs. In: Dale MM Foreman JC (eds) Textbook of immuno-pharmacology. Blackwell, London, pp 265–282.
Garland LG, Mongar JL (1974) Inhibition by cromoglycate of histamine release from rat peritoneal mast cells indued by mixtures of dextran, phosphatidyl serine and calcium ions. Br J Pharmacol 50:137–143
Garland LG, Mongar JL (1976) Differential histamine release by dextran and the ionophore A23187: the actions of inhibitors. Int Arch Allergy Appl Immunol 50:27–42
Garland LG Green AF, Hodson HF (1979) Inhibitors of the release of anaphylactic mediators. In: Vane JR Ferreira SH (eds) Anti-inflammatory drugs. Springer, Berlin Heidelberg New York, pp 467–530 (Handbook of experimental pharmacology, vol 50/II)
Godfrey S, Silverman M, Anderson SD (1973) Problems of interpreting exercise-induced asthma. J Allergy Clin Immunol 52:199–209
Golden HW, Jacuzio DA, Otterness IG (1987) Inhibition of C5a-induced basophil degranulation by disodium cromoglycate. Agents Actions 21:3–4
Graff-Lonnevig V, Hedlin G (1985) The effect of ketotifen on bronchial hyperreactivity in childhood asthma. J Allergy Clin Immunol 79:59–63
Greenwood C (1982) The pharmacology of ketotifen. Chest 82:45S–48S
Griffin MP, MacDonald N, McFadden ER (1983) Short-and long-term effects of cromolyn sodium on the airways reactivity of asthmatics. J Allergy Clin Immunol 71:331–338

Gushchin IS, Zebrev AI (1986) Ketotifen-induced histamine release, inhibition of histamine secretion and modulation of immune response. Agents Actions 18:92–95

Hachisuka H, Nomura H, Sakamoto F, Mori O, Okbo K, Sasai Y (1988) Effect of antianaphylactic agents on substance P-induced histamine release from rat peritoneal mast cells. Arch Dermatol Res 280:158–162

Holroyde MC, Altounyan REC, Cole M, Dixon M, Elliot EV (1981) Bronchoconstriction produced in man by leukotriene C and D. Lancet ii:17–18

Holtzman MJ, Fabbri LM, O'Bryne BD, Gold H, Aizawa EH, Walters SE, Aplert SE, Nadel JA (1983) Importance of airway inflammation for hyperresponsiveness induced by ozone. Am Rev Respir Dis 127:686–690

Huston DP, Bressler RB, Kaliner M, Sowell LK, Baylor MW (1986) Prevention of mast-cell degranulation by ketotifen in patients with physical urticarias. Ann Intern Med 104:507–510

Hutson PA, Holgate ST, Church MK (1988) The effect of cromolyn sodium and albuterol on early and late phase bronchoconstriction and airway leukocyte infiltration after allergen challenge of nonanesthetized guinea pigs. Am Rev Respir Dis 138:1157–1163

Jackson DM (1988) The effect of nedocromil sodium, sodium cromoglycate and codeine phosphate on citric acid-induced cough in dogs. Br J Pharmacol 93:609–612

Jackson DM, Eady RP (1986) Monkeys infected with *Ascaris suum* (a new in vivo model of airway disease): protective effect of nedocromil sodium and sodium cromoglycate against brochial antigen challenge. Eur J Respir Dis 69: [Suppl 147]: 202–205

Jackson DM, Eady RP (1988) Acute transient SO_2-induced airway hyperreactivity: effects of nedocromil sodium. J Appl Physiol 65:1119–1124

Jackson DM, Richards IM (1977) The effects of sodium cromoglycate on histamine aerosol-induced reflex bronchoconstriction in the anaesthetized dog. Br J Pharmacol 61:257–262

Jackson DM, Eady RP, Farmer JB (1986a) The effect of nedocromil sodium on non-specific bronchial hyperreactivity in the dog. Eur J Respir Dis 69 [Suppl 147]:217–219

Joseph M, Tonnel, AB, Capron A, Dessaint JP (1981) The interaction of IgE antibody with human alveolar macrophages and its participation in the inflammatory processes of lung allergy. Agents Actions 11:619–622

Joseph M, Capron A, Thorel T, Tonnel AB (1986) Nedocromil sodium inhibits JgE-dependent activation of rat macrophages and platelets as measured by schistosome killing, chemiluminscence and enzyme release. Eur J Respir Dis 69 [Suppl 147]:220–222

Joseph M, Marquette CH, Lassalle P, Vorng H, Tonnel AB, Capron A (1988) Inhalation of nedrocromil sodium, but not of sodium cromoglycate, inhibits the abnormal in vitro responses to aspirin of platelets from aspirin-sensitive asthmatics. Am Rev Respir Dis 137:29

Juniper EF, Latimer KM, Morris MM, Roberts RS, Hargreave FE (1986) Airway responses to hyperventilation of cold dry air: duration of protection by cromolyn sodium. J Allergy Clin Immunol 78:387–391

Juniper EF, Kline PA, Morris MM, Hargreave FE (1987) Airway constriction by isocapic hyperventilation of cold, dry air: comparison of magnitude and duration of protection by nedocromil sodium and sodium cromoglycate. Clin Allergy 17:523–528

Kakuta Y, Kato T, Sasaki H, Takishima T (1988) Effect of ketotifen on human alveolar macrophages. J Allergy Clin Immunol 81:469–474

Karlsson G, Pipkorn U (1984) Effect of disodium cromoglycate on changes in nasal airway resistance induced by platelet activating factor. Eur J Clin Pharmacol 27:371–373

Kato T, Terui T, Tagami H (1985) Effects of HC 20-511 (ketotifen) on chemiluminescence of human neutrophils. Inflammation 9:45–51
Kay AB, Walsh GM, Moqbel R, MacDonal AJ, Nagakura T, Carroll MP, Richerson HB (1987) Disodium cromoglycate inhibits activation of human inflammatory cells in vitro. J Allergy Clin Immunol 80:1–8
Kenakin TP (1987) Pharmacologic analysis of drug-receptor interaction. Raven, New York
Kennedy JD, Hasham F, Clay MJD, Jones RS (1980) Comparison of actions of disodium cromoglycate and ketotifen on exercise-induced brochoconstriction in childhood asthma. Br Med J 281:1458
Koenig JQ, Marshall SG, van Belle G, McManus MS, Bierman CW, Shapiro GG, Furukawa CT, Pierson WE (1988) Therapeutic range cromolyn dose-response inhibition and complete obliteration of SO_2-induced bronchoconstriction in atopic adolescents. J Allergy Clin Immunol 81:897–901
Konig P, Hordvik NL, Kreutz C (1987) The preventive effect and duration of action of nedocromil sodium and cromolyn sodium on exercise-induced asthma (EIA) in adults. J Allergy Clin Immunol 79:64–68
Kraemer R, Sennhauser F, Reinhardt M (1987) Effects of regular inhalation of belcomethasone dipropionate and sodium cromoglycate on bronchial hyperreativity in asthmatic children. Acta Paediatr Scand 76:119–123
Laitinen LA, Venho K, Poppius H (1986) A controlled study on the effect of treatment with cromolyn sodium pressurized aerosol on bronchial reactivity in patients with asthma. Ann Allergy 56:270–273
Latimer KM, O'Bryne PM, Morris MM, Roberts R, Hargreave FE (1983) Bronchoconstriction stimulated by airway cooling. Am Rev Respir Dis 128:440–443
Leitch AG, Lumb EM, Kay AB (1984) Disodium cromoglycate relieves symptoms in symptomatic young smokers. Allergy 39:211–215
Leung KBP, Flint KC, Brostoff J, Hudspith BN, Johnson NMcl, Pearce FL (1986) A comparison of nedocromil sodium and sodium cromoglycate on human lung mast cells obtained by bronchoalveolar lavage and by dispersion of lung fragments. Eur J Respir Dis 69 [Suppl 147]:223–226
Lisboa C, Moreno R, Cruz E, Barja S, Sanchez I, Moran J, Ferretti R (1985) Acute effect of ketotifen on the dose-response curve of histamine and methacholine in asthma. Br J Dis Chest 79:235–243
Loftus BG, Price JF (1987) Long-term, placebo-controlled trial of ketotifen in the management of preschool children with asthma. J Allergy Clin Immunol 79:350–355
Loftus BG, Price JF, Heaton R, Costello JF (1985) Effects of ketotifen on *in vitro* bronchoconstricton. Clin Allergy 15:465–471
Lowhagen O, Rak S (1985a) Modification of bronchial hyperreactivity after treatment with sodium cromoglycate during pollen season. J Allergy Clin Immunol 75: 460–467
Lowhagen O, Rak S (1985b) Bronchial hyperreactivity after treatment with sodium cromoglycate in atopic asthmatic patients not exposed to relevant allergens. J Allergy Clin Immunol 75(3):343–347
Lucas AM, Shuster S (1987) Cromolyn inhibition of protein kinase C activity. Biochem Pharmacol 36:562–565
Martelli NA, Usandivaras G (1977) Inhibition of aspirin-induced bronchoconstriction by sodium cromoglycate inhalation. Thorax 32:684–690
Martin U, Romer D (1978) The pharmacological properties of a new, orally active antianaphylactic compound: ketotifen, a benzocycloheptathiophene. Arzneimittelforschung 28:1–12
Mattoli S, Foresi A, Corbo GM, Polidori G, Ciappi G (1986) Protective effect of disodium cromoglycate on allergen-induced bronchoconstriction and increased hyperresponsiveness: a double-blind placebo-controlled study. Ann Allergy 57:295–300

Mazurek N, Berger G, Pecht I (1980a) A binding site on mast cells and basophils for the anti-allergic drug cromolyn. Nature 286:722–723
Mazurek N, Geller-Bernstein C, Pecht I (1980b) Affinity of calcium ions to the anti-allergic drug, dicromoglycate. FEBS Lett 111:194–196
Mazurek N, Bashkin P, Pecht I (1982) Isolation of a basophilic membrance protein binding the anti-allergic drug cromolyn. EMBO J 1:585–590
Mazurek N, Bashkin P, Loyter A, Pecht I (1983a) Restoration of Ca^{2+} influx and degranulation capacity of variant RBL-2H3 cells upon implantation of isolated cromolyn binding protein. Proc Natl Acad Sci USA 80:6014–6018
Mazurek N, Bashkin P, Petrank A, Pecht I (1983b) Basophil variants with impaired cromoglycate binding do not respond to an immunological degranulation stimulus Nature 303:528–530
Mazurek N, Schindler H, Schurolz T, Pecht I (1984) The cromolyn binding protein constitutes the Ca^{2+} channel of basophils opening upon immunological stimulus. Proc Natl Acad Sci USA 81:6841–6845
Mazzoni L, Morley J, Page CP, Sanjar S (1985) Induction of airway hyperreactivity by platelet activating factor in the guinea pig. J Physiol (Lond) 365:107P
Medici TC, Radielovic P, Morley J (1989) Prophylactic treatment of extrinsic bronchial asthma: a multicentre controlled double-blind study. Eur Respir J 2 [Suppl 5]:400s
Miller PD, Ainsworth D, Lam HF, Amdur MO (1988) Indomethacin and cromolyn sodium alter ozone-induced changes in lung function and plasma eiconsanoid concentrations in guinea pigs. Toxicol Appl Pharmacol 93:175–186
Mitchell I (1985) Sodium cromoglycate-induced changes in the dose-response curve of inhaled methacholine in cystic fibrosis. Ann Allergy 54:233–235
Moqbel R, Walsh GM, MacDonald AJ, Kay B (1986a) Effect of disodium cromoglycate on activation of human eosinophils and neutrophils following reversed (anti-IgE) anaphylaxis. Clin Allergy 16:73–83
Moqbel R, Walsh GM, Kay AB (1986b) Inhibition of human granulocyte activation by nedocromil sodium. Eur J Respir Dis 69 [Suppl 147]:227–229
Morley J, Sanjar S (1987) Isoprenaline induces increased airway reactivity in the guinea pig. J Phyiol (Lond) 390:180
Morley J Page CP, Mazzoni L, Sanjar S (1986) Effects of ketotifen upon responses to platelet activating factor: a basis for asthma prophylaxis. Ann Allery 56:335–339
Morley J, Chapman ID, Sanjar S, Schaeublin E (1989) Actions of ketotifen on PAF-induced airway hyperreactivity in the anaesthetised guinea-pig. Br J Pharmacol 96:76P
Morley J, Sanjar S, Boubekeur K, Aoki S, Kristersson A (1981) Pharmacological evaluation of prophylactic anti-asthma drugs by reference to the pathological sequelae of exposure to allergen or platelet activating factor. Agents Actions 23 [Suppl]:187–194
Morton AR, Fitch KD (1974) Sodium cromoglycate BP in the prevention of exercise-induced asthma. Med J Aust 2:158–162
Myers DJ, Bigby BG, Boushey HA (1986a) The inhibition of sulfur dioxide-induced bronchoconstriction in asthamtic subjects by cromolyn is dose dependent. Am Rev Respir Dis 133:1150–1153
Myers DJ Bigby BG, Calvayrac P, Sheppard D, Boushey HA (1986b) Interaction of cromolyn and a muscarinic antagonist in inhibiting bronchial reactivity to sulfur dioxide and to eucapnic hyperpnea alone. Am Rev Respir Dis 133:1154–1158
Nadel JA, Salem H, Tamplin B, Tokiwa Y (1965) Mechanism of bronchoconstriction during inhalation of sulfur dioxide. J Appl Physiol 20:164–167
Nemeth A, Magyar P, Herceg R, Huszti Z (1987) Potassium-induced histamine release from mast cells and its inhibition by ketotifen. Agents Actions 20:149–152
Ottenhof M, Ufkes JGR Van Rooij HV. (1985) The effect of prednisolone and ketotifen on the antigen-induced bronchoconstriction and mediator release in rat isolated lungs. Br J Pharmacol 86:627–636

Page CP, Guerreiro D, Sanjar S, Morley J (1985a) Platelet activating factor (Paf-acether) may account for late-onset reactions to allergen inhalation. Agents Actions 16:30–32
Page CP, Tomiak RHH Sanjar S, Morley J (1985b) Suppression of Paf-acether response: an anti-inflammatory effect of anti-asthma drugs. Agents Actions 16:33–35
Page CP Mazzoni L, Sanjar S (1985c) Induction of non-specific hyperreactivity by platelet activating factor (PAF) and its inhibition by anti-asthma drugs. Am Rev Respir Dis 131:A44
Papageorgiou N, Carrol M, Durham SR, Lee TH, Walsh GM, Kay AB (1983) Complement receptor enhancement as evidence of neutrophil activation after exercise-induced asthma. Lancet 2:1220–1223
Patel KR, Wall T (1986) Dose-duration effect of sodium cromoglycate aerosol in exercise-induced asthma. Eur J Respir Dis 69:256–260
Paterson JW, Yellin RH, Tarala RA (1983) Evaluation of ketotifen (HC20–511) in bronchial asthma. Eur J Clin Pharmacol 25:187–193
Pauwels R, Van der Straeten M (1986) The effects of nedocromil sodium on adenosine-induced bronchoconstriction in rats. Eur J Respir Dis 69 [Suppl 147]:235–237
Pauwels R Van Der Straeten M (1988) The effect of ketotifen on bronchial beta-adrenergic tachyphylaxis in normal human volunteers. J Allergy Clin Immunol 81:674–680
Pearce FL (1983) Mast cell heterogeneity. TIPS 4:165–167
Pearce FL, Truneh A (1981) Inhibition of histamine release from rat peritoneal mast cells treated with the ionophore A23187. Implications for the mode of action of anti-allergic compounds. Agents Actions 11:44–50
Pearce FL, Rafii-Tabor E (1983) Inhibition by disodium cromoglygate of anaphylactic histamine secretion from rat peritoneal mast cells in the presence of phosphatidylserine. Agents Actions 13:212–215
Pearce FL, Ali H, Barett KE, Befus AD, Bienenstock J, Brostoff J, Ennis M, Flint KC, Hudspith B, Johnson NM, Leung KBP, Peachell PT (1985) Functional characteristics of mucosal and connective tissue mast cells of man, the rat and other animals. Int Arch Allergy Appl Immunol 77:274–276
Pepys J, Hargreave FE, Char M, McCarthy DS (1968) Inhibitory effects of disodium cromoglycate on allergen-inhalation tests. Lancet ii:134
Phillips GD, Richards R, Scott VL, Holgate ST (1988) Sodium cromoglycate and nedrocromil sodium inhibit broncho-constriction provoked by adenosine 5'-monophosphate in atopic and non-atopic asthma. Am Rev Respir Dis 137:29
Phillips MJ, Meyrick Thomas RH, Moodley I, Davies RJ (1983) A comparison of the *in vivo* effects of ketotifen, clemastine, chlorpheniramine and sodium cromoglycate on histamine and allergen induced weals in human skin. Br J Clin Pharmacol 15:277–286
Pichurko BM, McFadden ER, Frederick Bowman H, Solway J, Burns S, Dowling N (1984) Influence of cromolyn sodium on airway temperature in normal subjects. Am Rev Respir Dis 130:1002–1005
Podleski WK, Panaszek BA, Schmidt JL, Burns RB (1984) Inhibition of eosinophil degranulation by ketotifen in a patient with milk allergy manifested as bronchial asthma – an electron microscopic study. Agents Actions 15:177–181
Polson JB, Krzanowski JJ, Anderson WH, Szentivanyi A (1982) Effects of ketotifen, a benzocycloheptathiophene, on methacholine- and acetylcholine-induced contractions of canine respiratory smooth muscle. Immunopharmacology 4:69–72
Rackham A, Brown CA, Chandra K, Ho P, Hoogerwerf PE, Kennedy RJ, Knight A, Langer H, Milne J, Moote DW, Nickerson GN, Rosen L, Stephenson H, Broadhead M, Lalonde Y, St Pierre JP (1989.) A Canadian multicenter study with zaditen (ketotifen) in the treatment of bronchial asthma in children age 5–17 years. J Allergy Clin Immunol 84(3):286–296

Rafferty P, Fergusson RJ, Tweedale PM, Biggs BA, Grant IWB (1985) Effects of verapamil and sodium cromoglycate on bronchoconstriction induced by isocapnic hyperventilation. Clin Allergy 15:531–534

Rafferty P, Tweedale PM, Ferguson RJ, Biggs BA,Grant IWB (1987) Does regular treatment with ketotifen inhibit broncho-constriction induced by isocapnic hyperventilation? Br J Clin Pharmacol 24:100–102

Read GW, Knoohuizen M, Goth A (1977) Relationship between phosphatidylserine and cromolyn in histamine release. Eur J Pharmacol 42:171–177

Rengo F, Trimarco B, Ricciardelli B, Volpe M, Violini R, Sacca L, Chiariello M (1979) Effects of disodium cromoglycate on hypoxic pulmonary hypertension in dogs. J Pharmacol Exp Ther 211:686–689

Reinhardt D, Ludwig J, Braun D, Kusenbach G, Griese M (1988) Effects of the antiallergic drug ketotifen on bronchial resistance and beta-adrenoceptor density of lymphocytes in children with exercise-induced asthma. Dev Pharmacol Ther 11:180–188

Richards IM, Dixon M, Jackson DM, Vendy K (1986) Alternative modes of action of sodium cromoglycate. Agents Actions 18:3–4

Roberts JA, Rodger IW, Thompson NC (1986) Effect of verapamil and sodium cromoglycate on leukotriene D_4 induced bronchoconstriction in patients with asthma. Throax 41:753–758

Robertson DN, Page CP (1987) The effect of platelet agonists on intrathoracic platelet accumulation and airways hyperresponsiveness. Br J Pharmacol 92:105–112

Ryo UY, Bann K, Townely RG (1976) Cromolyn therapy in patients with bronchial asthma. JAMA 236:927–931

Sagi-Eisenberg R (1985) Possible role for a calcium-activated, phospholipid-dependent protein kinase in mode of action of DSCG. TIPS 6:198–200

Sanjar S, Smith D, Kristersson A (1989) Incubation of platelets with PAF produces a factor which causes airway hyperreactivity in guinea-pigs. Br J Pharmacol 96: 75P

Shaw RJ, Kay AB (1986) Nedocromil sodium, a mucosal and connective tissue mast cell stabilizer, inhibits exercise-induced asthma. Eur J Respir Dis 69 [Suppl 147]:294–296

Sheppard D, Scott Wong W, Uehara CF, Nadel JA, Boushey HA (1980) Lower threshold and greater bronchomotor responsiveness of asthmatic subjects to sulfur dioxide. Am Rev Respir Dis 122:873–878

Sheppard D, Nadel JA, Boushey HA (1981) Inhibition of sulfur dioxide-induced bronchoconstriction by disodium cromoglycate in asthmatic subjects. Am Rev Respir Dis 124:257–259

Skedinger MC, Augustine NH, Morris EZ, Nielson DW, Zimmerman GA, Hill HR (1987) Effect of disodium cromoglycate on neutrophil movement and intracellular calcium mobilization. J Allergy Clin Immunol 80:573–577

Smith D, Sanjar S, Morley J (1989) Platelet activation and PAF-induced airway hyperreactivity in the anaesthetised guinea-pig. Br J Pharmacol 96:74P

Spry CJF, Kumaraswami V, Tai P-C (1986) The effect of nedocromil sodium on secretion from human eosinophils. Eur J Respir Dis 69 [Suppl 147]:241–243

St Pierre JP, Kobric M, Rackham A (1985) Effect of ketotifen treatment on cold-induced urticaria. Ann Allergy 55:840–843

Stafford WP, Mansfield LE, Yarbrough J (1984) Cromolyn modifies histamine bronchial reactivity in symptomatic seasonal asthma. Ann Allergy 52:401–405

Suschitzky JL, Sheard P (1984) The search for antiallergic drugs for the treatment of asthma – problems in finding a successor to sodium cromoglycate. Prog Med Chem 21:1–61

Szczeklik A, Serwonska M (1979) Inhibition of idiosyncratic reactions to aspirin in asthmatic patients by clemastine. Thorax 34:654–657

Tamura G, Mue S, Takishima T (1986) Protective effect of ketotifen on allergen-induced bronchoconstriction and skin weal. Clin Allergy 16:535–541

Tan WC, Crip E, Douglas N, Sudlow MF, (1982) Protective effect of drugs on bronchoconstriction induced by sulphur dioxide. Thorax 37:671–676
Tan WC, Lim TK (1987) Double-blind comparison of the protective effect of sodium cromoglycate and ketotifen on exercise-induced asthma in adults. Allergy 42:315–317
Theoharides TC, Sieghart W, Greengard P, Douglas WW (1980) Antiallergic drug cromolyn may inhibit histamine secretion by regulating phosphorylation of a mast cell protein. Science 207:80–82
Thompson NC, Roberts JA (1986) Nedocromil sodium attenuates exercise-induced asthma. Eur J Respir Dis 69 [Suppl 147]:297–298
Tinkleman DG, Moss BA, Bukantz SC, Sheffer AL, Dobken JH, Chodosh S, Cohen BM, Rosenthal RR, Rappaport I, Buckley CE, Chusid EL, Deutsch AJ, Settipane GA, Burns RBP (1985) A multicentre trial of the prophylactic effect of ketotifen, theophylline. and placebo in atopic asthma. J Allergy Clin Immunol 76:487–497
Tjwa MKT, Smeets J, Maesen F (1986) Ketotifen and methacholine-induced bronchospasm. Allergy 41:551–555
Truneh A, Pearce FL (1984) Effect of cyclic AMP, disodium cromoglycate and other antiallergic drugs on histamine secretion from rat mast cells stimulated with the calcium ionophore ionomycin. Agents Actions 14:179–184
Truneh A, White JR, Pearce FL (1982) Effect of ketotifen and oxatomide on histamne secretion from mast cells. Agents Actions 12:206–209
Tsicopoulos A, Lassale P, Joseph M, Tonnel AB, Thorel T, Dessaint JP, Capron A (1988) Effect of disodium cromoglycate on inflammatory cells bearing the Fc epsilon receptor type II (Fc epsilon RII). Int J Immunopharmacol 10:227–236
Van den Brink FG (1973a) The model of functional interaction. 1. Development and first check of a new model of functional synergism and antagonism. Eur J Pharmacol 22:270–278
Van den Brink FG (1973b) The model of functional interaction. II. Experimental verification of a new model: the antagonism of β-adrenoceptor stimulants and other agonists. Eur J Pharmacol 22:279–286
Vargaftig BB, Lefort J, Chignard ML (1980) Platelet activating factor induces a platelet-dependent bronchoconstriction unrelated to the formation of prostaglandin derivatives. Eur J Pharmacol 65:185–192
Vargaftig BB, Lefort J, Wal F, Chignard M, Medeiros MC (1982) Nonsteroidal anti-inflammatory drugs if combined with anti-histamine and anti-serotonin agents interfere with the bronchial and platelet effects of platelet activating factor (Paf-acether). Eur J Pharmacol 80:212–230
Wells A Taylor B (1979) A placebo-controlled trial of ketotifen (HC 20–511, Sandoz) in allergen induced asthma and comparison with disodium cromoglycte. Clin Allergy 9:237–240
Wells E, Jackson CG, Harper ST, Mann J, Eady RP (1986) Characterization of primate bronchoalveolar mast cells. J Immunol 137:3941–3945
Wells E, Mann J (1983) Phosphorylation of a mast cell protein in response to treatment with anti-allergic compounds. Biochem Pharmacol 32:837–842
White JR, Pearce FL (1983) Effects of anti-allergic and cyclic AMP-active drugs on histamine secretion from rat mast cells treated with the novel calcium ionophore chlortetracycline. Int Arch Allergy Appl Immunol 71:352–356
White JR, Ishizaka T, Ishizaka K, Sha'Afi RI (1984) Direct demonstration of increased intracellular concentration of free calcium as measured by quin-2 in stimulated rat peritoneal mast cell. Proc Natl Acad Sci USA 81:3978–3982
Woenne R, Kattan M, Levison H (1979) Sodium cromoglycate-induced changes in the dose-response curve of inhaled methacholine and histamine in asthmatic children. Am Rev Respir Dis 119:927–932

CHAPTER 10

Pathophysiology and Pharmacology of Aspirin-Induced Asthma

A. SZCZEKLIK, C. VIRCHOW, and M. SCHMITZ-SCHUMANN

A. Introduction

Soon after its introduction into therapy in 1899, aspirin became an extremely successful analgesic and anti-inflammatory agent. Recently, its clinical applications have increased dramatically and include coronary artery disease and several related cardiovascular disorders. In patients with unstable angina or transient ischemic attacks aspirin favourably affects the course of disease, probably through its anti-platelet action (REILLY and FITZGERALD 1988). In acute myocardial infarction it significantly lowers mortality (ISIS-2 COLLABORATIVE GROUP 1988). Indeed, aspirin is now recognized as the most popular drug in the world.

The majority of people tolerate aspirin well. Asthmatics, however, are an exception. There is a special type of asthma, characterized by intolerance to aspirin and other analgesics. For pharmacologists, biochemists and clinicians, it is a remarkable model for studying the pathogenesis of asthma.

B. Definition and Main Clinical Features

In about 10% of adults with asthma, but rarely in asthmatic children, aspirin and other nonsteroidal anti-inflammatory drugs (NSAID) precipitate asthma attacks. This distinct clinical syndrome is called aspirin-induced asthma (AIA). The majority of patients have a negative family history. When HLA typing was performed, no difference in class I HLA-A,B and C antigens or in HLA-DR antigens were found (JONES et al. 1984, MULLARKEY et al. 1986). However, there was a significant increase in HLA-Dqw2 in AIA as compared to other types of asthma with good tolerance of aspirin (MULLARKEY et al. 1986). In a recent study (SCHMITZ-SCHUMANN 1989) 142 patients with AIA were typed for MHC-1 and D antigens including the D locus. An HLA association of HLA-A2, B8, DR3 has been found with the relative risk of 1:8.

The course of the disease and its clinical picture are very characteristic (SAMTER and BEERS 1968; VIRCHOW 1976; SETTIPANE 1983; SZCZEKLIK and GRYGLEWSKI 1983; STEVENSON 1984). In individual patients the onset of symptoms before puberty or after the age of 60 has been well documented. In the majority of patients, however, the first symptoms appear during the third or fourth decade of life. The typical patient starts to experience intense vasomotor rhinitis characterized by intermittent and profuse watery rhinorrhea. Over a period of months, chronic nasal congestion appears and physical examination reveals nasal polyps. Bronchial asthma and intolerance to aspirin develop during subsequent stages of the illness. The intolerance presents itself as a unique picture: within an hour following ingestion of aspirin, acute asthmatic attacks develop, often accompanied by rhinorrhea, conjunctival irritation and scarlet flush of the head and neck. These reactions are dangerous; indeed, a single theraputic dose of aspirin or other anti-cyclooxygenase agent can provoke violent bronchospasm, shock, unconsciousness and respiratory arrest.

Asthma runs a protracted course, despite the avoidance of aspirin and cross-reactive drugs. The eosinophil count is elevated. Skin tests with common aero-allergens are often negative, and those with aspirin are always negative.

I. Major Offenders and Safe Alternatives

In patients with aspirin-induced asthma, not only aspirin but several other anti-inflammatory drugs precipitate bronchoconstriction. These include: indomethacin, mefenamic, flufenamic and cyclofenamic acids, ibuprofen, fenoprofen, ketoprofen, naproxen, diclofenac sodium, piroxicam, sulindac, tiaprofenic acid, aminopyrine, noramidopyrine, sulfinpyrazone, phenylbutazone, zomepirac sodium, tolmetin sodium, diflunisal and fenflumizole. All these drugs are contraindicated in patients with AIA. Not all of them produce adverse symptoms with the same frequency. This depends on both a drug's anti-cyclooxygenase potency and dosage as well as on the individual sensitivity of a patient (SZCZEKLIK et al. 1977a; STEVENSON 1984). If necessary, patients with AIA can take safely: sodium salicylate, salicylamide, choline magnesium trisalicylate, dextropropoxyphene, benzydamine, guacetisal (guaiacolic ester of acetylsalicylic acid) and chloroquine. The majority of patients will also tolerate paracetamol well (vide infra).

II. Salicylates

The exact mechanism of salicylate action is unknown (ROTH 1988). Salicylic acid has almost no effect on cyclooxygenase of gastric mucosa, macrophages, ram seminal vesical microsomes or thrombocytes. Platelet function is not influenced by salicylic acid either (RAINSFORD 1984; ROSENKRANZ et al. 1986). Sodium salicylate, in a therapeutic dose used for symptomatic treatment of rheumatic diseases, does not affect cyclooxygenase activity or prostanoid formation in healthy volunteers (ROSENKRANZ et al. 1986). In contrast, under certain experimental conditions, salicylic acid might become concentrated to such an extent that it inhibits the synthesis of prostaglandin (PG) E_2 (HIGGS et al. 1987; VANE 1987). On the other hand, SAGONE and HUSNEY (1987) proposed that the anti-inflammatory properties of salicylates might relate to their ability to alter the release of hydroxyl radicals from stimulated phagocytic cells.

In 1968 SAMTER and BEERS, in their classical paper, concluded that "intolerance to acetylsalicylic acid is certainly not an intolerance to salicylates". This notion was subsequently adopted by most clinicians, though some recent reports questioned its validity (STEVENSON et al. 1988). Differences in tolerance of salicylates by aspirin-sensitive asthmatics (Fig. 1) might be related to activity of these drugs towards cyclooxygenase. None of the salicylates approaches the inhibitory potency of aspirin but some are mild inhibitors. It is therefore interesting to note that diflunisal, a moderately active reversible cyclooxygenase inhibitor, produced adverse respiratory symptoms in half of 30 aspirin-intolerant patients (SZCZEKLIK and NIZANKOWSKA 1985; NIZANKOWSKA 1988). Salasate, a somewhat weaker inhibitor, precipitated adverse reactions in 2 of 10 such patients (STEVENSON et al. 1988). On the other hand, salicylates without anti-cyclooxygenase activity, like sodium salicylate (ROSENKRANZ et al. 1986; SAMTER and BEERS 1968), salicylamide (NIZANKOWSKA and SZCZEKLIK 1979) or guaiacolic ester of acetylsalicylic acid (BIANCO et al. 1979), were very well-tolerated by aspirin-sensitive asthmatics in controlled studies. Choline magnesium trisalicylate can be safely added to the latter group. This non-acetylated salicylate

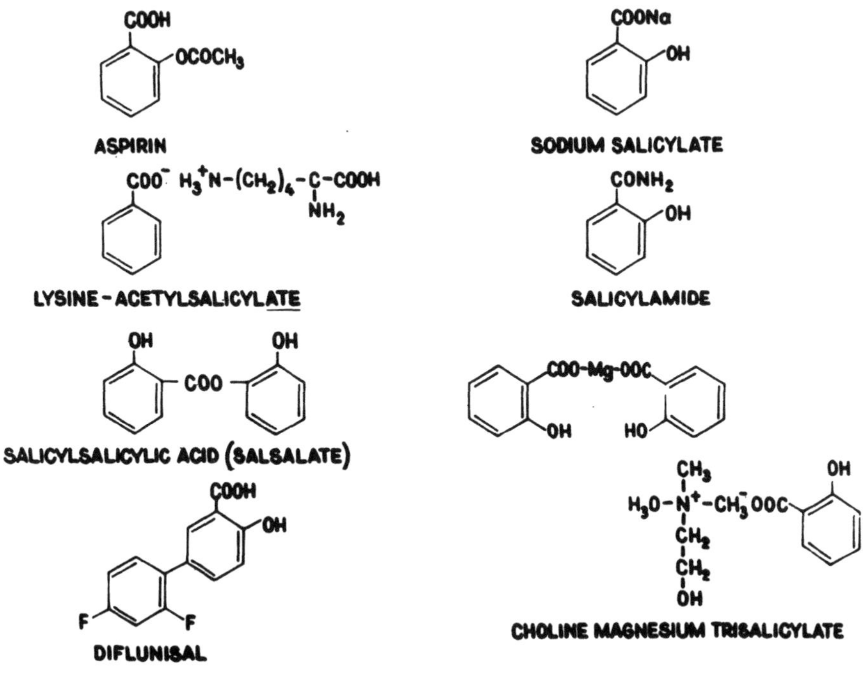

Fig. 1. Chemical structure of common salicylates. Those on the left precipitate asthmatic attacks; those on the right can be taken with impunity by patients with AIA

with strong anti-inflammatory and analgesic activity was shown in a controlled study to be perfectly well-tolerated by aspirin-sensitive asthmatics, even after prolonged administration in high therapeutic doses (SZCZEKLIK et al. 1989).

III. Paracetamol

Paracetamol can be taken with impunity by the majority of patients, However, it is safer to give half a tablet first and then observe a patient for 2 – 3 h for symptoms. We have found symptoms with paracetamol in 4% of Polish patients, a percentage similar to that reported by others. In a recent study, BARLES et al. (1988) observed mild, though definite, bronchoconstriction in 5 of 32 Spanish patients (15%). Our patients, in whom paracetamol produced obstruction to airflow, experienced the same effect following ingestion of phenacetin which is metabolised to paracetamol. SETTIPANE and STEVENSON (1988) desensitised to aspirin three patients known to be intolerant both to aspirin and to paracetamol. Following desensitisation all three tolerated well 1.0 g of paracetamol; however, when a dose of 2.0 g was given to one of them, adverse symptoms became clearly evident.

The mechanism of rare adverse reactions produced by paracetamol in aspirin-sensitive patients is unknown. In vivo, paracetamol does not affect cyclooxygenase function. In vitro, under standard assay conditions with purified enzyme, the drug has no anti-cyclooxygenase activity. However, if the level of lipid peroxides becomes drastically reduced in vitro, paracetamol might then become a cyclooxygenase inhibitor (LANDS 1981). Whether this unique property of the drug could thus account for the rare adverse symptoms is unknown. The reactions deiscussed here should be distinguished from acute hypersensitivity to paracetamol, which is not associated with aspirin-intolerance and is presumably mediated immunologically (STRICKER et al. 1985; ELLIS et al. 1988).

IV. Pyrazolones

Pyrazolones can precipitate adverse symptoms ranging from urticaria and angio-oedema to asthma and anaphylactic shock. Patiens with these symptoms do not form a homogeneous population but can be clearly divided into two groups with different disease pathogeneses.

In the first group, the mechanism responsible for the reactions appears to be allergic. In these patients: (1) Noramidopyrine and aminophenazone induce anaphylactic shock and/or urticaria; (2) Skin tests with these drugs are highly positive; (3) Phenylbutazone, sulfinpyrazone and several other cyclooxygenase inhibitors, including aspirin, can be taken with impunity; (4) Chronic bronchial asthma is present only in about one-fourth of the subjects.

In the second group the symptoms precipitated do not depend on antigen-antibody reactions but are due to inhibition of cyclooxygenase. Thus, in these patients: (1) Noramidopyrine, aminophenazone, sulfinpyrazone, phenylbutazone and several other inhibitors of cyclooxygenase, including aspirin, lead to asthmatic attacks; (2) Skin tests with pyrazolone drugs are negative; (3) All patients have chronic asthma.

The existence of these two groups, first noticed in 1977 (SZCZEKLIK et al. 1977a) and described in detail in later years (CZERNIAWSKA-MYSIK and SZCZEKLIK 1981; SZCZEKLIK 1986), has been confirmed by several authors (VOIGTLAENDER 1985; VIRCHOW et al. 1986; FABRO et al. 1987).

V. Tartrazine

Tartrazine, a yellow azo dye, is used for colouring foods, drinks, drugs and cosmetics (Fig. 2). Contrary to aspirin-like drugs, tartrazine does not inhibit cyclooxygenase activity (GERBER et al. 1979). In some aspirin-sensitive subjects, bronchoconstriction similar to that caused by aspirin has been observed following tartrazine ingestion. Older reports suggested that such a reactions could be quite common, affecting up to 40% of aspirin-intolerant asthmatics. A critical evaluation of the studies reporting a high incidence of tartrazine intolerance was published by SIMON (1984), who pointed out that these studies were either totally non-placebo controlled or the criteria for positive reactions were subjective and vaguely defined. Furthermore, many of the reactions considered positive were merly a product of clinical lability in these patients from whom all symptomatic treatment was usually withheld at the beginning of the test.

In a multicentre, international study we found tartrazine hypersensitivity to be rare. Of 168 Italian, Polish and Swiss patients with well-documented AIA, only 4 had positive reactions when challeged with tartrazine (VIRCHOW et al. 1988). Similarly, WEBER et al. (1979) were unable to confirm a single case of tartrazine intolerance among 44 patients. MORALES et al. (1985) observed one questionable case in 47 patients, and SIMON (1984) found none among 125 patients.

VI. Glucocorticosteroids

Corticosteroids are widely used in treatment of allergic or idiosyncratic reactions; therefore, it is surprising that they themselves may provoke

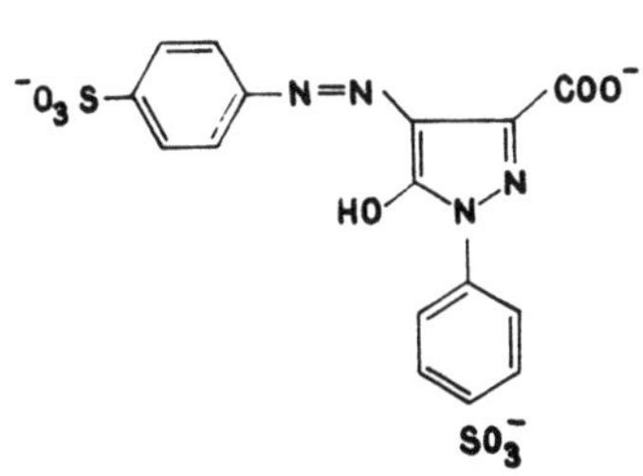

Fig. 2. Chemical structure of tartrazine

adverse responses and aggravate the conditions treated. Such paradoxical responses are very rare. They were observed after local administration of steroids and consited usually of urticaria and contact dermatitis (KOUNIS 1976); anaphylactoid reactions with bronchospasm were exceptional (KING 1960; MEDELSON et al. 1974; SHU CHAN et al 1984).

Recently, attention was drawn to acute idiosyncratic reactions to hydrocortisone in patients with AIA. PARTRIDGE and GIBSON (1978) described severe airflow obstruction in two patients within a few minutes after injection of 100 or 200 mg of hydrocortisone; one of the reactions was almost fatal. DAJANI et al. (1981), in 3 of 11 patients with AIA observed dyspnea and a fall in spirometric values beginning 3–5 min after intravenous injection of 100 mg hydrocortisone but not after saline or hydrocortisone solvent. SZCZEKLIK et al. (1985) studied the effects to intravenous injection of hydrocortisone succinate as compard to its solvent in 31 patients with AIA. Mean FEV1 (forced expiratory volume in first second [litre]) fell significantly 5 min after an intravenous bolus of 300 mg hydrocortisone but not after the solvent and returned to the initial values later. Only 3 of these 31 patients displayed clinical signs of increased impairment to airflow that resolved spontaneously. Neither in these three patients nor in another patient, already known to respond with bronchoconstriction to 30 mg hydrocortisone, did intravenous injections of 20 mg methylprednisolone, 4mg dexamethasone or 4mg betamethasone produce any signs of bronchial obstruction. In the whole group of patients, mean FEV1 increased significantly 3–5 h after hydrocortisone injection (Fig. 3).

The mechanism of these reactions is unknown. Under experimental conditions corticosteroids produce several potentially anti-asthmatic effects, of which the reduction of PG biosynthesis appears parthicularly interesting because of the likely involvement of prostanoids in AIA. Steroids inhibit release of arachidonic acid (GRYGLEWSKI et al. 1985) through lipocortins, a family of anti-phospholipase proteins (DIROSA et al. 1984; PASH and BAILEY 1988).

This action, however, is not limited to hydrocortisone, but encompasses prednisone, prednisolone, betamethasone and other glucocorticosteroids. Therefore, it is unlikely that the reactions discussd here could be mediated via lipocortins. An interesting explanation was recently suggested by TANIGUCHI and SATO (1988). They confirmed a previous report (SZCZEKLIK et al. 1985) that succinate is well tolerated by patients who respond with bronchoconstriction to hydroncortisone succinate. They challenged a group of 20 aspirin-sensitive asthmatics with a variety of glucocorticosteroid esters and observed cross-sensitivity with succinate esters. Accordingly, bronchoconstriction could be precipitated by succinate salts of both hydrocortisone and methylprednisolone but not by the phosphate salts.

Irrespective of its mode of action, hydrocortisone therapy should be used with caution in patients with AIA. In several patients it proudces a slight, transient and clinically irrelevant impairment of airflow, but it can also lead to

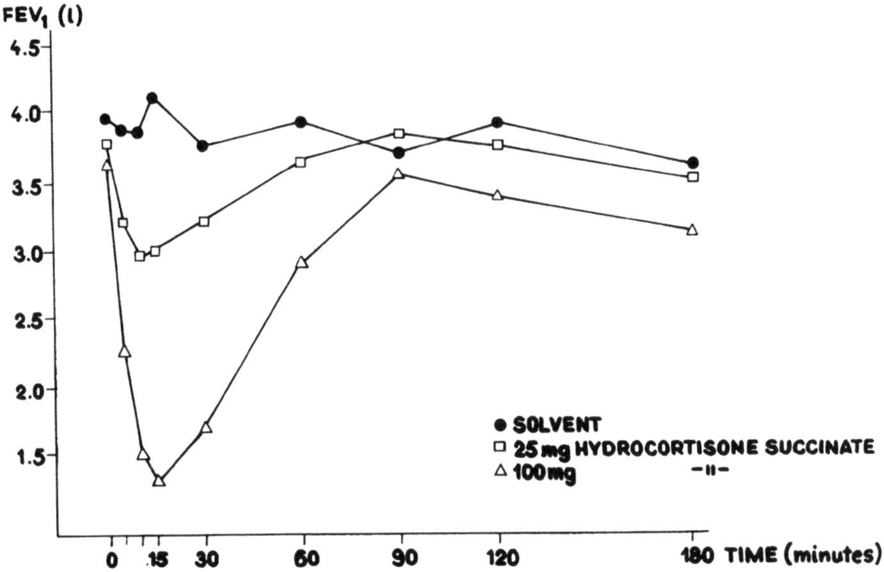

Fig. 3. Effects of intravenous injection of hydrocortisone succinate or its solvent on FEV1 in a 48 years old man with AIA. Only the 100 mg dose produced transient shortness of breath and wheezing. The symptoms resolved spontaneously without any pharmacological intervention

over-bronchoconstriction. It is therefore advisable to use other steroids in patients with AIA, preferably non-succinate salts.

C. Pathogenesis

I. Allergic Mechanisms

Clinical symptoms precipitated by aspirin in sensitive patients with asthma are reminiscent of immediate-type reactions. Therefore, an underlying antigen-antibody mechanism has been suggested. However, skin tests with aspirin are negative, and numerous attempts to demonstrate specific antibodies against aspirin or its derivatives were unsuccessful (see SCHLUMBERGER 1980). Furthermore, in patients with AIA asthmatic attacks can be precipitated not only by aspirin but by several other analgesics with different chemical structures, which makes immunological cross-reactivity most unlikely.

II. Abnormal Reactivity of Bradykinin Receptors

SAMTER and BEERS (1968) realized that aspirin-intolerance developed as a sequela of pre-existing disease. They hypothesized that a long-lasting pathological process might leta to altherations in bradykinin receptors resulting in their paradoxical response to aspirin. Aspirin would become an agonist of bradykinin receptors and by stimulating them would precipitate vascular permeability, bronchoconstriction and secretion of mucus. Though interesting, the hypothesis remains purely speculative.

III. Acetylation of Proteins

FARR (1970) proposed that aspirin-intolerance is the end stage of progressive acetylation of biologically active proteins by small doses of aspirin. Acetylation of receptors would result in their unusual drug responsivness. This concept is contradicted by the following clinical observations:

1. Aspirin-sensitive patients with asthma are intolerant to many other analgesics which have no acetylating properties.
2. Intolerance reactions have been reported after the first ingestion of aspirin.
3. Intolerance to aspirin continues after an extended period of strict avoidance of the drug.

IV. Complement Involvement

YURCHAK et al. (1970) suggested a direct activation of the complement system by aspirin as the pathogenic mechanism of aspirin-intolerance. The adverse reactions could be mediated by complement-derived split products, particularly the anaphylatoxins C3a and C5a.

Aspirin, indeed, activates complement both in vivo and in vitro. Such activation can be demonstrated in healthy subjects and in aspirin-intolerant patients (SCHLUMBERGER 1980). The significane of the above finding is not clear. VOIGTLAENDER et al. (1981) did not find any significant complement deficiencies in 40 aspirin-intolerant patients with either asthma or urticaria. These authors also noted that in vitro sodium salicylate induced about tenfold higher complement consumption than aspirin, yet sodium salicylate is very well-tolerated by patients with AIA.

V. The Cyclooxygenase Theory

1. Formulation of the Theory

A hypothesis was put forward (SZCZEKLIK et al. 1975) that, in sensitive patients, precipitation of asthma attacks by certain analgesics results from inhibition of cyclooxygenase, leading to an imbalance of prostanoids in the respiratory tract. Cyclooxygenase, an atypical lipoxygenase, is present in most human tissues, including the lungs. It introuduces two molecules of

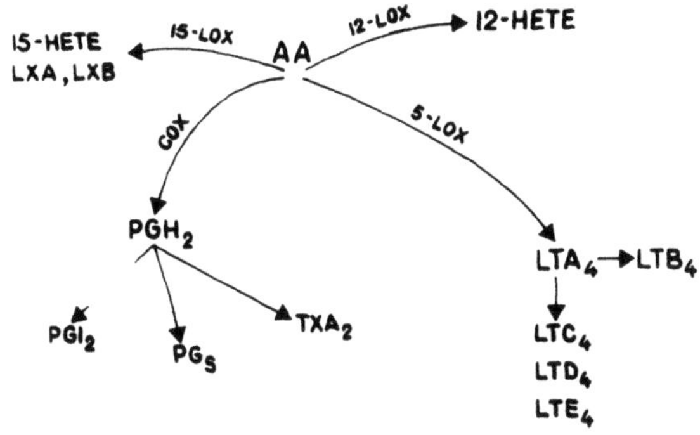

Fig. 4. Arachidonic acid liberated from phospholipid membranes can undergo several transfromations, depending on the action of the specific lipoxygenases. In this simplified diagram four main metabolic pathways are presented. *AA*, arachidonic acid; *COX*, cyclooxygenase; *LOX*, lipoxygenase; PGH_2, cyclic endoperoxide H_2; PGI_2, prostacyclin; *PGs*, prostaglandins; TXA_2, thromboxane A_2; LTA_4, leukotriene A_4; LTB_4, leukotriene B_4; LTC_4, leukotriene C_4; LTD_4, leukotriene D_4; LTE_4; leukotriene E_4; *12-HETE*, 12-hydroperoxyeicosatetraenoic acid; *15-HETE*, 15-hydroperoxyeicosatraenoic acid; *LXA*, lipoxin A; *LXB*, lipoxin B

oxygen into arachidonic acid, converting it into PG peroxides (Fig. 4). This is the beginning of a metabolic pathway leading to formation of PG, thromboxane and prostacyclin. Cyclooxygenase is inhibited by aspirin and several other analgesics, a phenomenon which might explain their pharmacological action (VANE 1971).

In the early 1970s allergic mechanisms, as an explanation for aspirin intolerance, were vigorously pursued. Contrary to these concepts, the cyclooxygenase, theory proposed that precipitation of asthmatic attacks by aspirin is not based on antigen-antibody reactions but stems from the pharmacological action of the drug. The original observation (SZCZEKLIK et al. 1975, 1977a), that the drug intolerance could be predicted on the basis of its in vitro inhibition of cyclooxygenase, has been consistenly reaffirmed during the ensuing years (STEVENSON and LEWIS 1987). Evidence in favour of the cyclooxygenase theory (SZCZEKLIK 1986) can be summarized as follows: (1) Analgesics with anti-cyclooxygenase activity invariably precipitate bronchoconstriction in aspirin-sensitive patients; (2) Analgesics not affecting cyclooxygenase are devoid of bronchospastic properties in these patients; (3) There is a positive correlation between the potency of analgesics to inhibit cyclooxygenase in vitro and their potency to induce asthmatic attacks in sensitive patient; (4) The degree of enzymatic inhibition that is sufficient to precipitate bronchoconstriction is an individual hallmark (thus, if the threshold dose for any anti-cyclooxygenase in a particular patients is known, one can predict the threshold doses for other analgesics in that patient); (5) In vitro anti-cyclooxygenase inhibitors cause platelets to release cytotoxic mediators in aspirin-sensitive asthmatics but not in atopic asthmatics or

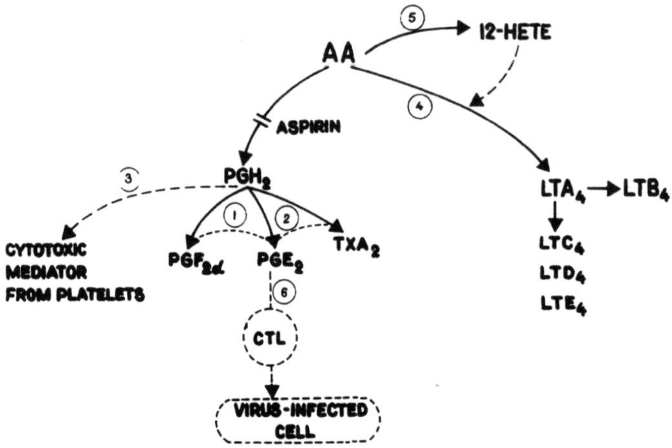

Fig. 5. AIA: Hypothetical alterations in arachidonic acid metabolism following inhibition of cyclooxygenase by aspirin. All abbreviations as in Fig. 4. *1*. Depression of PGE_2 and increase in $PGF_{2\alpha}$ (TOOGOOD 1977); *2*. Selective deficiency of bronchial PGE_2 and unchanged parenchymal TXA_2 (GRYGLEWSKI 1989); *3*. Undefined cyctotoxic mediator generated by platelets (AMEISEN et al. 1986); *4*. Overproduction of LTs as a result of increased availability of AA (SAMUELSON 1983); *5*. Stimulation of LT production by increased concentrations of 12-HETE (MACLOUF et al. 1982); *6*. Lack of PGE_2 of macrophage origin liberates specific cytotoxic lymphocytes which attack virus-infected cells of the respiratory tract (SZCZEKLIK 1988)

healthy subjects (AMEISEN et al. 1986; CAPRON et al. 1985); (6) In patients with AIA inhibition of thromboxane (TX) A2 formation, which occurs after the cyclooxygenase step in the arachidonic acid cascade, neither precipitates asthmatic attacks nor alters pulmonary function (SZCZEKLIK et al. 1987); (7) After aspirin desensitisation, cross-desensitisation to other analgesics which inhibit cyclooxygenase also occurs (STEVENSON and LEWIS 1987). Thus, the inhibition of bronchial cyclooxygenase by aspirin-like drugs appears to set off a chain reaction leading to asthma attacks in aspirin-intolerant patients. What follows at the biochemical level remains largely unknown (Fig. 5).

2. Early Explanations

At the time of publication of the cyclooxygenase hypothesis (SZCZEKLIK et al. 1975), the only eicosanoids known to be produced by the cells of respiratory tract were PGE_2 and $PGF_{2\alpha}$. It was presumed (TOOGOOD 1977) therefore that although cyclooxygenase was inhibited by aspirin, a selective deficiency among the cyclooxygenase products must also occur, such that bronchodilator PGE_2 was decreased relative to bronchoconstrictor PGF_2. This idea was not universally appealing because, although the efficacy of aspirin and certain other NSAID inhibitors of cyclooxygenase was clearly demonstrated, relative effects on PGH_2-PGE_2 isomerase and PGH_2-$PGF_{2\alpha}$ isomerase were not.

An early explanation (SZCZEKLIK et al. 1975), that aspirin-sensitive asthmatics might rely more on PGE_2 than on the β-adrenergic system to keep their bronchi unobstructed, seems unlikely today. These patients, indeed, respond better to inhaled PGE_2 than other asthmatics (SZCZEKLIK et al. 1977b), but clinical practice leaves no doubt that their lung function improves substantially following inhalation of beta-mimetics.

3. Participation of LTs

In 1979, when LTs were discovered, it became apparent that arachidonic acid could be diverted from the cyclooxygenase pathway to the 5-lipoxygenase pathway if cyclooxygenase was inhibited. The definition of LTs suggested that they would provide potent mediation of neutrophil influx into the tissue via the action of LTB_4 and a potent stimulation for bronchoconstriction, mucosal permeability and mucus secretion by the actions of LTC_4, LTD_4 and LTE_4. The explanation for AIA was then postulated as simply being caused by shunting of arachidonic acid from generation of PGs to the biosyntheis of LTs. In a simplified way, it would be a redirection of flow in a bifurcated vessel: from the blocked arm to an open one. More sophisticated explanations have been put forward. Biosyntheisis of LTs could be enhanced by overproduction of 12-hydroxyeicosatetraenoic acid (12-HETE) (MACLOUF et al. 1982) or removal of inhbiting control of PGE_2/PGI_2 (KUEHL et al. 1984). Both posibilities are a logical consequence of cyclooxygenase pathway inhibition.

There is some experimental support for the concept of a shift in arachidonic acid metabolism, though clinical evidence is still lacking. In a guinea-pig model of antigen-iuduced anaphylaxis, pretreatment of animals with indomethacin resulted in augmentation of the pulmonary mechanical responses to intravenous antigen and this was accompanied by increased generation of LTB_4 (LEE et al. 1986). Pretreatment of passively sensitised human airways with indomethacin resulted in an increased release of LTs from human bronchi in response to both antigen and anti-IgE stimulation (UNDEM et al. 1987). However, others have found (VIGANO et al. 1988) that in normal human lung parenchyma, an anti-IgE challenge in the presence of indomethacin does not produce a shift towards LT formation.

Two groups studied the release of LTs into the nasal cavity following aspirin administraction to patients with AIA. ORTOLANI et al. (1987) noticed an increase in the mean LTC_4 concentration in nasal washings of seven aspirin-sensitive asthmatics following nasal spray provocation with aspirin. However, clinical symptoms occurred within 1-2min after the challenge, while LTC_4 increase was observed 60 min later. FERRERI et al. (1988) used oral aspirin to provoke clinical symptoms in five intolerant patients. During the provoked reactions, LTC_4 increased in three patients. In two of five patients, a fall in PGE_2 preceded the appearnace of clinical symptoms. In control subjects, ingestion of higher doses of aspirin (650 mg) resulted in a distinct fall in PGE_2 without the release of LTC_4 into nasal washings. A recent study by BISYGAARD et al. (1988) casts serious doubts on the validity of mediator measurment in nasal lavage in relation to symptoms following local nasal challenge.

It is not clear in which cells of the respiratory tract alterations in arachidonic acid metabolism might occur. Leukocytes, especially eosinophils, presents in large amounts in nasal and bronchial tissue of aspirin-sensitive asthmatics (GODARD et al. 1982; SZCZEKLIK 1986), could be considered as a source of LTs. GOETZL et al. (1986) suggested a generalized abnormality in the regulation of arachidonic acid oxidative pathways in peripheral blood leukocytes of patients with AIA. Two recent studies do not support this idea. NIZANKOWSKA et al. (1988) studies production by polymorphonuclear leukocytes of 5-HETE and LTB_4 in ten aspirin-sensitive asthmatics and ten matched healthy controls. Blood samples were obtained before administration of threshold doses of aspirin and during the aspirin-induced reactions. Initial levels of eicosanoids did not differ between the two groups and remained unchanged following aspirin challenge. TSUDA et al. (1988) measured the production of LTB_4 and LTC_4 in peripheral blood leukocytes stimulated by calcium ionophore A23187. They compared four groups (controls, AIA, atopic and intrinsic asthma) before and after indomethacin challenge. All three asthmatic groups produced more LTC_4 than the healthy controls, but there was no difference between aspirin-intolerant patients and atopic or intrinsic ones. LTB_4 production (as well as PGE_2 and TXB_2) was similar in all four groups. Indomethacin did not affect LT generation in any of the groups studied.

The concept of arachidonic acid shunting needs the additional assumption that the airways of aspirin-intolerant patients are more sensitive to LTs than those of other patients with asthma (SZCZEKLIK and GRYGLEWSKI 1983). If not, all asthmatic patients would react with bronchoconstriction in response to aspirin-like drugs. Three research groups addressed this problem. VAGHI et al. (1985; see also BIANCO 1986) measured bronchial response to LTC_4 in ten aspirin-sensitive asthmatics and ten controls. They were unable to find any significant difference. SAKAKIBARA et al. (1988a) studied airway responsivnesss to methacholine, histamine and LTD_4 in 12 patients with AIA, 13 patients with extrinsic asthma and 12 patients with intrinsic asthma. There were no significant differences in either the concentrations of any of the agents producing a 20% fall in FEV1 or the slope of the FEV1 changes among the groups studied. The only positive finding was a somewhat delayed recovery in FEV1 following challenge with LTD_4 in the aspirin-intolerant group as compared to the others. These two studies do not support the concept of increased bronchial reactivity to LTC_4 or LTD_4. However, results of ARM et al. (1989) suggest a selective increase in airway responsiveness to LTE_4. They measured a 35% fall in the specific airway conductance (provocation dose 35, PD35) following histamine and LTE_4 inhalation in 5 subjects with aspirin-induced asthma and in 15 asthmatics without aspirin sensitivity. The airways of aspirin-intolerant patients had a significant, 13-fold increase in responsivness to LTE_4 relative to histamine when compared to control asthmatics. Interestingly, this hyperresponsivness to LTE_4 was abolished after aspirin-desensitisation.

The concept of diversion of arachidonic acid metabolism from prostanoids to LTs is hard to accept in view of the likely compartmentalization of arachidonic acid in the lung (GRYGLEWSKI et al. 1976). This concept still awaits testing with a powerful, specific LT inhibitor. In a recent trial (NIZANKOWSKA et al. 1987) pretreatment of aspirin-intolerant asthmatics with a LT inhibitor failed to prevent aspirin-precipitated bronchospasm. The bioavailability of the inhibitor, administered by inhalation, remained uncertain.

4. Platelet Involvement

In the last few years increasing attention has been paid to the possible participation of platelets in the pathogenesis of bronchial asthma (MORLEY et al. 1985; SZCZEKLIK et al. 1986), particularly in aspirin-induced asthma (MACLOUF et al. 1982; AMEISEN et al. 1986). In patients with AIA, aspirin challenge may lead to activation of peripheral blood platelets, which parallels the time course of the bronchospastic reaction (SCHMITZ-SCHUMANN et al. 1987). In contrast to platelet activation, the detection of endogenous PAF release has not been a consistent finding. Aspirin-induced bronchoconstriction seems not to be secondary to the release of PAF (SCHMITZ-SHUMANN et al. 1987).

In 1982 MACLOUF et al. noticed that platelet 12-hydroperoxyeicosatetraenoic acid (12-HPETE) stimulated the generation of LTB_4 and 5-HETE in a mixed platelet-leukocyte suspension. These authors hypothesized that administration of aspirin to intolerant patients with asthma may lead to increased generation of platelet 12-HPETE (BORGEAT et al. 1983) because of an impaired cyclooxygenase and 12-lipoxygenase balance, or because of inhibition of peroxidase activity in platelets. The released 12-HPETE could activate the 5-lipoxygenase of circulating blood leukocytes and pulmonary macrophages; the generated LTs would then precipitate asthma.

Capron's group (AMEISEN et al. 1986; CAPRON et al. 1985) reported that platelets isolated from patients with aspirin-induced asthma reacted abnormally in vitro to aspirin and other cyclooxygenase inhibitors by generating cytocidal molecules that could kill parasitic larvae. Aspirin-like drugs had no similar effect on platelets from normal donors or allergic asthmatics. This abnormality, according to the authors (JOSEPH et al. 1987), appears to be associated with the inhibiting properties of analgesics on the cyclooxygenase pathway, leading to a defect of the binding of PG endoperoxide PGH_2 to its receptor on the platelet membrane.

NIZANKOWSKA et al. (1988) measured 12-HETE production by platelets in ten aspirin-sensitive asthmatics and ten matched healthy controls before and after administration of threshold doses of aspirin. Initial levels of 12-HETE did not differ between the two groups. Following aspirin challenge, 12-HETE increased to similar levels in both groups. These data do not support the concept of a generalized abnormality in arachidonic acid oxidative pathways in platelets of aspirin-sensitive asthmatics. A lack of protective effect of prostacyclin infusions on aspirin challenge also raises doubts about participation of platelets in the discussed reactions. (NIZANKOWSKA et al. 1987).

5. Compartmentalization of Eicosanoids in the Lungs

An interestintg hypothesis was recently proposed by GRYGLEWSKI (1989). It is based on the idea that arachidonic acid metabolism in the lungs is compartmentalized (GRYGLEWSKI et al. 1976). Thus, PGE_2 is generated by smooth muscle of large airways and TXA_2 by contractile elements of lung parenchyma. Their generation is stimulated by non-specific spasmogens, e.g. histamine, LTs or high concentrations of arachidonic acid. Prostacyclin is produced by vascular endothelium in response to the stimulation of receptors by angiotensin II, adenosine diphosphate (ADP) or low concentrations of arachidonic acid. LTs are released by leukocytes residing in the lungs and by fibroblasts. All these compartments can be stimulated simultaneously by an immunological reaction.

The cyclooxygenase hypothesis postulates that in patients with AIA, cyclooxygenase of large airways is more susceptible to pharmacological inhibition than that of lung parenchyma. As a consequence, an ingestion of an anti-cyclooxygenase drug casuses an increase in the TXA_2/PGE_2 ratio. The most likely explanation for an augmented selective susceptibility of bronchial cyclooxygenase to analgesics is chronic viral infection of the upper airways of patients with AIA (GRYGLEWSKI et al. 1977; SZCZEKLIK 1988). This chronic infection might either change the biochemical characteristics of cyclooxygenase in upper airways or make it easily accesible to analgesics. If the hypothesis is right, then simultaneous pretreatment of patients with a TXA_2 synthetase inhibitor and a TXA_2/PGE_2 receptor antagonist should protect them against aspirin-induced bronchoconstriction.

6. Viral Infection

Viruses have been implicated in the pathogenesis of asthma (WELLIVER and OGRA 1988; FRICK and BUSSE 1988), including AIA (GRYGLEWSKI et al. 1977; SCHLUMBERGER 1980), though in the latter case no explanation was offered as to how virus infection could be linked with a cyclooxygenase dependent mechanism. Such an explanation has been given by a recent hypothesis (SZCZEKLIK 1988) stating that AIA results from chronic viral infection. In response to a virus, long after the initial exposure, specific cytotoxic lymphocytes are prodcued. Their activity is suppressed by PGE_2 synthesizd by pulmonary alveolar macrophages. Anti-cyclooxygenase analgesics block

PGE_2 production and allow cytotoxic lymphocytes to attack and kill their target cells, i.e. virus-infected cells of the respiratory tract. During this reaction, toxic oxygen intermediates, lysosomal enzymes and mediators are released, which precipitate attacks of asthma. These acute attacks can be prevented by avoidance of all drugs with anti-cyclooxygenase activity; however, asthma, continues to run a protracted course because of chronic viral infection (Fig. 6).

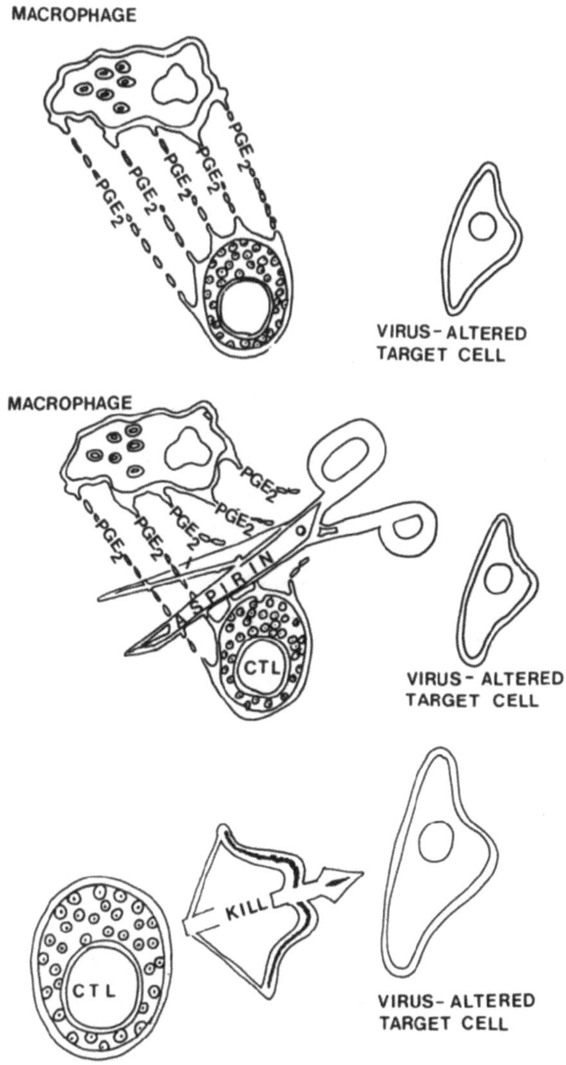

Fig. 6. Three stages of aspirin-precipitated reaction of asthma. *CTL*, cytotoxic T lymphocytes; *PGE_2*, prostaglandin E_2 (from SZCZEKLIK 1989, with permission)

The hypothesis is based on the following concepts: (1) The clinical course of AIA is reministcent of viral infection (SZCZEKLIK 1989; PRIETO et al. 1988); (2) Latency or semi-latency of viruses is increasingly being recognized (PODGWAITE and MAZZONE 1986; ANONYMOUS 1989). In humans, a notable example is Epstein-Barr virus, which causes an infection that persists for life and is subject to reactivation. Interestingly, some of the clinical manifestations of acute Epstein-Barr virus infection, such as Guillain-Barré syndrome, hepatitis or suppression of hematopoesis may be caused by secondary immune responses to latentlty infected lymphocytes (PAGANO 1988); (3) Cytotoxic T lymphocytes form a part of the human immune system in the respiratory tract. They increase in response to viral infections and are highly specific; (4) Lung macrophages produce PGE_2 which suppresses the immune response (MORLEY 1974), including cytotoxic activity of lymphocytes (HERBERMAN 1986; ROITT et al. 1985). This inhibition can be overcome by anti-cyclooxygenase analgesics, which deprive macrophages of PGE_2 (GOODWIN 1985).

D. Diagnosis

While a patient's clinical history might raise the suspicion of AIA the diagnosis can be established with certainty only by aspirin challenge. There are no in vitro tests suitable for routine clinical diagnosis.

Patients are challenged when their asthma is in remission and their FEV1 is greater than 70% of the predicted value. They continue regular medication, including corticosteroids, but stop sympathomimetics and methylxanthines for 10 h and antihistamines for 48 h prior to the challenge. Regular intake of sodium cromoglycate and ketotifen should be interrupted 72 before the challenge. Aspirin reactions can be blocked by ketotifen (DELANEY 1983; SZCZEKLIK et al. 1980, WUETRICH and FABRO 1981) and modified by H1 anti-histamines (SZCZEKLIK and SERWONSKA 1979). Sodium cromoglycate can alleviate aspirin-induced reactions (BASOMBA et al. 1976, MARTELLI and USANDIVARAS 1977), though this has not been a uniform experience (DAHL 1981; HOLLIGNSWORTH et al. 1984).

Glucocrticosteroids attenuate aspirin-precipitated reactions (NIZANKOWSKA and SZCZEKLIK 1989). A clinician trying to establish diagnosis of aspirin idiosyncrasy using challenge tests should be aware of the possibitity of false-negative results related to corticotherapy. This might be particularly relevant in patients with a mild degree of idiosyncrasy (threshold dose of aspirin > 200 mg). On the other hand, adverse reactions precipitated by aspirin in intolerant patients treated with steroids might be expected to become more accentuated once the steroids are stopped.

Oral challenge tests are most commonly performed. All challenges are carried out in the morning with a physician present and emergency treatment available. Single-blind placebo challenges with serial spirometric measurements are conducted on the first day (FEV1 variation should be less than 15%). On the next day, aspirin is administered in progressive, incremental doses. The starting dose is 20 mg. If there is no reaction, the dose is gradually increased until the reaction becomes positive or a dose of 400 mg is reached.

The reaction is considered positive if a decrease in FEV1 (or PEF) greater than 15% occurs, accompanied by symptoms of bronchial obstruction and irritation of nose or eyes. In the absence of these clinical findings, the reaction is considered positive only if a fall in FEV1 or PEF (peak expiratory flow rate) >25% is obtained. For relief of adverse symptoms, a β-2-adrenoceptor agonists is given by inhalation and, if necessary, aminophylline and steroids can be administered intravenously.

A modification of the challenge tests has been developed in which, instead of giving oral aspirin, an aerosol of lysine acetylsalicylate is administered by inhalation (Bianco et al. 1977a; Schmitz-Schumann et al. 1982, 1985; Sakakibara et al. 1988b).

In inhalation challenge tests the aspirin dose is increased every 30 min, and the test is completed in one morning. It is therefore faster than oral challenge which often takes 2–3 days. On the other hand, in oral tests there is no need for special equipment and the chances that aspirin will evoke other than bronchial symptoms are higher.

In a few patients only naso-ocular responses are provoked by orally administered aspirin. These patients usually exhibit bronchial asthma at other times. Pleskow et al. (1983) recorded this type of reaction in 3 of 50 patients with documented aspirin idiosyncrasy and chronic respiratory tract disease.

In the majority of patients, once aspirin intolerance is developed, it remains for the rest of the patient's life. Repeated aspirin challenges are therefore positive, though some variability in intensity and spectrum of symptoms occurs. However, we and others (Simon et al. 1984) have observed an occasional patient in whom a positive aspirin challenge became negative after a period of a few years.

AIA should be clearly differentiated from other forms of aspirin intolerance. For example, urticaria aggravated by aspirin is something quite different from AIA, and the two syndromes very rarely occur in the same patient (Settipane 1981; Szczeklik et al. 1977a; Szczeklik 1986).

E. Prevention and Therapy

Patients with AIA should avoid aspirin, all products containing it and other analgesics which inhibit cyclooxygenase. Nonetheless, sooner or later, almost every one of us needs aspirin or a related drug for fever, arthritis, chronic headache or other so-called minor discomforts. What can aspirin-intolerant asthmatics do in that case? They are left with two options. First, they can take safely, even for prolonged periods, certain agents which do not inhibit cyclooxygenase, e.g. choline magnesium trisalicylate, dextropropoxyphene or benzydamine. Most patients will also tolerate paracetamol well. Second, they can undergo desensitisation. This can be achieved by giving 4–8 incremental doses of aspirin every 2–3 h under careful observation. The procedure takes 2 days. It starts usually with a dose producing mild adverse symptoms and ends with 600 mg of aspirin which is then well tolerated. The state of aspirin tolerance (refractory period) lasts 2–5 days, but in most patients it can be extented over months if aspirin is administered regularly at a daily dose of 600 mg. At the same time patients can take other anti-cyclooxygenase drugs without any adverse effects. Refractoriness to aspirin following indomethacin administration has also been obsrved (SIMON et al. 1984). Desensitisation is possible in most (BIANCO et al. 1977a, b; STEVENSON et al. 1984) but not all (BALDOCCHI et al. 1983) aspirin-intolerant patients with asthma.

Some patients who were studied at the Scripps Clinic noted improvement in their underlying chronic respiratory symptoms (naso-ocular responses without asthma) after aspirin desensitisation and during maintenance of the desensitised state (LUMRY et al. 1983). Similarly, CHIU (1983) observed that 6 of 12 asthmatic patients, rendered aspirin tolerant and maintained on aspirin, improved when measured by such criteria as the need for anti-asthmatic medication. However, BIANCO et al. (1982) noted no such improvement in their patients. DOR et al. (1985), after inducing aspirin tolerance in nine corticosteroid-dependent asthmatics, noted deterioration of lung function in all patiens taking aspirin for more than 1 month. According to STEVENSON et al. (1984) desensitisation, followed by prolonged administration of aspirin, leads to significant improvement in nasal symptoms but not in the clinical manifestations and the course of asthma.

Despite being desensitised to aspirin, all patients remain asthmatic. Methacholine and histamine inhalation challenges before and after aspirin desensitisation are unchanged and the patients remain hyperreactive to these agents (SIMON et al. 1984). In addition, patients maintained on aspirin in a desensitised state still develop asthmatic relapses from all of their prior provoking factors, execept for aspirin.

The mechanism of desensitisation is unknown. KOWALSKI et al. (1984) offered the following explanation: Aspirin blocks cyclooxygenase irreversibly, and a new enzyme must be synthesized to replace molecules inactivated by the drug. After initial cyclooxygenase inhibition by aspirin, the original regulatory mechanisms are removed, and the functional balance in the bronchi is based on a prostanoid-independent regulatory system. Hypersensitivity to aspirin would then reoccur with the return of tissue capacity to generate PGs after aspirin withdrawal.

One can also think of another explanation. Incremental doses of aspirin are quickly biotransformed to salicylate which has been shown to protect cyclooxygenase against pharmacological inhibition by aspirin (VARGAFTIG 1978). Thus, pretreatment with sodium salicylate could theoretically prevent attacks of AIA. However, a note of caution is necessary. Protection of cyclooxygenase by salicylate against inhibition by analgesics has been shown in platelets (VARGAFTIG 1987), and this finding might not apply to bronchial smooth muscle.

Whether desensitised or not, most patients with AIA need regular therapy to control symptoms of their disease. The therapy does not differ from that of other types of asthma. Long-term treatment with systemic corticosteroids is necessary in at least half of the patients. Dietary fish oil in large quantities should be avoided because it might worsen the airflow obstruction, possibly through inhibition of the cyclooxygenase pathway (PICADO et al. 1988).

References

Ameisen JC, Capron A, Joseph M, Tonnel AB (1986) Platelets and aspirin induced asthma. In: Kay AB (ed) Clinical pharmacology and therapeutic progress in asthma. Blackwell, London, pp 226–236

Anonymous (1989) Herpes simplex virus latency. Lancet 1:194–195

Arm JP, O'Hickey SP, Spur BW, Lee TW (1989) Airways responsivness to histamine and leukotriene E_4 in subjects with aspirin-induced asthma. Am Rev Respir Dis 140:148–153

Baldocchi G, Vervloet D, Charpin J (1983) Acetyl-salicylic acid therapy in aspirin-sensitive asthmatics. J Allergy Clin Immunol 71:148–153

Barles PG, Garcia FD, Olmo JRP, Aznar JP, Alarma JLE (1988) Adverse reaction to acetaminophen as an alternative analgesic in AAS triad. Allergol Immunopathol (Madr) 16:321–325

Basomba A, Romar A, Pelaez A, Villamanzo IG, Campos A (1976) The effect of sodium cromoglycate in preventing aspirin-induced bronchospasm. Clin Allergy 6:269–275

Bianco S (1986) Asthme et medicaments anti-inflammatories non-steroidiens. In: Charpin J (ed) Allergologie. Flammarion, Paris, p 683

Bianco S, Robuschi M, Petrigni G (1977a) Aspirin-induced tolerance in aspirin-asthma detected by a new challenge technique. IRCS J Med Sci 5:129–130

Bianco S, Robuschi M, Petrigni G, Allegra L (1977b) Respiratory effects due to aspirin (ASA): ASA-induced tolerance in ASA-asthmatic patients. Bull Eur Physioplathol Respir 13:123–124

Bianco S, Petrigni C, Felisi R, Robuschi M (1979) Tolerance of guaiacolic ester of acetlsalicylic acid by patients with aspirin-induced asthma. Scand J Respir Dis 60:350–354

Bianco S, Robuschi M, Petrigni G (1982) Treatment of aspirin idiosyncrasy. J Allergy Clin Immunol 70:222.

Bisgaard H, Robinson C, Roemling F, Mygind N, Church M, Holgate S (1988) Leukotriene C_4 and histamine in early allergic reactions in the nose. Allergy 43:219–227

Borgeat P, Fruteau de Laclos B, Maclouf J (1983) New concepts in the modulation of leukotriene synthesis. Biochem Pharmacol 32:381–387

Capron A, Ameisen JC, Joseph M, Auriault C, Tonnel AB, Caen J (1985) New function for platelets and their pathological implications. Int Arch Allergy Appl Immunol 77:107–114

Chiu JT (1983) Improvement in aspirin-sensitive asthmatic subjects after rapid aspirin desensitization and aspirin maintenance (ADAM) treatment. J Allergy Clin Immunol 71:560–564

Czerniawska-Mysik G, Szczeklik A (1981) Idiosyncrasy to pyrazolone drugs. Allergy 36:381–384

Dahl R (1981) Oral and inhaled sodium cromoglycate in challenge test with food

allergens or acetylsalicylic acid. Allergy 36:161–165
Dajani BM, Sliman NA, Shubair KS, Hamzeh YS (1981) Bronchospasm caused by intravenous hydrocortisone sodium succinate (Solu-Cortef) in aspirin-sensitive asthmatics. J All Clin Immunol 68:201–205
Delaney JC (1983) The effect of ketotifen on aspirin-induced asthmatic reactions. Clin Allergy 13:247–251
Di Rosa M, Flower RJ, Hirata F, Parente L, Russo-Marie F (1984) Antiphospholipase proteins. Prostaglandins 28:441–442
Dor PJ, Vervloet D, Baldocchi G, Charpin J (1985) Aspirin intolerance and asthma induction of tolerance and long-term monitoring. Clin Allergy 15:37–42
Ellis M, Haydik I, Gillman S, Cummins L, Cairo M (1988) Immediate adverse reactions to acetaminophen. J Allergy Clin Immunol 81:180 (abstract 348)
Fabro L, Wuetrich B, Walti M (1987) Acetylsalicylic acid allergy and pyrazole allergy or pseudoallergy? Results of the skin tests and antibody determinations in a multicenter study. Z Hautkr 62:470–478
Farr RS (1970) The need to re-evaluate acetylsalicylic acid (aspirin). J Allergy 45:321–326
Ferreri NR, Howland WC, Stevenson DD, Spielberg HL, (1988) Release of leukotrienes, prostaglandins and histamine into nasal secretions of aspirin-sensitive asthmatics during reaction to aspirihn. Am Rev Respir Dis 137:847–854
Frick WE, Busse WW (1988) Respiratory infections: their role in airway responsivness and pathogenesis of asthma. Clin Chest Med 9:539–549
Gerber JG, Payne NA, Oelz O, Nies AS, Oates JA (1979) Tartrazine and the prostaglandin system. J Allergy Clin Immunol 63:289–294
Godard P, Chaintreuil J, Damon F, Coupe M, Flandre O, Crestes de Paulet A, Michel FB (1982) Functional assessment of alveolar macrophages: comparison of cells from asthmatic and normal subjects. J Allergy Clin Immunol 70:88–93
Goetzl EJ, Valacer DJ, Payan DG, Wong MYS (1986) Abnormal responses to aspirin of leukocyte oxygenation of arachidonic acid in adults with aspirin intolerance. J Allergy Clin Immunol 77:693–698
Goodwin JS (1985) Immunologic effects of nonsteroidal anti-inflammatory agents. Med Clin North Am 69:793–804
Gryglewski RJ (1989) Eicosanoids in aspirin-induced asthma. In: Schmitz-Schumann M, Menz G, Costabel U, Page CP (Eds) Intrinsic Asthma. Agents & Actions Suppl., Vol 28, pp 113–122
Gryglewski RJ, Dembinska-Kiec A, Grodzinska L, Panczenko B (1976) Differential generation of substances with prostaglandin-like and thromboxane-like activities by guinea-pig trachea and lung strips. In: Bouhuyus A (ed) Lung cells in disease. Amsterdam, pp 209–307
Gryglewski RJ, Szczeklik A, Nizankowska E (1977) Aspirin-sensitive asthma: its relationship to inhbition of prosaglandin biosynthesis. In: Berti F, Samuelson B, Velo GP (eds) Prostaglandins and thromboxanes. Plenum, New York , pp 191–203 (NATO Advance Study Institute series A: Life science)
Gryglewski RJ, Panczenko B, Korbut R, Grodzinska L, Ocetkiewicz A (1985) Corticosteroids inhibit prostaglandin release from perfused mesenteric blood vessels of rabbit and from perfused lungs of sensitised guinea-pig. Prostaglandins 10:343–350
Herberman RB (1986) Natural killer cells in lungs and other tissues and regulation of their activity by suppressor cells. J Allergy Clin Immunol 78:566–570
Higgs GA, Salmon JA, Henderson B, Vane JR (1987) Pharmacokinetics of aspirin and salicylate in relation to inhibition of arachidonate cyclooxygenase and antiinflammatory activity. Proc Natl Acad Sci USA 84:1417–1420
Hollingsworth HM, Downin ET, Braman SS, Glassroth J, Binder R, Center DM (1984) Identificantin and characterization of neutrophil chemotactic acitivty in aspirin-induced asthma. Am Rev Respir Dis 130:377–379
ISIS-2 Collaborative Group (1988) Randomised trial of intravenous streptokinase, oral aspirin, both, or neither among 17187 cases of suspected acute myocardial

infarction: ISIS-2. Lancet 2:154–176
Jones DH, May AG, Condemi JJ (1984) HLA DR typing of aspirin-sensitive asthmathics. Ann Allergy 52:87–89
Joseph M, Capron A, Ameisen JC, Martinot JB, Tonnel AB (1987) Plaquettes sanguines et asthma a l'aspirine. Allerg Immunol [Suppl] 19:7–10
King RA (1960) A severe anaphylactoid reaction to hydrocortisone. Lancet 2:1093–1095
Kounis NG (1976) Untoward reactions to corticosteroids: intolerance to hydorocortisone. Ann Allergy 36:203–207
Kowalski ML, Grzelewska-Rzymowska J, Rozniecki J, Szmidt M (1984) Aspirin tolerance induced in aspirin-sensitive asthmatics. Allergy 39:171–178
Kuehl FA, Dougherty HW, Ham EA (1984) Interactions between prostaglandins and leukotrienes. Biochem Pharmacol 33:1–5
Lands WEM (1981) Actions of anti-inflammatory drugs. Trends Pharmacol Sci, March: 78-80
Lee TH, Drazen JM, Leitch AG, Ravalese J, Corey EJ, Robinson DR, Lewis RA, Austen KF (1986) Enhancement of plasma levels of biologically active leukotriene B compounds during anaphylaxis of guinea-pig pretreated by indomethacin or a fish-oil enriched diet. J Immunol 136:2575–2582
Lumry WR, Curd JG, Zeiger RS, Pleskow WW, Stevenson DD (1983) Aspirin-sensitive rhinosinusitis: the clinical syndrome and effects of aspirin administration. J Allergy Clin Immunol 71:580–587
Maclouf J, Fruteau de Laclos B, Borgeat P (1982) Stimulation of leukotriene biosynthesis in human blood leukocytes by platelet-derived 12-hydroperoxyeicosatetraenoic acid. Proc Natl Acad Sci USA 79:6042–6046
Martelli NA, Usandivaras G (1977) Inhibition of aspirin-induced bronchoconstriction by sodium cromoglycate inhalation. Thorax 32:684–690
Mendelson LM, Melzer EO, Hamburger RN (1974) Anaphylaxis-like reactions to corticosteroid therapy. J Allergy Clin Immunol 54:125–129
Morales MC, Basomba A, Pelaez A, Villalmanzo IG, Campos A (1985) Challenge tests with tartrazine in patients with asthma associated with intolerant to analgesics (ASA-triad). Clin Allergy 15:55–59
Morley J (1974) Prostaglandins and lymphokines in arthritis. Prostaglandins 8:315–326
Morley J, Sanjar C, Page CP (1984) Platelets in asthma. Lancet 2:726–727
Mullarkey MF, Thomas PS, Hansen JA, Webb DR, Nisperos B (1986) Association of aspirin-sensitive asthma with HLA-DQw2. Am Rev Respir Dis 133:261–263
Nizankowska E (1988) Aspirin-sensitive asthma and arachidonic acid metabolism. Doctor's thesis. Copernicus Academy, Cracow
Nizankowska E, Szczeklik A (1979) Keine Bedenken gegen Solosin bei acetylsalicylsaure-emfindlichen Asthmatikern. Dtsch Med Wochenschr 104:1388–1389
Nizankowska E, Szczeklik A (1989) Glucocorticosteroids attenuate aspirin-precipitated adverse reactions in aspirin-intolerant patients with asthma. Ann Allergy 63:159–162
Nizankowska E, Sheridan AQ, Maile MH, Cross CJ, Czerniwska-Mysik G, Szczeklik A (1987) Pharmacological attempts to modulate leukotriene synthesis in aspirin-induced asthma. Agents Actions [Suppl] 21:203–213
Nizankowska E, Michalska Z, Wandzilak M, Radomski M, Marcinkiewicz E, Gryglewski R, Szczeklik A (1988) An abnormality of arachidonic acid metabolism is not a generalized phenomenon in patients with aspirin-induced asthma. Eicosanoids 1:45–48
Ortolani C, Mirone C, Fontana A, Folco GC, Miadonna A, Montalbetti N, Rinaldi M, Sala A, Tedeschi A, Valente D (1987) Study of mediators of anaphylaxis in nasal wash fluids after aspirin and sodium metabisulfite nasal provocation in intolerant rhinitic patients. Ann Allergy 59:106–112
Pagano JS (1988) Epstein-Barr virus. In: De Clerq E, Walker RT (eds) Anti-viral drug

development. A multidisciplinary approach. Plenum, New York, pp 81–90 (NATO Advance Study Institute series: Life science, Vol 143)
Partridge MR, Gibson GJ (1978) Adverse bronchial reactions to intravenous hydroncortisone in two aspirin-sensitive asthmatic patients. Br Med J 1:1521
Pash JM, Bailey M (1988) Inhibition by corticosteroids of epidermal growth factor-induced recovery of cyclooxygenase after aspirin inactivation. FASEB J 2:2613–2618
Picado C, Castillo JA, Schinca N, Pujades M, Ordinas A, Coronas A, Augusti-Vidal A (1988) Effects of a fish oil enriched diet on aspirin intolerant asthmatic patients: a pilot study. Thorax 43:93–97
Pleskow WW, Stevenson DD, Simon RA, Mathison DA, Schatz M, Zeiger RS (1983) Aspirin-sensitive rhino-sinusitis/asthma: spectrum of adverse reactions to aspirin. J Allergy Clin Immunol 71:574–580
Podgwaite JD, Mazzone HM (1986) Latency of insect virus. Adv Virus Res 31:293–317
Prieto L, Palop J, Castro J, Basomba A (1988) Aspirin-induced asthma in a patient with asthma previously improved by non-steroidal antiinflammatory drugs. Clin Allergy 18:629–632
Rainsford KD (1984) Aspirin and the salicylates. Butterworths, London
Reilly IAG, Fitzgerald GA (1988) Aspirin in cardiovascular disease. Drugs 35:154–176
Roitt IM, Brostoff J, Male D (1985) Immunology. Churchill Livingstone, London
Rosenkranz B, Fischer C, Messe CO, Froelich JC (1986) Effects of salicylic acid and acetylsalicylic acid alone and in combination on platelet aggregation and prostanoid synthesis in man. Br J Clin Pharmacol 21:309–317
Roth SH (1988) Salicylates revisited. Are they still the hall-mark of antiinflammatory therapy? Drugs 31:1–6
Sagone AL Jr, Huseny RM (1987) Oxidation of salicylates by stimulated granulocytes: evidence that these drugs act as free radical scavengers in biological systems. J Immunol 138:2177–2183
Sakakibara H, Suetsugu S, Saga T, Handa M, Suzuki M, Doizoe T, Minako T, Horiguchi T, Konishi Y, Umeda H (1988a) Bronchial hyperresponsivness in aspirin-induced asthma. J Jpn Thorac Soc 26:612–619
Sakakibara H, Tsuda M, Suzuki M, Handa M, Saga T, Umeda H, Suetsugu SD, Konishi Y (1988b) A new method for diagnosis of aspirin-induced asthma by inhalation test with water soluble aspirin (aspirine-D, L -lysine, Venopirine). J Jpn Thorac 26:275–283
Samter M, Beers RF Jr (1968) Intolerance to aspirin. Clinical studies and consideration of its pathogenesis. Ann Intern Med 68:975–983
Samuellsson B (1983) Leucotrienes and mediators of immediate hypersensitivity reactions and inflammation. Science 220:568-572
Schlumberger HD (1980) Drug-induced pseudo-allergic syndrome as exemplified by acetylsalicylic acid intolerance. In: Dukor P, Kallos P, Schlumberger HD, West GB (eds) Pseudo-allergic reactions. Involvement of drugs and chemicals. Karger, Basel, pp 125–203
Schmitz-Schumann M (1989) HLA-System and Analgetica-Asthma-Syndrom (AAS) Pneumologie 44:580–581
Schmitz-Schumann M, Schaub E, Virchow C (1982) Inhalative Provokation mit Lysin-Azetylsalicylsaeure bei Analgetika-Asthma-Syndrom. Prax Klin Pneumol 36:17–21
Schmitz-Schumann M, Juhl E, Costabel U, Ruehle K-H, Menz G, Virchow Chr, Mattys H (1985) Analgetikaprovokationsproben bei Analgetika-Asthma Syndrom. Atemwegs Lungenkrankh 11:479–486
Schmitz-Schumann M, Menz G, Schaufele A, von Felten A, Matthys H, de Souze V, Virchow C (1987) Evidence of PAF release and platelet activation in anaglesics-syndrome. Agents Actions [Suppl] 21;215–224

Settipane GA (1981) Adverse reactions to aspirin and related drugs. Arch Intern Med 141:328–332
Settipane GA (1983) Aspirin and allergic diseases: a review. Am J Med 74:102–109
Settipane RA, Stevenson DD (1988) Cross-sensitivity with acetaminophen in ASA-sensitive asthmatics. J Allergy Clin Immunol 81:180 (abstract 47)
Shu Chan C, Brown IG, Oliver WA, Zimmerman PV (1984) Hydrocortisone-induced anaphylaxis. Med J Aust 141:444–446
Simon RA (1984) Adverse reactions to drug additives. J Allergy Clin Immunol 74:623–630
Simon RA, Pleskow WW, Stevenson DD, Mathison DA (1984) Aspirin sensitivity: description of aspirin (ASA) respiratory sensitivity. In: Kornblat J, Wedner C (eds) Allergy: theory and practice. Grune and Stratton, New York, pp 435–452
Stevenson DD (1984) Diagnosis, prevention and treatment of adverse reactions to aspirin and nonsteroidal anti-inflammatory drugs. J Allergy Clin Immunol 74:617–622
Stevenson DD, Lewis RA (1987) Proposed mechanisms of aspirin sensitivity reactions. J Allergy Clin Immunol 80:788–790
Stevenson DD, Pleskow WW, Simon RA, Mathison DA, Lumry WR, Schatz M, Zeiger R (1984) Aspirin-sensitive rhinosinusitis asthma: a doule-blind cross-over study of treatment with aspirin. J Allergy Clin Immunol 73:500–507
Stevenson DD, Schrank PJ, Hougham AJ, Goldust MB, Wilson RR (1988) Salasate cross-sensitivity in aspirin-sensitive asthmatic. J Allergy Clin Immunol 81:181
Stricker BH, Meyboom RHB, Lindquist M (1985) Acute hypersensitivity reactions to paracetamol. Br Med J 291:938–939
Szczeklik A (1986) Analgesics, allergy and asthma. Drugs 32 [Suppl 4]:148–163
Szczeklik A (1988) Aspirin induced asthma as a viral disease. Clin Allergy 18:15–20
Szczeklik A (1990) Origin of aspirin-induced asthma. Agents Actions [Suppl] 28: 27–40
Szczeklik A, Serwonska M (1979) Inhibition of idiosyncratic reactions to aspirin in asthmatic patients by clemastine. Thorax 34:654–657
Szczeklik A, Gryglewski RJ (1983) Asthma and antiinflammatory drugs. Mechanisms and clinical patterns. Drugs 25:533–543
Szczeklik A, Nizankowska E (1985) The effect of diflunisal on pulmonary function in aspirin-sensitive asthma. J Allergy Clin Immunol 75:158 (abstract 216)
Szczeklik A, Gryglewski RJ, Czerniawska-Mysik G (1975) Relationship of inhibition of prostaglandin biosynthesis by analgesics to asthma attacks in aspirin-sensitive patients. Br Med J 1:67–69
Szczeklik A, Gryglewski RJ, Czerniawska-Mysik G (1977a) Clinical patterns of hypersensitivity to nonsteroidal antiinflammatory drugs and their pathogenesis. J Allergy Clin Immunol 60:276–284
Szczeklik A, Nizankowska E, Nizankowski R (1977b) Bronchial reactivity to prostaglandins F2alpha, E2 and histamine in different types of asthma. Respiration 34:323–331
Szczeklik A, Czerniawska-Mysik G, Serwonska M, Kuklinski P (1980) Inhibition by ketotifen of idiosyncratic reactions to aspirin. Allergy 35:421–424
Szczeklik A, Nizankowska E, Czerniawska-Mysik G, Sek S (1985) Hydrocortisone and airflow impairement in aspirin-induced asthma. J Allergy Clin Immunol 76:530–536
Szczeklik A, Milner PC, Birch J, Watkins J, Martin JF (1986) Prolonged bleeding time, reduced platelet aggregation, altered PAF-acether sensitivity and increased platelet mass are a trait of asthma and hay fever. Thromb Haemost 56:283–287
Szczeklik A, Nizankowska E, Dworski R, Splawinski J, Gajewski P, Splawinska B (1987) Effects of inhibition of thromboxane A2 synthesis in aspirin-induced asthma. J Allergy Clin Immunol 80:839–843
Szczeklik A, Nizankowska E, Dworski R (1990) Choline magnesium trisalicylate in patients with aspirin induced asthma. Eur Resp of 3: 535–539.

Taniguschi M, Sato A (1988) Aspirin-induced asthmatics (AIA) have cross-sensitivity with the steroid succinate esters. Abstr XIII Congress Allergy Clin Immunol, Montreux, abstract 358

Toogood JH (1977) Aspirin intolerance, asthma, prostaglandins and cromolyn sodium. Chest 72:35–37

Tsuda M, Sakakibara H, Kamidaira T, Saga T, Suetsugu S, Umeda H (1988) Arachidonic acid metabolism of peripheral blood leukocytes in aspirin-induced asthma. Abstr XIII Congress Allergy Clin Immunol, Montreux, abstract 755

Undem BJ, Pickett WC, Lichtenstein LM, Adams II GK (1987) The effect of indomethacin on imunologic release of histamine and sulfidopeptide leukotrienes from human bronchus and lung parenchuma. Am Rev Respir Dir 136:1183–1187

Vaghi A, Robuschi M, Simone P, Bianco S (1985) Bronchial response to leukotriene C_4 (LTC_4) in aspirin asthma. Abstracts SEP 4th Congress, Milano-Stresa, p 171

Vane JR (1971) Inhibition of prostaglandin synthesis as a mechanism of action for aspirin-like drugs. Nature 231:232–234

Vane JR (1987) The evolution of non-steroidal anti-inflammatory drugs and their mechanism of action. Drugs 33 [Suppl 1]:18–27

Vargaftig BB (1978) Salicylic acid fails to inhibit generation of thromboxane A_2 activity in platelets after in vivo administration to the rat. J Pharm Pharmacol 30:101–104

Vigano T, Toia A, Crivellari MT, Galli G, Mezzetti M, Folco GC (1988) Prostaglandin synthetase inhibition and formation of lipoxygenase products in immunologically challenged normal human lung parenchyma. Eicosanoids 1:73–77

Virchow C (1976) Analgetika-Intoleranz bei Asthmatikern (Analgetika-Asthma-Sydrom); vorläufige Mitteilung. Prax Pneumol 30:684–692

Virchow C, Schmitz-Schumann M, Juhl-Schaub E (1986) Pyrazolones and analgesics-induced asthma syndrome. Agents Actions [Suppl] 19:291–303

Virchow C, Szczeklik A, Bianco S, Schmitz-Schumann M, Juhl E, Robuschi M, Damonte C, Menz G, Serwonska M (1988) Intolerance to tartrazine in aspirin-induced asthma: results of a multicenter study. Respiration 53:20–23

Voigtlaender V (1985) Dermatologische Nebenwirkungen von pyrazolonen. In: Brune K, Lanz R (eds) 100 Jahre Phyrazolone. Urban Schwarzenberg, Munich, pp 261–266

Voigtlaender V, Haensch G, Rother U (1981) Acetylsalicylic acid intolerance: a possible role of complement. Int Arch Allergy Appl Immunol 66 [Suppl 1]:154–155

Weber RW, Hoffman M, Raine DA, Nelson HS (1979) Incidence of bronchoconstriction due to aspirin, azo dyes, non-azo dyes, and preservatives in a population of perennial asthmatics. J Allergy Clin Immunol 64:32–37

Welliver RC, Ogra PL (1988) Immunology of respiratory viral infections. Annu Rev Med 39:147–162

Wuetrich B, Fabro L (1981) Azetylsalicylsaeure and Lebensmittel additiva-Intoleranz bei Urtikaria, Asthma bronchiale und chronischer Rhinopathie. Schweiz Med Wochenschr 111:1445–1449

Yurchak AM, Wicher K, Arbesman CE (1970) Immunological studies on aspirin. Clinical studies with aspiryl-protein conjugates. J Allergy 46:245–251

Subject Index

Handbook of Experimental Pharmacology

Volume 56
Cardiac Glycosides
Part 1: **Experimental Pharmacology**
Part 2: **Pharmacokinetics and Clinical Pharmacology**

Volume 57
Tissue Growth Factors

Volume 58
Cyclic Nucleotides
Part 1: **Biochemistry**
Part 2: **Physiology and Pharmacology**

Volume 59
Mediators and Drugs in Gastrointestinal Motility
Part 1: **Morphological Basis and Neurophysiological Control**
Part 2: **Endogenous and Exogenous Agents**

Volume 60
Pyretics and Antipyretics

Volume 61
Chemotherapy of Viral Infections

Volume 62
Aminoglycoside Antibiotics

Volume 63
Allergic Reactions to Drugs

Volume 64
Inhibition of Folate Metabolism in Chemotherapy

Volume 65
Teratogenesis and Reproductive Toxicology

Volume 66
Part 1: **Glucagon I**
Part 2: **Glucagon II**

Volume 67
Part 1: **Antibiotics Containing the Beta-Lactam Structure I**
Part 2: **Antibiotics Containing the Beta-Lactam Structure II**

Volume 68
Antimalarial Drugs
Part 1: **Biological Background, Experimental Methods and Drug Resistance**
Part 2: **Current Antimalarias and New Drug Developments**

Volume 69
Pharmacology of the Eye

Volume 70
Part 1: **Pharmacology of Intestinal Permeation I**
Part 2: **Pharmacology of Intestinal Permeation II**

Volume 71
Interferons and Their Applications

Volume 72
Antitumor Drug Resistance

Volume 73
Radiocontrast Agents

Volume 74
Antieliptic Drugs

Volume 75
Toxicology of Inhaled Materials

Springer-Verlag
Berlin
Heidelberg
New York
London
Paris
Tokyo
Hong Kong
Barcelona

Handbook of Experimental Pharmacology

Volume 76
Clinical Pharmacology of Antiangial Drugs

Volume 77
Chemotherapy of Gastrointestinal Helminths

Volume 78
The Tetracyclines

Volume 79
New Neuromuscular Blocking Agents

Volume 80
Cadmium

Volume 81
Local Anesthetics

Volume 82
Radioimmunoassay in Basic and Clinical Pharmacology

Volume 83
Calcium in Drug Actions

Volume 84
Antituberculosis Drugs

Volume 85
The Pharmacology of Lymphocytes

Volume 86
The Cholinergic Synapse

Volume 87
Part 1: **Pharmacology of the Skin I**
Part 2: **Pharmacology of the Skin II**

Volume 88
Drugs for the Treatment of Parkinson's Disease

Volume 89
Antiarrhythmic Drugs

Volume 90
Part 1: **Catecholamines I**
Part 2: **Catecholamines II**

Volume 91
Microbial Resistance to Drugs

Volume 92
Insulin

Volume 93
Pharmacology of Antihypertensive Therapeutics

Volume 94
Part 1: **Chemical Carcinogenesis and Mutagenesis I**
Part 2: **Chemical Carcinogenesis and Mutagenesis II**

Volume 95
Part 1: **Peptide Growth Factors and Their Receptors I**
Part 2: **Peptide Growth Factors and Their Receptors II**

Volume 96
Chemotherapy of Fungal Diseases

Volume 97
Histamine and Histamine Antagonists

Springer-Verlag
Berlin
Heidelberg
New York
London
Paris
Tokyo
Hong Kong
Barcelona